AF736551

VOLUME FOUR HUNDRED AND SIX

INTERNATIONAL REVIEW OF CELL AND MOLECULAR BIOLOGY

Autophagy – Part C

INTERNATIONAL REVIEW OF CELL AND MOLECULAR BIOLOGY

VOLUME FOUR HUNDRED AND SIX

INTERNATIONAL REVIEW OF CELL AND MOLECULAR BIOLOGY

Autophagy – Part C

Edited by

JOSÉ MANUEL BRAVO-SAN PEDRO
Universidad Complutense de Madrid, Madrid, Spain

LORENZO GALLUZZI
Fox Chase Cancer Center, Philadelphia, PA, United States

ELSEVIER

ACADEMIC PRESS
An imprint of Elsevier

Academic Press is an imprint of Elsevier
125 London Wall, London, EC2Y 5AS, United Kingdom
50 Hampshire Street, 5th Floor, Cambridge, MA 02139, United States

First edition 2026

ISBN: 978-0-443-41437-4
ISSN: 1937-6448

For information on all Academic Press publications
visit our website at https://www.elsevier.com/books-and-journals

Publisher: Zoe Kruze
Acquisition Editor: Leticia M. Lima
Editorial Project Manager: Devwart Chauhan
Production Project Manager: James Selvam
Cover Designer: Muthukumaran Thangaraj

Typeset by MPS Limited, India

Contents

Contributors

Eloy Bejarano
Department of Biomedical Sciences, School of Health Sciences, Universidad Cardenal Herrera-CEU, CEU Universities, Valencia, Spain

Irene Carriles-Moar
Departamento de Biología Molecular, Instituto Universitario de Biología Molecular (IUBM) and Centro de Biología Molecular Severo Ochoa (CBM), CSIC-Universidad Autónoma de Madrid, Madrid, Spain

Sandra Cutanda-Tesouro
Department of Physiological Sciences, Universitat de Barcelona, Barcelona, Spain

Fatima Domenica Elisa De Palma
Team "Metabolism, Cancer & Immunity", Équipe labellisée par la Ligue contre le cancer, Centre de Recherche des Cordeliers, UMRS1138, Inserm, Université Paris Cité, Sorbonne Université, Paris, France; Department of Molecular Medicine and Medical Biotechnologies, University of Napoli Federico II, Naples, Italy

Angelica Di Matteo
Team "Metabolism, Cancer & Immunity", Équipe labellisée par la Ligue contre le cancer, Centre de Recherche des Cordeliers, UMRS1138, Inserm, Université Paris Cité, Sorbonne Université, Paris, France; Department of Precision Medicine, University of Campania "L. Vanvitelli", Naples, Italy

Mojgan Djavaheri-Mergny
Team "Metabolism, Cancer & Immunity", Équipe labellisée par la Ligue contre le cancer, Centre de Recherche des Cordeliers, UMRS1138, Inserm, Université Paris Cité, Sorbonne Université, Paris, France

Alicia Domenech-Bendaña
Department of Biomedical Sciences, School of Health Sciences, Universidad Cardenal Herrera-CEU, CEU Universities, Valencia, Spain

Rana El Nahas
Laboratory of Immunoregulation, Research Department, Sidra Medicine, Doha, Qatar

Meritxell Espino-Guarch
Laboratory of Immunoregulation, Research Department, Sidra Medicine, Doha, Qatar

Álvaro F. Fernández
Departamento de Bioquímica y Biología Molecular, Universidad de Oviedo; Instituto Universitario de Oncología del Principado de Asturias (IUOPA); Instituto de Investigación Sanitaria del Principado de Asturias (ISPA), Oviedo, Spain

Pablo M. Garcia-Roves
Department of Physiological Sciences, Universitat de Barcelona; Metabolism and Gene Therapy Group, Diabetes and Metabolism Program, Institut D'Investigació Biomèdica de Bellvitge (IDIBELL), Barcelona; Instituto de Salud Carlos III, Centro de Investigación Biomédica en Red Fisiopatología de La Obesidad y La Nutrición (CIBEROBN), Madrid, Spain

Dafne García-Mateos
Departamento de Biología Molecular, Instituto Universitario de Biología Molecular (IUBM) and Centro de Biología Molecular Severo Ochoa (CBM), CSIC-Universidad Autónoma de Madrid, Madrid, Spain

José Ignacio García-Muñoz
Department of Biomedical Sciences, School of Health Sciences, Universidad Cardenal Herrera-CEU, CEU Universities, Valencia, Spain

Lucia Gimeno-Mallench
Department of Biomedical Sciences, School of Health Sciences, Universidad Cardenal Herrera-CEU, CEU Universities, Valencia, Spain

Maho Hamasaki
Department of Genetics, The University of Osaka Graduate School of Medicine, Suita, Osaka; Graduate School of Frontier Biosciences, The University of Osaka, Japan

Carmen Hernández
Instituto de Biología Molecular y Celular de Plantas (IBMCP; CSIC-UPV) (Consejo Superior de Investigaciones Científicas-Universidad Politécnica de Valencia), Ciudad Politécnica de la Innovación, Calle Ingeniero Fausto Elio, Valencia, Spain

Manuela Lizarralde-Guerrero
Team "Metabolism, Cancer & Immunity", Équipe labellisée par la Ligue contre le cancer, Centre de Recherche des Cordeliers, UMRS1138, Inserm, Université Paris Cité, Sorbonne Université, Paris; Université Paris-Saclay, Faculté de Médecine, Kremlin-Bicêtre, France

Daniel López-López
Clinical Bioinformatics Area, Fundación Progreso y Salud (FPS), CDCA; Computational Systems Medicine, Institute of Biomedicine of Seville (IBIS), Hospital Virgen del Rocio; Bioinformatics in Rare Diseases (BiER), Centro de Investigación Biomédica en Red de Enfermedades Raras (CIBERER), FPS, Hospital Virgen del Rocío, Sevilla, Spain

Maria Chiara Maiuri
Team "Metabolism, Cancer & Immunity", Équipe labellisée par la Ligue contre le cancer, Centre de Recherche des Cordeliers, UMRS1138, Inserm, Université Paris Cité, Sorbonne Université, Paris; Université Paris-Saclay, Faculté de Médecine, Kremlin-Bicêtre, France

Guillermo Marcos-Zapatero
Departamento de Biología Molecular, Instituto Universitario de Biología Molecular (IUBM) and Centro de Biología Molecular Severo Ochoa (CBM), CSIC-Universidad Autónoma de Madrid, Madrid, Spain

Satoshi Minami
Department of Genetics, The University of Osaka Graduate School of Medicine, Suita, Osaka; Division of Health Sciences, Graduate School of Medicine, The University of Osaka, Japan

Jose Luis Nieto-Torres
Department of Biomedical Sciences, School of Health Sciences, Universidad Cardenal Herrera-CEU, CEU Universities, Valencia, Spain

Lucio Pastore
Department of Molecular Medicine and Medical Biotechnologies, University of Napoli Federico II; CEINGE-Biotecnologie Avanzate Franco Salvatore, Naples, Italy

Jonathan Pol
Team "Metabolism, Cancer & Immunity", Équipe labellisée par la Ligue contre le cancer, Centre de Recherche des Cordeliers, UMRS1138, Inserm, Université Paris Cité, Sorbonne Université, Paris, France

Alejandro Ponce-Mora
Department of Biomedical Sciences, School of Health Sciences, Universidad Cardenal Herrera-CEU, CEU Universities, Valencia, Spain

Miryam Pérez-Cañamás
Instituto de Biología Molecular y Celular de Plantas (IBMCP; CSIC-UPV) (Consejo Superior de Investigaciones Científicas-Universidad Politécnica de Valencia), Ciudad Politécnica de la Innovación, Calle Ingeniero Fausto Elio, Valencia, Spain

Natalia Reglero-Real
Departamento de Biología Molecular, Instituto Universitario de Biología Molecular (IUBM) and Centro de Biología Molecular Severo Ochoa (CBM), CSIC-Universidad Autónoma de Madrid; Instituto de Investigación Sanitaria La Princesa, Madrid, Spain

Filippo Scialó
Department of Molecular Medicine and Medical Biotechnologies, University of Napoli Federico II; CEINGE-Biotecnologie Avanzate Franco Salvatore, Naples, Italy

Laura Vipraio Tiberi
Team "Metabolism, Cancer & Immunity", Équipe labellisée par la Ligue contre le cancer, Centre de Recherche des Cordeliers, UMRS1138, Inserm, Université Paris Cité, Sorbonne Université, Paris, France; Department of Molecular Medicine and Medical Biotechnologies, University of Napoli Federico II, Naples, Italy

Maria Vitale
CEINGE-Biotecnologie Avanzate Franco Salvatore, Naples, Italy

CHAPTER ONE

Lysosomal function and homeostasis: From cellular adaptation to disease pathogenesis

Satoshi Minami[a,c] and Maho Hamasaki[a,b,*]
[a]Department of Genetics, The University of Osaka Graduate School of Medicine, Suita, Osaka, Japan
[b]Graduate School of Frontier Biosciences, The University of Osaka, Japan
[c]Division of Health Sciences, Graduate School of Medicine, The University of Osaka, Japan
*Corresponding author. e-mail address: hamasaki@gt.med.osaka-u.ac.jp

Contents

Abstract

Lysosomes were once considered terminal degradative organelles responsible for disposing of cellular waste. However, recent studies have revealed that lysosomes

International Review of Cell and Molecular Biology, Volume 406
ISSN 1937-6448, https://doi.org/10.1016/bs.ircmb.2026.04.002

serve as dynamic signalling and metabolic hubs at the center of diverse biological processes, including nutrient sensing, metabolic regulation, membrane trafficking, autophagy, inflammation, and cell death. To support this broad functional repertoire, lysosomes must possess robust mechanisms to maintain their integrity in the face of damage or stress.

In response to lysosomal membrane damage, cells engage multilayered adaptive mechanisms that act in coordination—membrane repair (Repair), selective removal of damaged organelles (Removal), and *de novo* biogenesis of lysosomes (Regeneration). These processes are mediated by a range of molecular pathways, including the ESCRT complex, the PITT pathway, lysophagy, and TFEB-dependent lysosomal regeneration. Notably, recent findings highlight the noncanonical autophagy-like pathway known as ATG8ylation (conjugation of ATG8s on single membranes), which is activated via the STING–V-ATPase–ATG16L1 axis and functions as a critical hub connecting multiple arms of the lysosomal damage response.

In this review, we systematically outline the molecular basis of lysosomal damage responses, including ATG8ylation, and explore how these networks are implicated in a broad spectrum of pathological conditions such as aging, neurodegeneration, cancer, obesity-related disorders, and immune dysfunction. Understanding these lysosomal quality control mechanisms not only sheds light on the fundamental principles of organelle homeostasis but also opens new avenues for therapeutic innovation.

1. Introduction

Lysosomes have long been regarded as the terminal degradative organelles of the cell, responsible for recycling macromolecules delivered via endocytosis, phagocytosis, and autophagy through the action of more than 60 acidic hydrolases. For decades, lysosomal dysfunction was considered relevant only to rare congenital disorders such as lysosomal storage diseases.

However, this view has undergone a profound transformation. Recent studies have revealed that lysosomes are not merely waste-processing compartments but function as dynamic signaling hubs that sense and integrate diverse environmental and intracellular cues—including nutrient status, metabolic stress, and pathogen exposure—to coordinate essential cellular programs such as growth, autophagy, immune response, and cell death (Ballabio & Bonifacino, 2020). Maintaining lysosomal homeostasis is thus fundamental to cellular integrity and organismal health.

Importantly, impaired lysosomal function has been implicated in a wide range of common diseases, including neurodegenerative disorders (e.g., Alzheimer's and Parkinson's diseases), metabolic syndromes (e.g., type 2 diabetes and fatty liver disease), cancer, and immune dysfunctions (Bonam et al., 2019). Moreover, mounting evidence indicates that lysosomal decline is a hallmark of aging, contributing to the progressive deterioration of cellular clearance and metabolic control across tissues. This age-related dysfunction may act as a key driver of chronic inflammation, cellular senescence, and reduced regenerative capacity, thereby accelerating systemic aging.

In this review, we begin by providing an overview of the lysosome's multifaceted functions in degradation, signal integration, and homeostatic regulation. We then focus on the emerging cellular responses to lysosomal membrane damage, highlighting the multilayered mechanisms by which cells repair, eliminate, and regenerate compromised lysosomes. Finally, we discuss how disruption of these damage-response pathways contributes to diverse pathological states, including aging, and consider how lysosome-centered strategies may open new therapeutic avenues.

2. Multifaceted roles and structural plasticity of the lysosome

Lysosomes are membrane-bound organelles responsible for the degradation and recycling of intracellular and extracellular materials delivered via endocytosis, phagocytosis, and autophagy. While their degradative function has long been appreciated, recent studies have significantly broadened our understanding of lysosomes, revealing their central role in coordinating metabolic responses, signaling pathways, and cellular homeostasis under diverse physiological and stress conditions (Ballabio & Bonifacino, 2020).

The lysosome's dynamic nature underpins its functional versatility. Once considered static compartments anchored within the cytoplasm, lysosomes are now recognized as highly motile and adaptable structures whose positioning and behavior are influenced by nutrient availability, cellular stress, and metabolic cues. Moreover, lysosomes form physical contact sites with various organelles—including mitochondria, endoplasmic reticulum, and endosomes—enabling the exchange of ions, metabolites, and signaling molecules. These inter-organelle interactions position the lysosome as an integrative platform in maintaining intracellular organization and adaptive homeostasis.

2.1 Nutrient sensing and mTORC1 signaling via lysosomes

Among the many regulatory functions attributed to lysosomes, their role in nutrient sensing and control of metabolic signaling has garnered particular attention. Central to this role is the regulation of mechanistic target of rapamycin complex 1 (mTORC1), a master regulator of cellular growth, protein synthesis, and autophagy (Saxton & Sabatini, 2017).

Activation of mTORC1 requires its translocation to the lysosomal membrane, a process mediated by Rag GTPases and the Ragulator complex (Bar-Peled et al., 2012; Sancak et al., 2010). Beyond serving as a mere scaffold, the lysosome actively participates in this process by detecting luminal amino acid concentrations—particularly arginine—via sensors such as v-ATPase and the amino acid transporter SLC38A9. These sensors modulate Ragulator–Rag signaling to control mTORC1 activation in a nutrient-dependent manner (Abu-Remaileh et al., 2017; Rebsamen et al., 2015; Wang et al., 2015a; Zoncu et al., 2011).

In addition to amino acid sensing, lysosomes integrate signals from other metabolic pathways. For instance, lysosomal cystine export has been shown to influence mTORC1 suppression during nutrient deprivation, suggesting a role for sulfur amino acid metabolism in growth regulation (Jouandin et al., 2022). Moreover, calcium efflux through lysosomal TRPML (MCOLN) channels has been implicated in the modulation of mTORC1 and activation of the transcription factor TFEB, further highlighting the role of lysosomal ion homeostasis in metabolic control (Huang et al., 2024a).

Recent findings also suggest that specific phosphoinositides, such as PI $(3,5)P_2$, regulate the spatial activation of mTORC1 on lysosomes, potentially differentiating lysosome-localized and non-lysosomal mTORC1 pools based on the origin and context of amino acid input (Ebner et al., 2025).

Taken together, these insights establish the lysosome not only as a structural platform for mTORC1 activation but also as an active metabolic hub capable of integrating diverse nutritional, ionic, and lipid signals to orchestrate context-specific cellular responses.

2.2 Lysosomes as integrators of intracellular stress sensing and response

In addition to their established role in degrading and recycling intracellular material, lysosomes have recently emerged as critical integrators of cellular stress signals. They respond to a variety of stress stimuli—including nutrient deprivation, oxidative stress, mechanical perturbation, hypoxia, and endoplasmic reticulum (ER) stress—by coordinating adaptive signaling cascades that restore cellular homeostasis.

A central player in this stress-sensing machinery is the lysosomal Ca^{2+} channel TRPML1 (Transient Receptor Potential Mucolipin 1). TRPML1 becomes activated under various stress conditions and mediates the release of Ca^{2+} from the lysosomal lumen into the cytosol (Medina et al., 2015; Wang et al., 2015b; Zhang et al., 2016). This Ca^{2+} signal triggers multiple downstream responses, including facilitation of autophagosome–lysosome fusion (Cheng et al., 2010), nuclear translocation of the transcription factor TFEB (Medina et al., 2015; Sardiello et al., 2009), and reformation of functional lysosomes from autolysosomes (Miller et al., 2015). Moreover, TRPML1-mediated Ca^{2+} efflux promotes lysosomal exocytosis, which contributes to plasma membrane repair (Reddy et al., 2001) and extrusion of cellular waste (Medina et al., 2011).

Lysosomes also play a regulatory role in inflammation. TRPML1-mediated import of Fe^{2+} has been shown to suppress NF-κB activation via prolyl hydroxylases (PHDs) (Xing et al., 2025). Similarly, the lysosomal membrane protein CD151 helps maintain proteostasis within endolysosomal compartments, thereby attenuating inflammatory signaling (Chen et al., 2024a). In addition, lysosomes act as negative regulators of the cGAS–STING pathway, preventing excessive innate immune activation in response to self-DNA accumulation (Xu & Wan, 2024).

Under hypoxic stress, lysosomes promote degradation of HIF-1α to limit unnecessary hypoxic responses (Hubbi et al., 2014). During ER stress, relocalization of V-ATPase to ER exit sites redirects COPII vesicles from the secretory to the lysosomal degradation pathway, thereby enhancing the clearance of misfolded proteins (Sun et al., 2023).

However, prolonged or excessive activation of these stress responses can overwhelm the lysosomal system. Recent studies have demonstrated that sustained lysosomal damage impairs cellular recovery from quiescence and may lead to irreversible cell cycle arrest (Murley et al., 2025). These findings underscore the importance of lysosomes not only as effectors of stress adaptation but also as critical checkpoints that preserve long-term cellular integrity.

Taken together, lysosomes orchestrate a broad spectrum of stress-related responses, ranging from autophagy and membrane repair to transcriptional regulation and inflammation suppression. Maintaining lysosomal functionality under stress conditions is therefore vital for cellular survival. The following section will discuss the molecular mechanisms by which cells detect and repair lysosomal damage to safeguard organelle integrity and ensure homeostasis.

2.3 Lysosomal interactions with other organelles

The maintenance of cellular homeostasis relies not only on the autonomous function of individual organelles but also on their dynamic interplay. Recent studies have revealed that lysosomes form membrane contact sites (MCSs) with various organelles, enabling the exchange of ions and metabolites, modulation of membrane dynamics, and integration of signaling pathways. These contacts position lysosomes as a central hub for inter-organelle communication that coordinates a wide array of cellular processes.

Among these, the physical and functional interaction between lysosomes and mitochondria has emerged as particularly critical. For example, at ER–lysosome–mitochondrion tri-junctions, phosphatidylinositol 4-phosphate [PI(4)P] supplied by lysosomes via ORP1L facilitates mitochondrial fission (Boutry & Kim, 2021). Lysosomal TRPML1 channels also release Ca^{2+} into the cytosol, which is then taken up by mitochondria to support mitochondrial function and energy metabolism (Peng et al., 2020). Disruption of these contacts has pathological implications. In neurodegenerative disease models, α-synuclein interferes with the formation of mitochondria–lysosome contacts, disturbing Ca^{2+} homeostasis and impairing TFEB-mediated lysosomal transcriptional activation (Giamogante et al., 2024).

The endoplasmic reticulum (ER) also forms functional contacts with lysosomes, playing a key role in calcium exchange and protein quality control. The lysosomal membrane protein DRAM1 interacts with the ER calcium sensor STIM1, promoting lysosomal Ca^{2+} uptake, ER stress adaptation, and ER-specific autophagy (ERphagy) (Wang et al., 2024b). Additionally, a selective lysosome-dependent degradation pathway has been identified in which TOLLIP mediates the clearance of misfolded ER membrane proteins, distinct from proteasome-dependent ER-associated degradation (ERAD) (Hayashi et al., 2023).

In the context of lipid metabolism, lysosomes are essential for the bidirectional trafficking of cholesterol and phospholipids, particularly between lysosomes and the ER. Disruption of this exchange can compromise autophagic flux and lipid homeostasis. For instance, respiratory syncytial virus (RSV) infection impairs ER–lysosome cholesterol trafficking, resulting in lysosomal cholesterol accumulation and autophagy suppression, which in turn supports viral replication (Chen et al., 2024b).

Interactions between lysosomes and the Golgi apparatus are also functionally significant. Lysosomes in proximity to the trans-Golgi network are involved in the degradation of aberrant or excess secretory proteins through a microautophagy-like mechanism. When this process is disrupted, abnormal accumulation of these proteins in autolysosomes has been observed (Zhou et al., 2023).

Recently, novel nuclear–lysosome crosstalk has also been described. The transcription factor OTX2, important for neurodevelopment, is exported from the nucleus via nuclear envelope budding and subsequently trafficked to lysosomes for degradation or secretion (Park et al., 2023).

Collectively, these findings redefine lysosomes not merely as degradative endpoints but as dynamic participants in cellular organization. Through membrane contact sites, lysosomes integrate and modulate organelle-specific processes such as calcium signaling, lipid exchange, stress adaptation, and organelle quality control. Specific membrane proteins and lipid regulators orchestrate these contacts, and their dysregulation can propagate dysfunction across multiple organelles, ultimately contributing to disease pathogenesis. Understanding these lysosome-centered networks offers novel insights into cellular systems biology and provides fertile ground for developing innovative therapeutic strategies.

3. Mechanisms for maintaining lysosomal homeostasis

Lysosomes are constantly exposed to a variety of potentially damaging stressors, including pathogens, crystals, lipids, pharmacological agents, and reactive oxygen species (ROS) (Bussi et al., 2018; Chauhan et al., 2016; Gabandé-Rodríguez et al., 2014; Maejima et al., 2013; Papadopoulos et al., 2017; Rajamäki et al., 2010; Terman et al., 2006). Upon injury, the release of lysosomal hydrolases—such as cathepsins—into the cytosol can trigger cell death programs, including apoptosis and necroptosis (Wang et al., 2018). In addition, lysosomal dysfunction impairs fundamental cellular processes such as autophagy, endocytosis, and phagocytosis, as well as metabolic signaling (Ballabio & Bonifacino, 2020). These consequences are increasingly recognized as key drivers in the pathogenesis of a wide range of diseases, including neurodegeneration, metabolic disorders, and cancer (Bonam et al., 2019). Therefore, preserving lysosomal integrity is essential for maintaining cellular and organismal homeostasis.

For a long time, the mechanisms by which cells monitor and restore lysosomal function after damage remained largely unknown. However, recent advances have revealed that cells possess a multilayered response system to cope with lysosomal injury. This system orchestrates membrane repair, selective removal of damaged lysosomes, and the regeneration of new lysosomal compartments (Fig. 1). Rather than functioning independently, these processes are believed to operate in an integrated and context-dependent manner, allowing cells to dynamically adapt to stress while maintaining the integrity of the lysosomal network.

In the following sections, we describe in detail the emerging mechanisms that collectively contribute to lysosomal homeostasis, including membrane repair, selective lysosome clearance, and TFEB-driven lysosomal biogenesis.

3.1 ATG8ylation as a central hub in the lysosomal damage response: An integrative role of noncanonical autophagy pathways

When cells encounter lysosomal damage, the restoration of homeostasis requires a tightly regulated molecular system capable of integrating sequential responses, ranging from damage sensing and repair to regeneration. Recent

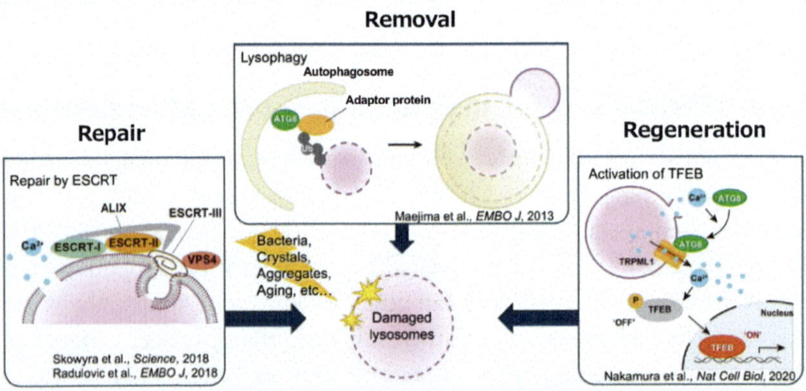

Fig. 1 Homeostatic mechanisms in response to lysosomal damage. When lysosomes are damaged, cells activate three principal homeostatic responses—repair, removal, and regeneration—in a coordinated and context-dependent manner. In cases of mild and localized damage to the membrane, the ESCRT complex or the PITT pathway attempts to repair. Concurrently, microlysophagy is engaged to selectively remove the damaged membrane region by inward invagination, thereby preserving overall lysosomal function. If the damage is extensive and deemed irreversible, macrolysophagy targets the entire lysosome for degradation through autophagosome-mediated pathways. Finally, to compensate for the loss of degradative capacity, activation of the transcription factor TFEB promotes the regeneration of new lysosomes. These multilayered and adaptive responses constitute an optimal strategy to maintain lysosomal homeostasis under various stress conditions.

studies have identified a pivotal platform in this process: ATG8ylation (the Conjugation of ATG8s to Single Membranes, a process involving the lipidation of ATG8 family proteins (such as LC3 and GABARAP) directly on damaged lysosomal membranes.

Unlike canonical autophagy, ATG8ylation proceeds independently of upstream factors such as the ULK1 complex and WIPI2, and it does not involve the formation of double-membrane autophagosomes. Instead, ATG8 lipidation occurs directly on single-membrane structures, specifically at the sites of lysosomal damage (Meng et al., 2024). Moreover, ATG8ylation is considered an evolutionarily ancient mechanism, likely predating the emergence of ULK1-related pathways, and has been characterized as a "primordial" membrane stress response system (Lv et al., 2024).

Rather than serving solely as a damage sensor, ATG8ylation functions as a multifaceted platform that orchestrates a broad range of events, including membrane repair, homeostatic recovery, and lysosomal regeneration. For example, the formation of ATG8ylation facilitates early recruitment of the ESCRT complex for membrane stabilization (Ogura et al., 2023), promotes stress granule assembly to seal ruptured lysosomal membranes (Meng et al., 2024), contributes to microlysophagy for selective removal of damaged lysosomal membranes (Ogura et al., 2023), and activates TFEB-mediated lysosomal biogenesis programs (Goodwin et al., 2021; Nakamura et al., 2020). Furthermore, ATG8ylation has been shown to recruit and activate LRRK2, which contributes to lysosomal morphological remodeling and membrane dynamics regulation (Kuwahara & Iwatsubo, 2024).

Collectively, these multilayered responses highlight ATG8ylation as a central signaling hub that coordinates the full spectrum of lysosomal damage responses—from early containment to late-stage organelle reconstruction.

One of the most critical triggers for ATG8ylation activation involves the STING–V-ATPase–ATG16L1 axis. Upon lysosomal injury, the cytosolic cGAS–STING pathway is activated, leading STING to induce localized environmental changes on the lysosomal membrane, such as proton efflux and calcium release (Lv et al., 2024; Meng et al., 2024). These events disrupt the membrane potential, causing conformational changes in the V-ATPase complex, which then promotes interaction with ATG16L1. This interaction results in the recruitment of the ATG5–ATG12–ATG16L1 complex to the site of damage, thereby initiating ATG8ylation formation (Eguchi et al., 2024).

In addition to this V-ATPase–ATG16L1–dependent mechanism, a parallel pathway has recently been identified in which SM proteins and the autophagy adaptor TECPR1 recruit the ATG5 complex to damaged membranes. This alternative route appears to operate independently of V-ATPase signaling and may be selectively engaged depending on the type or context of lysosomal injury (Florey, 2023).

Taken together, ATG8ylation operates as a modular, evolutionarily conserved response platform that integrates multiple upstream signals to execute a temporally ordered series of events—including early ESCRT recruitment, damage containment via stress granules, selective clearance through microlysophagy, homeostatic reprogramming through TFEB activation, and membrane remodeling via LRRK2 engagement. These coordinated functions reinforce the concept of ATG8ylation not merely as a variant of autophagy, but as a specialized and indispensable mechanism for preserving lysosomal integrity.

Notably, cells also utilize ATG8ylation-independent complementary mechanisms to stabilize damaged lysosomes and maintain homeostasis. These include localized structural reorganization and environmental adaptation mediated by Annexins, SLC12A9, and ATG9A. A detailed discussion of these auxiliary pathways will follow in the next Section (3.2) (Fig. 2).

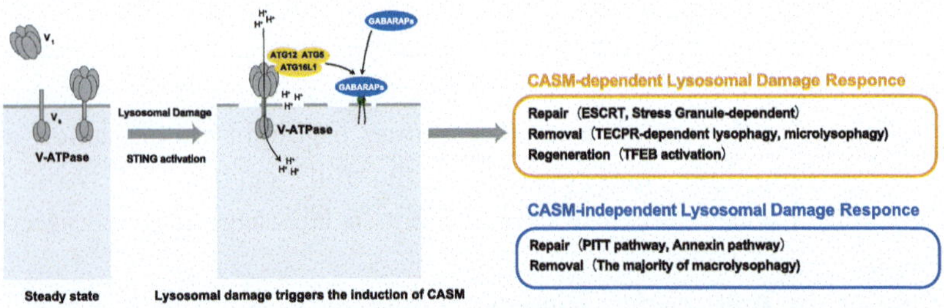

Fig. 2 ATG8ylation-dependent and -independent lysosomal damage response mechanisms. Upon lysosomal damage, activation of STING leads to proton leakage from the lysosomal membrane, resulting in changes in membrane potential, triggering a conformational change in V-ATPase and promoting its direct interaction with ATG16L1. Consequently, the ATG16L1–ATG5–ATG12 complex is recruited to the lysosomal membrane, leading to lipidation of ATG8. This process represents a non-canonical form of ATG8ylation, which is mechanistically distinct from classical autophagy. Induction of ATG8ylation serves as a trigger for subsequent homeostatic responses, including membrane repair via the ESCRT complex or stress granules, lysosomal clearance through TECPR1-mediated lysophagy and microlysophagy, and lysosome regeneration through TFEB activation.

3.2 Membrane sealing of damaged lysosomes

Lysosomes are highly acidic organelles packed with hydrolytic enzymes, and any compromise in their membrane integrity can trigger catastrophic cellular consequences. A variety of intrinsic and extrinsic stressors—such as aging, lipotoxicity, mechanical stress, and pathogen infection—can inflict damage on lysosomal membranes. A key event that signals such damage is the abrupt leakage of calcium ions (Ca^{2+}) from the lysosomal lumen into the cytosol, initiating a rapid and highly coordinated damage response.

One of the earliest steps in this response is the activation of protein kinase R (PKR), which leads to the phosphorylation of eukaryotic initiation factor 2α (eIF2α), thereby inducing a transient suppression of global translation. As a result, RNA-binding proteins such as G3BP1 and TIA-1 condense with stalled mRNAs and translation factors to form stress granules (SGs) (Duran et al., 2024; Jayabalan et al., 2025). While traditionally viewed as sites of translational repression, SGs are now recognized to play a direct role in lysosomal damage containment. Recent evidence shows that SGs are recruited to sites of lysosomal rupture, where they form "molecular patches" that physically plug membrane breaches prior to the recruitment of canonical repair machinery such as ALG-2 and ESCRT (Bussi et al., 2023).

These SG-based patches act as a sentinel response that limits the spread of lysosomal injury. Furthermore, lipidation of ATG8 family proteins (a process known as ATG8ylation) anchors SG components such as NUFIP2 and G3BP1 to the lysosomal membrane, suppressing local mTORC1 activity and promoting TFEB dephosphorylation and nuclear translocation (Jia et al., 2022). Thus, SGs play multifaceted roles beyond translation repression—they contribute to membrane stabilization and set the stage for lysosomal renewal programs.

In cases where membrane damage is too extensive to be contained by SGs alone, a more structurally robust and sustained sealing mechanism is required. The annexin family of Ca^{2+}-binding proteins provides this, particularly annexin A1 and A2. These proteins rapidly localize to damaged membranes via interactions with negatively charged phospholipids such as phosphatidylserine and phosphatidylinositol. Annexin A1, in particular, forms lateral arrays that generate a temporary "membrane cap" over rupture sites, maintaining local structural integrity and providing a scaffold for downstream repair processes (Yim et al., 2022).

Although both SGs and annexins respond to the same Ca^{2+} signal and localize to sites of lysosomal injury, they operate via distinct modalities. SGs serve as reversible, signaling-integrative structures that modulate translation and initiate early damage responses. In contrast, annexins offer mechanical stabilization through direct membrane association, functioning as physical supports until more definitive repair pathways are engaged.

Together, SGs and annexins form a tiered membrane-sealing system that adapts to the extent and context of lysosomal damage. By temporarily containing membrane rupture, this system prevents cytotoxic leakage and maintains cellular viability. Moreover, it creates a permissive environment for the activation of subsequent repair and regeneration systems, which will be discussed in the following sections. Understanding these early containment strategies not only elucidates the first line of defense against lysosomal damage but also highlights how membrane integrity is intricately tied to cellular homeostasis.

3.3 Repair of damaged lysosomes

Damage to the lysosomal membrane represents a critical cellular threat that compromises organelle integrity and overall homeostasis. In response, cells initiate a rapid and multi-tiered repair program aimed at restoring membrane continuity and reestablishing lysosomal function. Among the earliest and most conserved components of this repair response are two distinct molecular systems: the ESCRT (endosomal sorting complex required for transport) machinery and the PITT (phosphoinositide-initiated membrane tethering and lipid transport) pathway (López-Jiménez et al., 2018; Radulovic et al., 2018; Tan & Finkel, 2022). These systems act locally at sites of injury and precede lysophagic clearance or lysosome biogenesis, serving as the first line of defense against lysosomal membrane permeabilization (LMP). The following sections outline the distinct mechanisms and roles of these repair strategies.

3.3.1 ESCRT-mediated membrane repair: Reclosure of membrane discontinuities

The ESCRT machinery is a well-characterized, evolutionarily conserved membrane remodeling system originally implicated in intraluminal vesicle formation, cytokinesis, and viral budding. Recent studies have extended its functional repertoire to include membrane repair following lysosomal damage (Skowyra et al., 2018).

Upon LMP, calcium efflux from the lysosomal lumen into the cytosol acts as a damage signal, activating the calcium-binding protein ALG-2.

ALG-2, in turn, recruits ALIX and canonical ESCRT-I/II complexes, which nucleate the assembly of ESCRT-III subunits such as CHMP4B at the lesion site (Scheffer et al., 2014). ESCRT-III filaments physically bridge the edges of the ruptured membrane, enabling membrane reapproximation. This repair process is dynamically regulated by the AAA+ ATPase VPS4, which remodels ESCRT-III assemblies and facilitates membrane scission and closure (Shukla et al., 2022).

Notably, the Parkinson's disease-associated kinase LRRK2 has been shown to modulate this repair response by promoting the recruitment of CHMP4B via Rab8A, suggesting that ESCRT-mediated repair is tightly regulated by signaling pathways linking lysosomal stress to membrane restoration (Herbst et al., 2020).

3.3.2 PITT pathway: Lipid remodeling through ER–lysosome contacts

Parallel to the physical resealing by ESCRT, the PITT pathway contributes to membrane stabilization by orchestrating lipid exchange between the endoplasmic reticulum (ER) and the damaged lysosome (Tan & Finkel, 2022). This pathway is initiated by the rapid recruitment of PI4K2A to the lysosomal membrane, where it synthesizes phosphatidylinositol 4-phosphate (PI4P), generating a localized lipid signaling domain.

This PI4P domain recruits lipid transfer proteins of the OSBP-related protein (ORP) family—including ORP9, ORP10, ORP11, and OSBP—which establish membrane contact sites between the ER and the lysosome. These ORPs mediate the bidirectional exchange of PI4P with phosphatidylserine (PS) and cholesterol, reinforcing the damaged membrane by restoring its lipid composition. In addition, the lipid transfer protein ATG2 is recruited in a PS-dependent manner, further promoting the recovery of lysosomal membrane lipid homeostasis.

Unlike the ESCRT machinery, which restores membrane continuity through mechanical rejoining, the PITT pathway facilitates metabolic reinforcement of the membrane, thereby supporting the early recovery of lysosomal function.

3.3.3 Membrane and functional reorganization: Toward full lysosomal restoration

Following the initial repair phase, cells enter a more advanced stage aimed at the structural and functional reconstruction of the lysosome. This process ensures not only the re-establishment of membrane integrity but also the restoration of lysosomal degradative capacity and ionic homeostasis.

A central player in this phase is ATG9, a transmembrane protein known for its role in autophagy but now recognized for its ability to localize independently to damaged lysosomes and facilitate membrane supply and remodeling (Peng et al., 2025; Wang & Baehrecke, 2025). ATG9 function fine-tuned by ARFIP2 modulates the local PI4P concentration to create a lipid environment conducive to membrane restructuring (De Tito et al., 2025).

Additionally, the VPS13C–Rab10 axis emerges as a key regulatory module in lysosomal remodeling, particularly in neuronal cells. VPS13C interacts with Rab10 and, under the control of LRRK2-mediated phosphorylation, influences membrane dynamics essential for functional restoration (Schröder et al., 2024).

Beyond membrane repair, the maintenance of lysosomal osmotic and ionic balance is critical. The lysosomal transporter SLC12A9 plays a pivotal role in this context by mediating the efflux of NH_4^+ and Cl^-, thereby preventing pathological lysosomal swelling and ensuring functional recovery even after structural repair has been completed (Levin-Konigsberg et al., 2025).

Together, ATG9, ARFIP2, VPS13C, and SLC12A9 constitute a coordinated network that governs lysosomal reconstitution beyond physical repair. These factors act downstream or in parallel with ESCRT and PITT systems, enabling long-term maintenance of lysosomal homeostasis.

Should this reconstitution process prove insufficient, cells may transition to quality control mechanisms such as lysophagy or lysosomal exocytosis, or initiate TFEB-dependent lysosomal biogenesis to replenish the network. Thus, membrane repair and post-repair remodeling form a continuum that not only preserves organelle function but ultimately determines the fate of the damaged lysosome and the cell itself.

3.4 Removal of damaged lysosomes

As described above, lysosomal membrane damage (LMD) represents a serious cellular threat due to the leakage of hydrolytic enzymes and the activation of inflammatory and cell death pathways. To maintain lysosomal homeostasis, cells initiate a multilayered response that progresses from initial containment and membrane repair to selective elimination and eventual regeneration.

The initial response includes the recruitment of Annexins and stress granules to locally sequester the damaged site, followed by repair through the ESCRT and PITT complexes. If these efforts fail to restore lysosomal integrity, the cell transitions to a damage elimination phase, selectively removing irreparably damaged lysosomes from the intracellular environment. This elimination phase consists of three principal pathways: macrolysophagy, microlysophagy, and lysosomal exocytosis. Each pathway contributes uniquely to lysosomal quality control, and the degree of damage, repair feasibility, and cellular context determine their activation.

3.4.1 Removal of damaged lysosomes via macrolysophagy

Macrolysophagy is a selective form of autophagy in which damaged lysosomes are targeted for degradation via autophagosomes, thereby restoring cellular homeostasis. This process is tightly orchestrated through a multilayered cascade of molecular events, encompassing damage detection, tagging, sequestration, degradation, and ultimately organelle regeneration.

The most well-established trigger for lysophagy involves the cytosolic exposure of glycan moieties following lysosomal membrane rupture. These are detected by the galectin family, particularly Galectin-3 and Galectin-8, which initiate downstream signaling. Subsequently, E3 ubiquitin ligase complexes such as SCF^{FBXO2}, SCF^{FBXO27}, and CUL4A–DDB1 ubiquitinate lysosomal membrane proteins (e.g., LAMP2) with K48- or K63-linked ubiquitin chains, effectively marking the damaged lysosome for selective clearance (Koerver et al., 2019; Liu et al., 2020; Teranishi et al., 2022). FBXO2 and FBXO27, possessing glycan-recognition domains, are thought to confer specificity to this tagging process (Yoshida et al., 2017).

In parallel, Galectin-3 promotes the ubiquitination of ULK1 and ATG16L1 via TRIM16, thereby facilitating the initiation of isolation membrane formation (Chauhan et al., 2016). Thus, the interplay among galectins, ubiquitin ligases, and autophagy adaptors forms a pivotal regulatory axis in the early phase of lysophagy.

The composition and abundance of ubiquitin chains also critically influence pathway specificity. In particular, excess K48-linked chains may shift the fate of damaged lysosomes toward proteasomal degradation. To counteract this, the ELDR (endo-lysosomal damage response) complex, comprising the deubiquitinase YOD1 and the segregase VCP/p97, selectively removes K48 chains, thereby preserving autophagic progression (Papadopoulos et al., 2017).

Following ubiquitination, adaptor proteins such as p62/SQSTM1, NDP52, OPTN, and TAX1BP1 are recruited to the lysosomal surface. These adaptors bridge ubiquitinated cargo and LC3 on the isolation membrane, enabling selective encapsulation of damaged lysosomes. TAX1BP1, in particular, is activated via the TBK1–SCF^FBXO3–TMEM192 axis and serves as a central scaffold for autophagosome nucleation (Park et al., 2025).

Of note, CCDC50 has recently emerged as a novel lysophagy adaptor capable of simultaneously recognizing Galectin-3 and K63-linked ubiquitin chains (Jia et al., 2023). Unlike canonical adaptors such as p62 or OPTN, CCDC50 is structurally and functionally distinct, underscoring the diversity and cell-type specificity of the lysophagic machinery.

Significantly, the functionality of these adaptors is dynamically modulated by stress-responsive signaling. The p38 MAPK–HSP27–p62 axis, for example, responds to oxidative stress and mTORC1 inactivation by modulating the phase separation properties of p62, thereby enhancing its interaction with ubiquitin aggregates and autophagic membranes (Gallagher & Holzbaur, 2023; Gallagher et al., 2024). This axis exemplifies the crosstalk between stress responses and selective autophagy.

In terms of spatial regulation, the actin cytoskeleton also plays a crucial role in lysophagy. The actin-stabilizing protein CNN2 localizes to the surface of damaged lysosomes and promotes the nucleation of isolation membranes. CNN2 is subsequently ubiquitinated and removed via a VCP/p97-dependent mechanism, facilitating actin remodeling and membrane elongation (Kravić et al., 2022).

Beyond the canonical ubiquitin-dependent pathways, a distinct ubiquitin-independent mechanism for lysophagy has also been identified. The TECPR1–ATG5–ATG12 complex directly senses exposed membrane lipids such as glycosphingolipids and sphingomyelin on damaged lysosomes and triggers ATG8 lipidation without the need for upstream ubiquitination (Boyle et al., 2023). This alternative pathway enables the detection and clearance of subtle or atypical lysosomal injuries that may evade galectin- or ubiquitin-based sensors, thereby contributing to the robustness and redundancy of the lysophagy system.

Strikingly, recent findings have revealed that lysosomal damage may also be detected *prior* to ESCRT recruitment. SPART and SPG20, two factors traditionally unrelated to lysophagy, have been shown to recognize lipid packing defects on damaged membranes and activate the E3 ligase ITCH, thereby initiating K63-linked ubiquitination upstream of ESCRT assembly (Gahlot et al., 2024). This discovery challenges the long-held notion that ESCRT acts first in the damage response hierarchy and suggests a revised model of early detection and decision-making in lysosomal quality control.

Finally, lysosome clearance and regeneration are not discrete processes but are often executed in a sequential and coordinated manner. LIMP2, for instance, functions both as a lysophagy receptor and a regulator of spatial segregation by interacting with the Rab7 GAP TBC1D15. It also promotes the recruitment of autophagic lysosome reformation (ALR) components such as Dynamin2, Kinesin5B, and Clathrin (Bhattacharya et al., 2023). This dual functionality exemplifies an integrated strategy wherein damage removal is coupled with lysosomal biogenesis.

In summary, macro-lysophagy constitutes a sophisticated homeostatic mechanism that flexibly integrates multiple damage-sensing, tagging, sequestration, degradation, and regeneration pathways. By deploying such a multilayered and context-adaptive system, cells are able to finely tune their response to lysosomal injury, ensuring the preservation of organelle function and cellular viability.

3.4.2 Clearance of damaged lysosomes via micro-lysophagy

Micro-lysophagy involves the inward budding of the lysosomal membrane to form intralumenal vesicles (ILVs), which sequester and degrade membrane-associated components. This mechanism plays a central role not only in membrane protein turnover but also in the structural remodeling of lysosomes in response to stress, contributing to the preservation of lysosomal homeostasis.

The fundamental architecture of micro-lysophagy is structurally analogous to the formation of multivesicular bodies (MVBs) in endosomes. It is initiated by ubiquitination of membrane proteins and proceeds through the stepwise recruitment of ESCRT (Endosomal Sorting Complex Required for Transport) complexes, ultimately driving ILV formation. This evolutionarily conserved mechanism is broadly recognized as a quality control system for lysosomal membranes across species ranging from yeast to mammals (Li et al., 2015a; Li et al., 2015b; Zhang et al., 2021; Zhu et al., 2017).

Recent studies have revealed that micro-lysophagy functions not only as a basal maintenance mechanism but also as a flexible adaptive response to cellular stress. Under starvation, for example, calcium release from lysosomes triggers activation of ALG-2 at ER exit sites (ERES), which subsequently recruits membrane remodeling factors such as ALIX and ESCRT-III. This process suggests the activation of a dynamic micro-autophagy reaction at organelle contact sites and indicates that micro-lysophagy participates in membrane reorganization during lysosomal stress (Liao et al., 2024).

A particularly noteworthy advance is the emerging role of the autophagy-related protein ATG8 (LC3/GABARAP family) in micro-lysophagy. Traditionally regarded as essential for autophagosome formation during macroautophagy, ATG8 has now been shown to also play a pivotal role in micro-lysophagy. Through covalent conjugation with membrane lipids, a process known as ATG8ylation, ATG8 gains membrane localization and functions as a spatial and functional "code" for membrane identity and selection (Deretic & Lazarou, 2022).

While ATG8ylation typically targets phosphatidylethanolamine (PE), recent findings have demonstrated that it can also occur with phosphatidylserine (PS) (Durgan et al., 2021). This lipid conjugation versatility is thought to be dynamically regulated by the membrane environment and cellular stress, influencing the duration and specificity of ATG8 membrane association. In particular, PS conjugation may serve a compensatory and context-specific role under conditions such as lysosomal membrane damage, where PE content is reduced.

Moreover, ATG8ylation has been shown to promote the recruitment and activation of ESCRT complexes. Membrane-localized ATG8 facilitates the assembly of ESCRT-III subunits such as CHMP4B, thereby enhancing ILV formation (Ogura et al., 2023). This finding adds a new layer to the previously ubiquitin-dependent model of ESCRT activation, positioning ATG8ylation as a complementary or alternative membrane code that functions in concert with or independently from ubiquitin signaling.

Taken together, micro-lysophagy is far more than a passive protein degradation pathway—it represents a highly orchestrated system in which ubiquitination, ATG8ylation, and ESCRT activation are temporally and spatially integrated to achieve selective membrane clearance and repair. Especially under stress conditions, the flexible interplay of these pathways enables rapid and precise membrane remodeling, thus ensuring the functional integrity of lysosomes.

Future investigations are needed to elucidate the molecular determinants that govern the switch between ubiquitin- and ATG8ylation-based recognition systems and to determine how these signals cooperatively regulate micro-lysophagy in a context-dependent and spatiotemporally coordinated manner.

3.4.3 Removal of damaged lysosomes via lysosomal exocytosis

Lysosomal exocytosis serves as an alternative mechanism to remove irreversibly damaged lysosomes by expelling their contents into the extracellular space. Upon severe damage, initial responses such as membrane sealing of damaged lysosomes, ESCRT/PITT-mediated repair, and ATG9-dependent remodeling may be insufficient. In such cases, the cell redirects damaged lysosomes for exocytosis.

Lysosomal exocytosis involves the fusion of damaged lysosomes with the plasma membrane, enabling the expulsion of their contents into the extracellular space. While initially characterized as a calcium-dependent constitutive process involved in cellular homeostasis, recent studies suggest that this pathway is also activated in response to lysosomal injury, functioning as a damage-adaptive mechanism.

At the center of the transcriptional regulation of this pathway is TFEB (transcription factor EB). In response to lysosomal stress, TFEB translocates to the nucleus and activates the expression of a broad set of lysosomal genes through CLEAR (Coordinated Lysosomal Expression and Regulation) elements. Importantly, this transcriptional program includes factors involved in lysosomal exocytosis, indicating that TFEB governs not only the reinforcement of degradative capacity but also the selective extrusion of dysfunctional lysosomes from the cell (Spampanato et al., 2013).

In addition, Connexin43, traditionally known as a component of gap junctions, has recently emerged as a key effector of lysosomal exocytosis. Upon lysosomal damage, Connexin43 facilitates actin cytoskeleton remodeling, promoting the recruitment of damaged lysosomes to the plasma membrane and enhancing membrane fusion and content release (Domingues et al., 2024). This pathway functions independently of ESCRT-mediated repair and is particularly important for the clearance of lysosomes that are deemed irrecoverable based on the extent of damage or accumulation.

Thus, lysosomal exocytosis represents not merely a waste disposal system but a highly regulated and selective elimination mechanism governed by transcriptional and cytoskeletal factors such as TFEB and Connexin43. As an integral component of lysosomal quality control, this pathway enables the preservation of lysosomal homeostasis under stress conditions. Its dysregulation may contribute to pathological processes including inflammation, neurodegeneration, and cancer, highlighting its significance as a potential therapeutic target.

3.5 Lysosome biogenesis via TFEB activation

When cells encounter lysosomal damage or dysfunction, they must not only eliminate the defective organelles but also promptly and efficiently reconstruct new ones to maintain intracellular homeostasis. At the heart of this restorative process is the transcription factor TFEB (Transcription Factor EB), which acts as the central regulator of the lysosomal gene expression program, orchestrating adaptive responses from damage sensing to homeostatic recovery.

The earliest trigger of this pathway is the release of calcium (Ca^{2+}) from the lumen of damaged lysosomes, a process mediated by the activation of the TRPML1 channel on the lysosomal membrane. The resultant increase in cytosolic Ca^{2+} activates the phosphatase calcineurin, which in turn dephosphorylates cytosolic TFEB, enabling its translocation into the nucleus. Once in the nucleus, TFEB binds to CLEAR (Coordinated Lysosomal Expression and Regulation) motifs and promotes the expression of a broad array of lysosomal and autophagy-related genes (Medina et al., 2015; Sardiello et al., 2009). This transcriptional program triggers not only lysosome replenishment but also global metabolic and structural reprogramming.

Recent discoveries have highlighted the membrane-associated modification "ATG8ylation" as a key process tightly linked to TFEB activation: it involves lipid conjugation of ATG8 family proteins, especially GABARAPs, to the lysosomal membrane, triggered by lysosomal damage. Structural rearrangement of V-ATPase upon stress recruits ATG16L1, which catalyzes ATG8 lipidation. The membrane-tethered ATG8s serve as a molecular platform to enhance calcineurin activity and sequester the TFEB inhibitor complex FLCN–FNIP, thereby suppressing mTORC1 activity and promoting sustained TFEB activation and nuclear import (Goodwin et al., 2021; Kumar et al., 2020; Ogura et al., 2023). Thus, ATG8ylation provides spatial and functional specificity for TFEB activation at sites of lysosomal injury.

In addition to these signaling cascades, TFEB is also subject to metabolic regulation. For example, the immunometabolite itaconate, produced during bacterial infection, covalently modifies Cys212 of TFEB, disrupting its interaction with 14–3-3 proteins and thereby facilitating nuclear translocation (Zhang et al., 2022). Moreover, under chronic stress conditions such as hypoxia or persistent metabolic burden, intracellular lactate accumulation induces TFEB lactylation, a post-translational modification that impedes TFEB ubiquitination and degradation (Huang et al., 2024b). This results in prolonged TFEB stability and transcriptional activity, allowing cells to maintain lysosomal biogenesis over extended periods. Lactylation thus appears to represent a physiological adaptation that equips cells to endure sustained lysosomal stress by ensuring long-term homeostatic capacity.

Nevertheless, TFEB activity must be tightly regulated, as uncontrolled lysosomal biogenesis could lead to unnecessary energy expenditure or abnormal cell proliferation. To prevent such maladaptive outcomes, TFEB-induced transcription factors such as BHLHE40/41 form a negative feedback loop to suppress TFEB activity in the nucleus (Carey et al., 2020). Furthermore, completion of lysosomal reconstitution via ESCRT-I and reactivation of mTORC1 signaling serve to attenuate TFEB activation progressively (Wróbel et al., 2022).

In summary, TFEB activation is not a simple on-off switch, but a highly orchestrated and multi-layered cellular strategy. It integrates damage sensing through calcium signaling, amplification via ATG8ylation, metabolic modulation through lactylation, and finely tuned transcriptional control followed by feedback inhibition. This elaborate regulatory framework enables cells to execute lysosomal biogenesis with spatial and temporal precision, thereby restoring homeostasis and ensuring adaptive resilience under both acute and chronic stress conditions.

3.6 Homeostatic regulation through lysosomal fusion and fission

Lysosomes, as central organelles in cellular catabolism, are not static entities but dynamically remodel their number, size, and structure in response to environmental and metabolic demands. This dynamic plasticity is primarily mediated through two opposing yet coordinated processes: fusion and fission. These structural transformations are not merely physical events but represent tightly regulated adaptive strategies that enable the cell to maintain homeostasis under fluctuating conditions.

Lysosomal fusion occurs primarily through interactions with autophagosomes and endosomes, enabling the degradation of intracellular cargo such as misfolded proteins, damaged organelles, and exogenous particles. Fusion events not only allow lysosomes to expand their content capacity but also optimize their internal environment for enzymatic digestion. Under nutrient deprivation or increased catabolic stress, enhanced fusion facilitates the formation of larger, more active lysosomes capable of meeting the cell's increased degradative demands.

In contrast, lysosomal fission functions to regenerate smaller, functionally competent lysosomes from larger ones, particularly following prolonged autophagic flux. This process is exemplified by autophagic lysosome reformation (ALR), which replenishes the lysosomal pool after extensive autophagic activity, and is regulated by the mTORC1 pathway

(Yu et al., 2010). Fission ensures the maintenance of lysosomal capacity without excessive organelle enlargement, preserving cellular architecture and functional balance.

For many years, the molecular mechanisms controlling lysosomal fission remained elusive. However, recent studies have identified HPO-27, a HEAT repeat–containing protein, as a key regulator of lysosomal fission in response to nutritional cues (Li et al., 2024). Moreover, membrane tension across the lysosome can influence this remodeling process: increased tension leads to closure of the TRPML1 channel, reducing calcium efflux and ultimately suppressing TFEB activation and lysosomal biogenesis (Cai et al., 2024).

Importantly, fusion and fission do not operate in isolation and are tightly integrated with broader homeostatic networks such as autophagy, nutrient sensing, and stress responses. For instance, during lysophagy—the selective autophagic removal of damaged lysosomes—fission may facilitate the fragmentation of damaged compartments, improving the efficiency of recognition and degradation (Gahlot et al., 2024; Li et al., 2024; Toifl et al., 2025).

Recent findings have also revealed a mechanosensitive link between lysosomal morphology and transcriptional activation: transitions from tubular to spherical lysosomes can trigger TFEB activation, thereby promoting lysosomal biogenesis (Cai et al., 2024), suggesting that structural remodeling of lysosomes itself serves as a signal that tunes gene expression to reinforce lysosomal homeostasis.

Together, lysosomal fusion and fission constitute an adaptive morpho-functional feedback system, so the lysosomal network can flexibly expand or renew itself in response to cellular demands. Disruption of this balance may underlie the pathogenesis of lysosome-related disorders and aging-associated degenerative diseases, making the regulation of lysosomal dynamics an increasingly critical area of investigation and therapeutic interest.

4. The significance of lysosomes in human diseases

Cells face continuous environmental fluctuations and endogenous stress throughout their lifespan. The ability to maintain intracellular homeostasis under such conditions is a fundamental determinant of cellular integrity and survival. In this context, lysosomes have gained recognition not only as terminal degradative compartments but also as multifunctional organelles involved in diverse cellular processes, including metabolism, signal transduction, membrane trafficking, autophagy, and immune regulation.

Recent advances have highlighted the central role of lysosomes in integrating these processes, positioning them as key hubs in the cellular stress response. Among these functions, the maintenance of lysosomal quality—through mechanisms that enable repair, selective degradation, and regeneration of damaged lysosomes—has emerged as a critical factor influencing the onset and progression of various diseases.

Disruption of lysosomal homeostasis affects a wide range of pathological conditions, including aging, neurodegenerative diseases, cancer, metabolic syndromes, and immune disorders. In many of these contexts, failure to efficiently remove damaged lysosomes contributes to impaired proteostasis, metabolic imbalance, and sustained inflammatory signaling.

Importantly, lysosomal dysfunction is not merely a consequence but often a driving force in disease pathogenesis. For example, the accumulation of damaged lysosomes can interfere with autophagic flux, activate stress response pathways, and alter inter-organelle communication. These events may progressively exacerbate cellular dysfunction and tissue pathology.

Given this growing body of evidence, lysosomes have become promising therapeutic targets. Understanding the molecular mechanisms that govern lysosomal repair, turnover, and biogenesis may provide novel opportunities for disease intervention. The following sections explore how lysosomal integrity is maintained or compromised in specific disease contexts, and how modulating lysosomal function might contribute to restoring cellular and tissue homeostasis.

4.1 Importance of lysosomes in aging

Aging is a progressive biological process in which cells gradually lose their ability to maintain internal homeostasis, ultimately resulting in functional decline. In response to this loss, cells mobilize multilayered adaptive mechanisms to sustain survival. At the core of these mechanisms are systems such as autophagy, DNA repair, and protein quality control. In recent years, lysosomes have emerged as pivotal players in aging, functioning alongside these canonical pathways. Once regarded merely as passive degradative compartments, lysosomes have now become active regulatory organelles that both respond to and orchestrate the aging process.

For example, under stress conditions such as nutrient deprivation, lysosomes undergo morphological transformation from spherical to tubular forms, a change that enhances their degradative capacity and helps sustain autophagic flux (Villalobos et al., 2023). The formation of such "tubular lysosomes" is not a mere structural anomaly but an adaptive reconstruction linked to increased metabolic flexibility and extended lifespan. Among the key regulators of lysosomal adaptation are the MiT/TFE family transcription factors, particularly TFEB, which coordinate the expression of genes related to lysosomes and autophagy, thereby maintaining homeostasis. It has been shown, for instance, that the neurotrophic factor MANF localizes to lysosomes and activates TFEB, leading to improved proteostasis and lifespan extension (Taylor et al., 2024).

Moreover, pharmacological interventions that modulate lysosomal function have gained attention as realistic anti-aging strategies. A representative example is low-dose metformin, which activates the AMPK–TFEB axis by modulating V-ATPase activity through the lysosomal protein PEN2, thereby promoting longevity (Ma et al., 2022).

Within this context, a newly identified mechanism termed the lysosomal surveillance response has attracted considerable interest. In this response, cells sense V-ATPase dysfunction caused by aging or metabolic stress and activate transcription factors such as GATA4/6 and TFEB, promoting the regeneration of damaged lysosomes and restoration of lysosomal homeostasis (Li et al., 2025). This mechanism not only serves as a repair response to organelle damage but also contributes to the acquisition of stress resistance and longevity.

In addition, recent findings reveal that lysosomes may function as part of an endocrine network that regulates systemic aging. Specifically, long-chain fatty acids generated through lipid metabolism in adipose tissue lysosomes act as signaling molecules that influence the nervous system, extending organismal lifespan via NHR-49 and NLP-11 pathways (Savini et al., 2022). This discovery highlights a novel role for lysosomes beyond intracellular catabolism: they function as endocrine components that modulate whole-body aging.

On the other hand, lysosomes themselves are susceptible to structural and functional deterioration with advancing age. In senescent cells, despite an increase in lysosome number, their quality often declines, as evidenced by reduced acidity, decreased hydrolase activity, and compromised membrane integrity (Tan & Finkel, 2023).

Such lysosomal dysfunction contributes to the establishment and persistence of the senescence-associated secretory phenotype (SASP), which drives chronic inflammation and tissue aging. For example, stabilization of GATA4 during cellular senescence induces ABCA1 expression, leading to cholesterol accumulation within lysosomes. This accumulation aberrantly activates mTORC1 and perpetuates the production of pro-inflammatory cytokines and SASP factors (Roh et al., 2023).

Furthermore, lysosomes altered by aging actively provide catabolic products like amino acids, serving as nutritional resources for the biosynthesis and secretion of SASP components. Some studies also indicate that the lysosomal proteome itself may directly contribute to the composition of SASP, positioning lysosomes as potential "inducer platforms" for SASP formation (Rovira et al., 2022).

Notably, TFEB, while generally regarded as a protective factor, may paradoxically exacerbate SASP under certain conditions. For instance, TFEB nuclear translocation induced by CDK4 inhibition increases lysosome numbers, which unexpectedly promotes SASP production in senescent cells (Curnock et al., 2023). These findings emphasize the dualistic and dynamic nature of lysosomes, which can both suppress and drive aging processes depending on context.

Therefore, the essence of aging regulation lies not in simply increasing lysosome number or activity, but in ensuring their qualitative integrity. Of particular importance is the ability to accurately identify and selectively remove damaged lysosomes through lysophagy. This selective autophagic process, involving damage recognition via galectins and TRIM proteins, recruitment of ATG8, and engagement of ESCRT or CASM complexes, plays a critical role in maintaining cellular fitness and suppressing aging (Henn et al., 2025).

In addition, senescent cells may evade ferroptotic cell death by sequestering iron within lysosomes, which has been implicated in resistance to senolytic therapies (Feng et al., 2024). Thus, strategies that inhibit ferritinophagy or destabilize lysosomal membranes are being explored as novel approaches to eliminate senescent cells.

In summary, the fundamental key to preserving cellular youthfulness lies not in the quantity or activity of lysosomes alone, but in the precise regulation of their quality. Advancing our understanding of lysosomal dynamics, plasticity, and limitations—through spatiotemporal visualization and functional analysis—will open new avenues for preventing age-associated diseases and extending healthy lifespan.

4.2 The importance of lysosomes in neurodegenerative diseases

Neurodegenerative diseases are characterized by the irreversible loss of neuronal structure and function, often manifesting as progressive cognitive decline or motor deficits. Neurons are exceptionally susceptible to disruptions in intracellular homeostasis due to their post-mitotic nature, high metabolic demand, and complex morphology. Among the various organelles that safeguard cellular homeostasis, lysosomes have emerged as a central player.

Lysosomes not only degrade waste materials and dysfunctional organelles via autophagy and endocytosis but also coordinate organelle quality control, nutrient sensing, and intracellular signaling. In neurons, which must retrieve and process cellular components from distal axonal or dendritic terminals, precise spatial distribution and maturation of lysosomes are critical for maintaining cellular integrity (Nixon & Rubinsztein, 2024).

However, with aging or under pathological stress, lysosomal acidification, membrane permeability, and enzymatic function become compromised, leading to impairment of their degradative capacity, resulting in the accumulation of undegraded substrates and organelles, ultimately disrupting neuronal proteostasis. In Alzheimer's disease, for instance, metabolic products of the amyloid precursor protein (APP) impair v-ATPase function, leading to defective lysosomal acidification and failure to degrade β-amyloid (Aβ), thereby contributing to plaque formation (Im et al., 2023; Lee et al., 2022). Similarly, in Parkinson's disease, dysfunction of lysosomal ion channels such as TMEM175 disrupts proton homeostasis and promotes α-synuclein aggregation (Hu et al., 2022). Moreover, astrocytic lysosomal impairment caused by DJ-1 deficiency accelerates α-synuclein accumulation in neighboring neurons, compromising proteostasis across the midbrain organoid environment (Morrone Parfitt et al., 2024).

These disorders share a common pathological feature—accumulation of damaged lysosomes and subsequent failure of cellular clearance systems. Damaged lysosomes, particularly those with membrane rupture, pose a significant threat by leaking hydrolytic enzymes or redox-active metals, triggering secondary oxidative and inflammatory damage. In this context, lysophagy—a selective autophagic mechanism for removing compromised lysosomes—acts as a critical cellular defense. For example, α-synuclein aggregates can induce lysosomal membrane rupture, facilitating cytosolic propagation and aggregation, whereas activation of lysophagy can mitigate such spreading (Kakuda et al., 2024).

Importantly, lysosomal function in glial cells is equally vital for neural health. In microglia, the RagA–FLCN–TFEB axis negatively regulates excessive lysosomal biogenesis and modulates inflammatory responses and migratory behavior (Iyer et al., 2022). In astrocytes, lysosome-derived calcium release is essential for neuron–glia communication and synaptic plasticity (Sun et al., 2022). These findings highlight lysosomes as indispensable organelles not only for neuronal survival but also for maintaining the functional integrity of the neuroglial network.

Furthermore, restoring lysosomal function may offer therapeutic potential for neurodegenerative diseases. Small-molecule activators of TFEB, a master regulator of lysosomal biogenesis, have been shown to ameliorate cognitive symptoms in Alzheimer's disease models (Lin et al., 2024), suggesting that enhancing lysosomal renewal and damaged lysosome clearance may delay disease progression.

Neurodegeneration may not simply result from neuronal loss but instead originate from "lysosomal failure." The accumulation of dysfunctional lysosomes, collapse of repair and removal pathways, and breakdown of quality control mechanisms form a pathological cascade leading to irreversible neuronal degeneration. Advancing our understanding of lysosomal quality control and damage response will be essential for uncovering novel therapeutic targets and achieving meaningful intervention in neurodegenerative disease.

4.3 The significance of lysosomes in cancer

Cancer cells undergo extensive reprogramming of intracellular homeostatic networks to sustain their distinctive characteristics, such as uncontrolled proliferation, enhanced stress resistance, and invasive or metastatic potential. At the core of this reorganization lies the lysosome, an organelle essential for intracellular degradation and metabolic regulation. Beyond serving as the endpoint of autophagy, lysosomes act as signaling hubs by engaging in crosstalk with nutrient-sensing pathways such as mTORC1, thereby enabling cancer cells to flexibly adapt to fluctuations in nutrient availability and the extracellular environment.

Recent studies have shown that perturbation of lysosomal integrity disrupts the survival strategies of cancer cells. Cationic amphiphilic drugs (CADs), for instance, preferentially accumulate within lysosomes and induce lysosomal membrane destabilization through aberrant accumulation of lysoglycerophospholipids, ultimately leading to cell death (Nielsen et al., 2024). These compounds also promote proton leakage across lysosomal membranes and suppress tumor growth by inhibiting the STAT3 signaling pathway (Liu et al., 2023). Moreover, decreased lysosomal activity has been linked to impaired autophagic flux and abnormal autophagosome accumulation, contributing to competitive elimination of cancer cells under cellular competition (Akter & Kon, 2023).

Cellular responses to damaged lysosomes are also crucial in regulating the motility and invasiveness of cancer cells. Recent findings indicate that liquid-liquid phase separation (LLPS) condensates formed by p62 and NBR1 promote the degradation of the small GTPase Rac1 near damaged lysosomes, thereby attenuating cancer cell migration and invasion (Noguchi et al., 2023). Such mechanisms of lysosomal damage sensing, selection, and degradation may function as intrinsic suppressors of cancer progression.

In hepatocellular carcinoma (HCC), lysosomes exhibit multifaceted roles in tumor dynamics. The lysosomal Ca^{2+} channel TRPML1 regulates energy metabolism through functional interactions with mitochondria, and its altered expression significantly affects the metabolic balance of tumor cells (Siow et al., 2022). In hepatic stellate cells, p62 promotes the ubiquitination and activation of STING via the E3 ubiquitin ligase TRIM32, facilitating anti-tumor immune responses. Conversely, NBR1 facilitates lysosome-dependent degradation of STING, contributing to immune evasion (Nishimura et al., 2024).

These findings underscore how cancer cells strategically exploit the functional diversity and plasticity of lysosomes to maintain survival and adaptability. However, disruption of lysosomal structural or functional homeostasis compromises critical processes such as metabolism, signal transduction, cellular motility, and immune regulation. Thus, targeting lysosomal quality control systems and damage response pathways presents a promising avenue for the development of cancer-specific therapeutic strategies and may drive innovation in future cancer treatment paradigms.

4.4 Importance of lysosomes in obesity-related diseases

The pathogenesis and progression of obesity-related metabolic disorders are deeply rooted in intracellular stress caused by lipid overload and the resultant failure of organelle homeostasis. Among these organelles, lysosomes have emerged as frontline responders to lipotoxic stress. Once considered mere degradative compartments, lysosomes are now recognized as central regulators of metabolic homeostasis, capable of dynamic "adaptive remodeling" in response to pathological stimuli.

In hepatocytes, excessive fatty acids damage lysosomal membranes, disrupt intraluminal pH regulation and enzyme activity, and impair cargo degradation. This triggers a vicious cycle: undegraded substrates accumulate, leading to further lipid deposition, mitochondrial dysfunction, and inflammation, thereby accelerating the progression of metabolic-associated steatotic liver disease (MASLD) (Pu, 2022). Interestingly, suppression of the lysosomal membrane protein LAMTOR1 reduces mTORC1 activity and promotes lysosomal biogenesis, contributing to hepatocellular protection under lipotoxic conditions (Jang et al., 2025).

Pancreatic β-cells are similarly vulnerable to lipid-induced lysosomal dysfunction. Lipid overload disrupts insulin granule turnover and autophagic organelle remodeling, resulting in impaired insulin secretion. In this context, restoring lysosomal function could improve β-cell resilience and secretory capacity under metabolic stress (Arden et al., 2024).

Proximal tubular epithelial cells in the kidney also experience lysosomal stress under diabetic or dyslipidemic conditions. Saturated fatty acids accumulate within lysosomes, compromise autophagic flux, and drive dedifferentiation and fibrotic remodeling. TFEB-mediated lysosomal reprogramming serves a cytoprotective role, highlighting the centrality of lysosomal quality control in diabetic kidney disease (Pierre et al., 2024).

In macrophages involved in atherosclerosis, lysosomes are again a critical point of vulnerability. Uptake of oxidized LDL and cholesterol transforms macrophages into foam cells, where accumulated lipids disrupt lysosomal integrity. Lysosomal rupture leads to cytosolic leakage of contents, triggering NLRP3 inflammasome activation and amplifying chronic inflammation. While lysophagy can remove damaged lysosomes, inadequate clearance accelerates plaque progression and destabilization (Bonacina et al., 2025).

These findings collectively underscore that lysosomal damage under lipotoxic stress is a shared feature across metabolically active organs and immune cells. Despite differing cellular contexts, the fundamental response remains conserved: detection of lysosomal injury, selective removal of dysfunctional compartments, and de novo lysosome biogenesis. In particular, adaptive remodeling mechanisms involving TFEB and AMPK represent a cross-tissue protective axis that mitigates cellular damage and preserves organ function.

When the capacity for lysosomal quality control is overwhelmed, disease progression becomes inevitable. Thus, targeting the dynamic regulation of lysosomal homeostasis—beyond simple lipid clearance—represents a promising strategy for treating obesity-related diseases. At the heart of this approach lies the understanding of how cells withstand chronic metabolic stress by constantly sensing, repairing, and renewing their lysosomal network.

4.5 The central role of lysosomes in immune regulation

The immune system is a highly regulated network that precisely distinguishes "non-self" entities such as pathogens and tumor cells from "self," eliminating the former while preserving the latter and preventing excessive inflammation. This remarkable balance relies heavily on the integrity of intracellular organelle function within immune cells. Among these organelles, lysosomes have garnered renewed attention not only as degradative compartments but also as hubs of metabolic control, antigen processing, cell death regulation, and immune signaling.

Recent findings have highlighted that lysosomal membrane permeabilization (LMP), which leads to lysosomal damage, can drastically reshape the metabolic and functional state of immune cells, including macrophages and T cells. In macrophages, LMP-induced mitochondrial remodeling promotes a pro-inflammatory phenotype (Bussi et al., 2022). Simultaneously, the leakage of double-stranded DNA from damaged lysosomes into the cytosol can chronically activate the cGAS–STING pathway, contributing to age-related sterile inflammation (inflammaging) (Wang et al., 2024a). In T cells, LMP results in the cytosolic release of mitochondrial DNA, disrupting immune homeostasis and fueling chronic inflammation (Jin et al., 2023).

Thus, LMP-mediated lysosomal damage represents a critical stress signal that reshapes the metabolic and functional identity of immune cells and may serve as a trigger for chronic inflammation and immune disorders. In response to such damage, cells deploy a multi-step quality control system, including lysosomal repair, selective clearance, and regeneration. These mechanisms are orchestrated in a coordinated and context-dependent manner, allowing immune cells to cope with varying degrees of lysosomal stress. The concerted action of these pathways is thought to be essential for preserving immune cell homeostasis and ensuring proper immune responses.

Interestingly, a recent study has revealed a noncanonical mechanism for the clearance of damaged lysosomes that bypasses LC3 conjugation. TAX1BP1 and FIP200 can recognize and target damaged lysosomes independently of LC3, recruiting a caspase-8 activation complex that mitigates TNFα-induced cell death (Huyghe et al., 2022). This mechanism suggests the existence of a lysosome-based immune regulatory axis and highlights the organelle's flexible capacity to adapt to inflammatory contexts.

When lysosomal homeostasis is disrupted, the consequences extend beyond innate and adaptive immunity, influencing tumor immunity and autoimmunity. Of particular interest is the emerging role of lysosomes in shaping modes of cell death. Inducing lysosomal membrane damage in cancer cells leads to the release of tumor antigens and damage-associated molecular patterns (DAMPs) such as ATP and HMGB1, which activate dendritic cells and elicit antitumor T cell responses. This cascade promotes immunogenic cell death—a form of cell death that enhances tumor clearance by the immune system. Notably, the antidepressant sertraline induces immunogenic cell death in cancer cells through lysosomal cholesterol accumulation and activation of LC3 lipidation and TFEB (Alvarez-Valadez et al., 2024). Conversely, mTOR–TFEB signaling has also been linked to impaired PD-L1 degradation, facilitating immune evasion in the tumor microenvironment (Wang et al., 2025). These findings underscore lysosomes as pivotal modulators of tumor–immune interactions.

Beyond cell death, lysosomal integrity also influences antigen presentation and the maintenance of immune memory. In tissue-resident $CD8^+$ T cells, TFEB-driven nutrient-sensing pathways are critical for cell differentiation and long-term persistence. Molecules such as Flcn and the Ragulator/Rag complex suppress nuclear translocation of TFEB, thereby acting as brakes on memory T cell formation (Raynor et al., 2024). These observations suggest that lysosomes function as metabolic niches that calibrate immune memory.

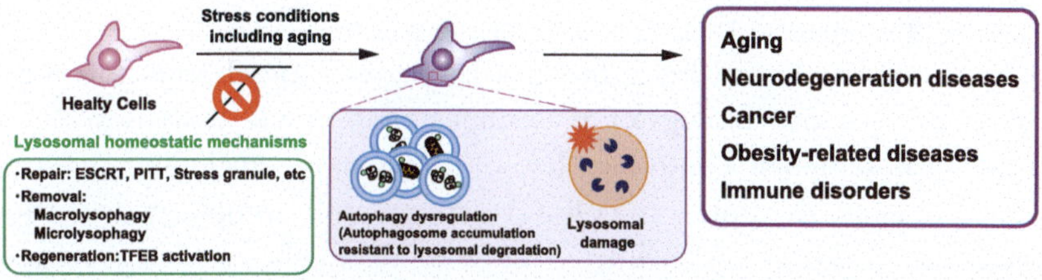

Fig. 3 Pathological relevance of lysosomal damage and its homeostatic responses. Disruption of lysosomal homeostasis is a fundamental pathogenic mechanism in a broad spectrum of diseases, including aging-associated decline, neurodegenerative diseases, cancer, obesity-related diseases, and immune disorders. Cellular stressors—particularly those associated with aging—compromise lysosomal integrity and function, resulting in the accumulation of dysfunctional lysosomes and impaired autophagic flux. To counteract these pathological changes, cells need to activate a sequence of tightly regulated responses: damage sensing, repair, selective removal of irreparably damaged organelles, and biogenesis of new lysosomes. The proper execution of these homeostatic mechanisms is essential for maintaining cellular function and preventing disease progression across diverse biological contexts.

Taken together, the maintenance of lysosomal homeostasis in immune cells determines the threshold of inflammatory activation, the efficacy of antitumor immunity, and the durability of immune memory. Far from being mere digestive compartments, lysosomes serve as central nodes in the integrated regulation of immune function. Advancing our understanding of lysosome biology promises to unlock new therapeutic strategies for a spectrum of immune-related diseases, including inflammation, cancer, and infection (Fig. 3).

5. Recent advances in lysosome research and therapeutic applications

Lysosomes were once regarded as passive terminal compartments responsible for degrading and disposing of cellular waste. However, this view has dramatically shifted. It is now well established that lysosomes are not merely "cellular garbage bins" but central organelles that integrate nutrient sensing, regulate energy metabolism, coordinate membrane trafficking, execute autophagy, and even determine cell fate through death and differentiation. This paradigm shift highlights lysosomes as dynamic regulatory hubs and underscores their crucial role in maintaining cellular homeostasis. Importantly, lysosomal dysfunction plays a role in a wide range of diseases—including neurodegeneration, lipid storage disorders, cancer, and aging—establishing lysosomes as emerging therapeutic targets in modern medicine.

Recent technological breakthroughs have greatly accelerated this redefinition of lysosomal biology. For example, optogenetic tools now allow precise spatiotemporal control of lysosomal functions, including pH regulation, calcium dynamics, mTORC1 activity, and autophagy induction (Zeng et al., 2024a). In Alzheimer's disease models, light-based activation of lysosomes reduces amyloid-β accumulation, offering compelling proof that lysosomal plasticity can be therapeutically harnessed (Zeng et al., 2024b).

Concurrently, technologies for dissecting lysosomal composition and function at the molecular level are advancing rapidly. One notable innovation is the LysoIP (Lysosome ImmunoPrecipitation) method, which enables rapid, high-purity isolation of lysosomes for integrated proteomic, lipidomic, metabolomic, and transcriptomic analysis (Laqtom, 2023). In CLN3-deficient mice, LysoIP uncovered the pathological accumulation of glycerophosphodiesters within lysosomes and their leakage into cerebrospinal fluid, identifying potential new biomarkers (Laqtom et al., 2022). Additional breakthroughs—such as single-lysosome metabolite profiling (Zhu et al., 2021) and phospho-proteomic mapping of lysosome-localized proteins (Zi et al., 2022) —are providing multi-faceted insights into the lysosomal state.

These advances have laid the foundation for novel lysosome-targeted therapeutic strategies. One promising approach is therapeutic modulation of lysosomal pH. In Parkinson's disease models, administration of acidic nanoparticles restores lysosomal acidity, suppresses α-synuclein accumulation, and prevents neurodegeneration (Arotcarena et al., 2022). Similarly, in NAFLD models, re-acidification of lysosomes enhances autophagy and restores metabolic function (Zeng et al., 2023), reinforcing the concept that the lysosomal microenvironment critically influences cellular fate.

From a proteostasis perspective, smart liposomes exploit the acidic lysosomal environment and cathepsin B activity for the selective degradation of pathogenic proteins (Jeon et al., 2025). These autonomous, degradative nanostructures—designed to mimic lysosomal functionality—represent a conceptual leap beyond conventional drug delivery systems, pointing toward a new class of degradation-inducing therapeutics.

In parallel, small molecules that activate TFEB, a master regulator of lysosomal biogenesis and function, are emerging as promising therapeutic candidates. For instance, lysosome-enhancing compounds (LYECs) inhibit CDK9, a kinase that suppresses TFEB nuclear translocation, thereby boosting lysosomal function and ameliorating pathology in Alzheimer's models (Yin et al., 2023). Similarly, arimoclomol restores lysosomal function via TFEB activation and reduces neurodegeneration in Niemann–Pick type C models (Shammas et al., 2025).

More recently, attention has turned to the subcellular positioning of lysosomes as a new axis of regulation. In Parkinson's disease models, compounds that activate the JIP4–TRPML1 axis promote lysosomal clustering near the microtubule-organizing center, enhancing autophagy and exerting neuroprotective effects (Date et al., 2024). These findings reveal that the physical positioning of lysosomes is not merely a consequence of cellular architecture but a tunable determinant of function and a potential therapeutic target.

Taken together, lysosomes are no longer viewed as static, uniform compartments. They exhibit remarkable plasticity, are subject to dynamic regulation, and can become selectively vulnerable in disease contexts. The next frontier lies in delineating disease-specific mechanisms of lysosomal dysfunction and developing personalized lysosome-targeted interventions. Therapeutic strategies that integrate lysosomal structure, function, environment, and spatial dynamics will define the cutting edge of lysosome-based medicine and may revolutionize the treatment of diverse human diseases.

Acknowledgments

A Grant-in-Aid for Scientific Research supported this work from the Ministry of Education, Culture, Sports, Science and Technology of Japan (23K07718 to S.M.); the Japan Agency for Medical Research and Development (AMED) (25ek0210214h0001 and 25ek0310027h0001 to S.M.); and AMED-CREST (JP22gm1410014 to S.M. and M.H.).

References

Abu-Remaileh, M., Wyant, G. A., Kim, C., Laqtom, N. N., Abbasi, M., Chan, S. H., ... Sabatini, D. M. (2017). Lysosomal metabolomics reveals V-ATPase- and mTOR-dependent regulation of amino acid efflux from lysosomes. *Science (New York, N. Y.), 358*(6364), 807–813.

Akter, E., & Kon, S. (2023). Autophagic perturbation caused by reduced lysosomal activity positively regulates cell competition. *Autophagy, 19*(6), 1874–1875.

Alvarez-Valadez, K., Sauvat, A., Diharce, J., Leduc, M., Stoll, G., Guittat, L., ... Ferret, L. (2024). Lysosomal damage due to cholesterol accumulation triggers immunogenic cell death. *Autophagy*, 1–23.

Arden, C., Park, S. H., Yasasilka, X. R., Lee, E. Y., & Lee, M. S. (2024). Autophagy and lysosomal dysfunction in diabetes and its complications. *Trends in Endocrinology and Metabolism: TEM.*

Arotcarena, M. L., Soria, F. N., Cunha, A., Doudnikoff, E., Prévot, G., Daniel, J., ... Crauste-Manciet, S. (2022). Acidic nanoparticles protect against α-synuclein-induced neurodegeneration through the restoration of lysosomal function. *Aging Cell, 21*(4), e13584.

Ballabio, A., & Bonifacino, J. S. (2020). Lysosomes as dynamic regulators of cell and organismal homeostasis. *Nature Reviews. Molecular Cell Biology, 21*(2), 101–118.

Bar-Peled, L., Schweitzer, L. D., Zoncu, R., & Sabatini, D. M. (2012). Ragulator is a GEF for the rag GTPases that signal amino acid levels to mTORC1. *Cell, 150*(6), 1196–1208.

Bhattacharya, A., Mukherjee, R., Kuncha, S. K., Brunstein, M. E., Rathore, R., Junek, S., ... Dikic, I. (2023). A lysosome membrane regeneration pathway depends on TBC1D15 and autophagic lysosomal reformation proteins. *Nature Cell Biology, 25*(5), 685–698.

Bonacina, F., Zhang, X., Manel, N., Yvan-Charvet, L., Razani, B., & Norata, G. D. (2025). Lysosomes in the immunometabolic reprogramming of immune cells in atherosclerosis. *Nature Reviews Cardiology, 22*(3), 149–164.

Bonam, S. R., Wang, F., & Muller, S. (2019). Lysosomes as a therapeutic target. *Nature Reviews. Drug Discovery, 18*(12), 923–948.

Boutry, M., & Kim, P. K. (2021). ORP1L mediated PI(4)P signaling at ER-lysosome-mitochondrion three-way contact contributes to mitochondrial division. *Nature Communications, 12*(1), 5354.

Boyle, K. B., Ellison, C. J., Elliott, P. R., Schuschnig, M., Grimes, K., Dionne, M. S., ... Randow, F. (2023). TECPR1 conjugates LC3 to damaged endomembranes upon detection of sphingomyelin exposure. *The EMBO Journal, 42*(17), e113012.

Bussi, C., Heunis, T., Pellegrino, E., Bernard, E. M., Bah, N., Dos Santos, M. S., ... Fearns, A. (2022). Lysosomal damage drives mitochondrial proteome remodelling and reprograms macrophage immunometabolism. *Nature Communications, 13*(1), 7338.

Bussi, C., Mangiarotti, A., Vanhille-Campos, C., Aylan, B., Pellegrino, E., Athanasiadi, N., ... Šarić, A. (2023). Stress granules plug and stabilize damaged endolysosomal membranes. *Nature, 623*(7989), 1062–1069.

Bussi, C., Peralta Ramos, J. M., Arroyo, D. S., Gallea, J. I., Ronchi, P., Kolovou, A., ... Schwab, Y. (2018). Alpha-synuclein fibrils recruit TBK1 and OPTN to lysosomal damage sites and induce autophagy in microglial cells. *Journal of Cell Science, 131*(23), jcs226241.

Cai, R., Scott, O., Ye, G., Le, T., Saran, E., Kwon, W., ... Saric, A. (2024). Pressure sensing of lysosomes enables control of TFEB responses in macrophages. *Nature Cell Biology, 26*(8), 1247–1260.

Carey, K. L., Paulus, G. L. C., Wang, L., Balce, D. R., Luo, J. W., Bergman, P., ... Singh, S. (2020). TFEB transcriptional responses reveal negative feedback by BHLHE40 and BHLHE41. *Cell Reports, 33*(6), 108371.

Chauhan, S., Kumar, S., Jain, A., Ponpuak, M., Mudd, M. H., Kimura, T., ... Bruun, J. A. (2016). TRIMs and galectins globally cooperate and TRIM16 and galectin-3 co-direct autophagy in endomembrane damage homeostasis. *Developmental Cell, 39*(1), 13–27.

Chen, J., Ding, Y., Jiang, C., Qu, R., Wren, J. D., Georgescu, C., ... Giles, C. B. (2024a). CD151 maintains endolysosomal protein quality to inhibit vascular inflammation. *Circulation Research, 134*(10), 1330–1347.

Chen, L., Zhang, J., Xu, W., Chen, J., Tang, Y., Xiong, S., ... Liu, Z. (2024b). Cholesterol-rich lysosomes induced by respiratory syncytial virus promote viral replication by blocking autophagy flux. *Nature Communications, 15*(1), 6311.

Cheng, X., Shen, D., Samie, M., & Xu, H. (2010). Mucolipins: Intracellular TRPML1-3 channels. *FEBS Letters, 584*(10), 2013–2021.

Curnock, R., Yalci, K., Palmfeldt, J., Jäättelä, M., Liu, B., & Carroll, B. (2023). TFEB-dependent lysosome biogenesis is required for senescence. *The EMBO Journal, 42*(9), e111241.

Date, Y., Sasazawa, Y., Kitagawa, M., Gejima, K., Suzuki, A., Saya, H., ... Hattori, N. (2024). Novel autophagy inducers by accelerating lysosomal clustering against Parkinson's disease. *Elife, 13*, e98649.

De Tito, S., Almacellas, E., Dai Yu, D., Millard, E., Zhang, W., de Heus, C., ... Panagi, I. (2025). ATG9A and ARFIP2 cooperate to control PI4P levels for lysosomal repair. *Developmental Cell, S1534-5807*(25), 00318-1.

Deretic, V., & Lazarou, M. (2022). A guide to membrane atg8ylation and autophagy with reflections on immunity. *The Journal of Cell Biology, 221*(7), e202203083.

Domingues, N., Catarino, S., Cristovao, B., Rodrigues, L., Carvalho, F. A., Sarmento, M. J., ... Correia-Rodrigues, A. (2024). Connexin43 promotes exocytosis of damaged lysosomes through actin remodelling. *The EMBO Journal, 43*(17), 3627–3649.

Duran, J., Salinas, J. E., Wheaton, R. P., Poolsup, S., Allers, L., Rosas-Lemus, M., ... Salemi, M. (2024). Calcium signaling from damaged lysosomes induces cytoprotective stress granules. *The EMBO Journal, 43*(24), 6410–6443.

Durgan, J., Lystad, A. H., Sloan, K., Carlsson, S. R., Wilson, M. I., Marcassa, E., ... Wakelam, M. J. (2021). Non-canonical autophagy drives alternative ATG8 conjugation to phosphatidylserine. *Molecular Cell, 81*(9), 2031–2040.e2038.

Ebner, M., Puchkov, D., López-Ortega, O., Muthukottiappan, P., Su, Y., Schmied, C., ... Dornan, G. L. (2025). Nutrient-regulated control of lysosome function by signaling lipid conversion. *Cell, 188*(9), 2560.

Eguchi, T., Sakurai, M., Wang, Y., Saito, C., Yoshii, G., Wileman, T., ... Iwatsubo, T. (2024). The V-ATPase-ATG16L1 axis recruits LRRK2 to facilitate the lysosomal stress response. *The Journal of Cell Biology, 223*(3), e202302067.

Feng, Y., Wei, H., Lyu, M., Yu, Z., Chen, J., Lyu, X., & Zhuang, F. (2024). Iron retardation in lysosomes protects senescent cells from ferroptosis. *Aging (Albany NY) 16(9), 7683-7703.*

Florey, O. (2023). TECPR1 helps bridge the CASM during lysosome damage. *The EMBO Journal, 42*(19), e115210.

Gabandé-Rodríguez, E., Boya, P., Labrador, V., Dotti, C. G., & Ledesma, M. D. (2014). High sphingomyelin levels induce lysosomal damage and autophagy dysfunction in Niemann Pick disease type A. *Cell Death and Differentiation, 21*(6), 864–875.

Gahlot, P., Kravic, B., Rota, G., van den Boom, J., Levantovsky, S., Schulze, N., ... Meyer, H. (2024). Lysosomal damage sensing and lysophagy initiation by SPG20-ITCH. *Molecular Cell, 84*(8), 1556–1569.e1510.

Gallagher, E. R., & Holzbaur, E. L. F. (2023). The selective autophagy adaptor p62/SQSTM1 forms phase condensates regulated by HSP27 that facilitate the clearance of damaged lysosomes via lysophagy. *Cell Reports, 42*(2), 112037.

Gallagher, E. R., Oloko, P. T., Fitch, T. C., Brown, E. M., Spruce, L. A., & Holzbaur, E. L. F. (2024). Lysosomal damage triggers a p38 MAPK-dependent phosphorylation cascade to promote lysophagy via the small heat shock protein HSP27. *Current Biology: CB, 34*(24), 5739–5757.e8.

Giamogante, F., Barazzuol, L., Maiorca, F., Poggio, E., Esposito, A., Masato, A., ... Brini, M. (2024). A SPLICS reporter reveals [Formula: see text]-synuclein regulation of lysosome-mitochondria contacts which affects TFEB nuclear translocation. *Nature Communications, 15*(1), 1516.

Goodwin, J. M., Walkup, W. G. T., Hooper, K., Li, T., Kishi-Itakura, C., Ng, A., ... Fletcher, K. (2021). GABARAP sequesters the FLCN-FNIP tumor suppressor complex to couple autophagy with lysosomal biogenesis. *Science Advances, 7*(40), eabj2485.

Hayashi, Y., Takatori, S., Warsame, W. Y., Tomita, T., Fujisawa, T., & Ichijo, H. (2023). TOLLIP acts as a cargo adaptor to promote lysosomal degradation of aberrant ER membrane proteins. *The EMBO Journal, 42*(23), e114272.

Henn, D., Yang, X., & Li, M. (2025). Lysosomal quality control review. *Autophagy, 21*(7), 1413–1432.

Herbst, S., Campbell, P., Harvey, J., Bernard, E. M., Papayannopoulos, V., Wood, N. W., ... Gutierrez, M. G. (2020). LRRK2 activation controls the repair of damaged endomembranes in macrophages. *The EMBO Journal, 39*(18), e104494.

Hu, M., Li, P., Wang, C., Feng, X., Geng, Q., Chen, W., ... Reid, W. (2022). Parkinson's disease-risk protein TMEM175 is a proton-activated proton channel in lysosomes. *Cell, 185*(13), 2292–2308.e2220.

Huang, P., Dong, R. Y., Wang, P., Xu, M., Sun, X., & Dong, X. P. (2024a). MCOLN/TRPML channels in the regulation of MTORC1 and autophagy. *Autophagy, 20*(5), 1203–1204.

Huang, Y., Luo, G., Peng, K., Song, Y., Wang, Y., Zhang, H., ... Liu, X. (2024b). Lactylation stabilizes TFEB to elevate autophagy and lysosomal activity. *The Journal of Cell Biology, 223*(11), e202308099.

Hubbi, M. E., Gilkes, D. M., Hu, H., Kshitiz, Ahmed, I., & Semenza, G. L. (2014). Cyclin-dependent kinases regulate lysosomal degradation of hypoxia-inducible factor 1α to promote cell-cycle progression. *Proceedings of the National Academy of Sciences of the United States of America, 111*(32), E3325–E3334.

Huyghe, J., Priem, D., Van Hove, L., Gilbert, B., Fritsch, J., Uchiyama, Y., ... Bertrand, M. J. M. (2022). ATG9A prevents TNF cytotoxicity by an unconventional lysosomal targeting pathway. *Science (New York, N. Y.), 378*(6625), 1201–1207.

Im, E., Jiang, Y., Stavrides, P. H., Darji, S., Erdjument-Bromage, H., Neubert, T. A., ... Nixon, R. A. (2023). Lysosomal dysfunction in down syndrome and Alzheimer mouse models is caused by v-ATPase inhibition by Tyr(682)-phosphorylated APP βCTF. *Science Advances, 9*(30), eadg1925.

Iyer, H., Shen, K., Meireles, A. M., & Talbot, W. S. (2022). A lysosomal regulatory circuit essential for the development and function of microglia. *Science Advances, 8*(35), eabp8321.

Jang, Y., Ko, M., Lee, J. Y., Kim, J. Y., Lee, E. W., & Kwon, H. J. (2025). Inhibition of lysosomal LAMTOR1 increases autophagy by suppressing the MTORC1 pathway to ameliorate lipid accumulations in MAFLD. *Autophagy,* 1–17.

Jayabalan, A. K., Ayeni, A., & Jia, J. (2025). Calcium release from damaged lysosomes triggers stress granule formation for cell survival. *Autophagy, 21*(6), 1390–1392.

Jeon, M., Kim, D. E., Choi, S. Y., Kim, S., Kim, S., Lee, H., & Kim, H. (2025). Lysosomal targeting of liposomes with acidic pH and Cathepsin B induces protein aggregate clearance. *Cell Communication and Signaling: CCS, 23*(1), 296.

Jia, J., Wang, F., Bhujabal, Z., Peters, R., Mudd, M., Duque, T., ... Behrends, C. (2022). Stress granules and mTOR are regulated by membrane atg8ylation during lysosomal damage. *The Journal of Cell Biology, 221*(11), e202207091.

Jia, P., Tian, T., Li, Z., Wang, Y., Lin, Y., Zeng, W., ... Pan, J. (2023). CCDC50 promotes tumor growth through regulation of lysosome homeostasis. *EMBO Reports, 24*(10), e56948.

Jin, J., Mu, Y., Zhang, H., Sturmlechner, I., Wang, C., Jadhav, R. R., ... Goronzy, J. J. (2023). CISH impairs lysosomal function in activated T cells resulting in mitochondrial DNA release and inflammaging. *Nature Aging, 3*(5), 600–616.

Jouandin, P., Marelja, Z., Shih, Y. H., Parkhitko, A. A., Dambowsky, M., Asara, J. M., ... Perrimon, N. (2022). Lysosomal cystine mobilization shapes the response of TORC1 and tissue growth to fasting. *Science (New York, N. Y.), 375*(6582), eabc4203.

Kakuda, K., Ikenaka, K., Kuma, A., Doi, J., Aguirre, C., Wang, N., ... Badawy, S. M. M. (2024). Lysophagy protects against propagation of α-synuclein aggregation through ruptured lysosomal vesicles. *Proceedings of the National Academy of Sciences of the United States of America, 121*(1), e2312306120.

Koerver, L., Papadopoulos, C., Liu, B., Kravic, B., Rota, G., Brecht, L., ... Ehrmann, M. (2019). The ubiquitin-conjugating enzyme UBE2QL1 coordinates lysophagy in response to endolysosomal damage. *EMBO Reports, 20*(10), e48014.

Kravić, B., Bionda, T., Siebert, A., Gahlot, P., Levantovsky, S., Behrends, C., & Meyer, H. (2022). Ubiquitin profiling of lysophagy identifies actin stabilizer CNN2 as a target of VCP/p97 and uncovers a link to HSPB1. *Molecular Cell, 82*(14), 2633–2649.e2637.

Kumar, S., Jain, A., Choi, S. W., da Silva, G. P. D., Allers, L., Mudd, M. H., ... Lazarou, M. (2020). Mammalian Atg8 proteins and the autophagy factor IRGM control mTOR and TFEB at a regulatory node critical for responses to pathogens. *Nature Cell Biology, 22*(8), 973–985.

Kuwahara, T., & Iwatsubo, T. (2024). CASM mediates LRRK2 recruitment and activation under lysosomal stress. *Autophagy, 20*(7), 1692–1693.

Laqtom, N. N. (2023). Studying lysosomal function and dysfunction using LysoIP. *Nature Reviews. Molecular Cell Biology, 24*(11), 773.

Laqtom, N. N., Dong, W., Medoh, U. N., Cangelosi, A. L., Dharamdasani, V., Chan, S. H., ... Tang, R. (2022). CLN3 is required for the clearance of glycerophosphodiesters from lysosomes. *Nature, 609*(7929), 1005–1011.

Lee, J. H., Yang, D. S., Goulbourne, C. N., Im, E., Stavrides, P., Pensalfini, A., ... Berg, M. J. (2022). Faulty autolysosome acidification in Alzheimer's disease mouse models induces autophagic build-up of Aβ in neurons, yielding senile plaques. *Nature Neuroscience, 25*(6), 688–701.

Levin-Konigsberg, R., Mitra, K., Spees, K., Nigam, A., Liu, K., Januel, C., ... Kundaje, A. (2025). An SLC12A9-dependent ion transport mechanism maintains lysosomal osmolarity. *Developmental Cell, 60*(2), 220–235.e227.

Li, L., Liu, X., Yang, S., Li, M., Wu, Y., Hu, S., ... Zhang, J. (2024). The HEAT repeat protein HPO-27 is a lysosome fission factor. *Nature, 628*(8008), 630–638.

Li, M., Koshi, T., & Emr, S. D. (2015a). Membrane-anchored ubiquitin ligase complex is required for the turnover of lysosomal membrane proteins. *The Journal of Cell Biology, 211*(3), 639–652.

Li, M., Rong, Y., Chuang, Y. S., Peng, D., & Emr, S. D. (2015b). Ubiquitin-dependent lysosomal membrane protein sorting and degradation. *Molecular Cell, 57*(3), 467–478.

Li, T. Y., Gao, A. W., Yang, R., Sun, Y., Lei, Y., Li, X., ... Morales, K. (2025). A lysosomal surveillance response to stress extends healthspan. *Nature Cell Biology, 27*(7), 1083–1097.

Liao, Y. C., Pang, S., Li, W. P., Shtengel, G., Choi, H., Schaefer, K., ... Lippincott-Schwartz, J. (2024). COPII with ALG2 and ESCRTs control lysosome-dependent microautophagy of ER exit sites. *Developmental Cell, 59*(11), 1410–1424.e1414.

Lin, J., Yuan, Y., Huang, C., Zi, J., Li, L., Liu, J., ... Li, Y. (2024). TFEB agonist clomiphene citrate activates the autophagy-lysosomal pathway and ameliorates Alzheimer's disease symptoms in mice. *The Journal of Biological Chemistry, 300*(12), 107929.

Liu, B., Chen, R., Zhang, Y., Huang, J., Luo, Y., Rosthøj, S., ... Jäättelä, M. (2023). Cationic amphiphilic antihistamines inhibit STAT3 via Ca(2+)-dependent lysosomal H (+) efflux. *Cell Reports, 42*(2), 112137.

Liu, E. A., Schultz, M. L., Mochida, C., Chung, C., Paulson, H. L., & Lieberman, A. P. (2020). Fbxo2 mediates clearance of damaged lysosomes and modifies neurodegeneration in the Niemann-Pick C brain. *JCI Insight, 5*(20), e136676.

López-Jiménez, A. T., Cardenal-Muñoz, E., Leuba, F., Gerstenmaier, L., Barisch, C., Hagedorn, M., ... Soldati, T. (2018). The ESCRT and autophagy machineries cooperate to repair ESX-1-dependent damage at the Mycobacterium-containing vacuole but have opposite impact on containing the infection. *PLoS Pathogens, 14*(12), e1007501.

Lv, B., Dion, W. A., Yang, H., Xun, J., Kim, D. H., Zhu, B., & Tan, J. X. (2024). A TBK1-independent primordial function of STING in lysosomal biogenesis. *Molecular Cell, 84*(20), 3979–3996.e3979.

Ma, T., Tian, X., Zhang, B., Li, M., Wang, Y., Yang, C., ... Yu, Y. (2022). Low-dose metformin targets the lysosomal AMPK pathway through PEN2. *Nature, 603*(7899), 159–165.

Maejima, I., Takahashi, A., Omori, H., Kimura, T., Takabatake, Y., Saitoh, T., ... Isaka, Y. (2013). Autophagy sequesters damaged lysosomes to control lysosomal biogenesis and kidney injury. *The EMBO Journal, 32*(17), 2336–2347.

Medina, D. L., Di Paola, S., Peluso, I., Armani, A., De Stefani, D., Venditti, R., ... Forrester, A. (2015). Lysosomal calcium signalling regulates autophagy through calcineurin and TFEB. *Nature Cell Biology, 17*(3), 288–299.

Medina, D. L., Fraldi, A., Bouche, V., Annunziata, F., Mansueto, G., Spampanato, C., ... Sardiello, M. (2011). Transcriptional activation of lysosomal exocytosis promotes cellular clearance. *Developmental Cell, 21*(3), 421–430.

Meng, Y., Li, X., & Xu, H. (2024). TFEB links the cGAS-STING pathway to lysosome biogenesis. *Trends in Cell Biology, 34*(12), 983–985.

Miller, A., Schafer, J., Upchurch, C., Spooner, E., Huynh, J., Hernandez, S., ... Fares, H. (2015). Mucolipidosis type IV protein TRPML1-dependent lysosome formation. *Traffic (Copenhagen, Denmark), 16*(3), 284–297.

Morrone Parfitt, G., Coccia, E., Goldman, C., Whitney, K., Reyes, R., Sarrafha, L., ... Crary, J. F. (2024). Disruption of lysosomal proteolysis in astrocytes facilitates midbrain organoid proteostasis failure in an early-onset Parkinson's disease model. *Nature Communications, 15*(1), 447.

Murley, A., Popovici, A. C., Hu, X. S., Lund, A., Wickham, K., Durieux, J., ... Genuth, N. R. (2025). Quiescent cell re-entry is limited by macroautophagy-induced lysosomal damage. *Cell, 188*(10), 2670–2686.e2614.

Nakamura, S., Shigeyama, S., Minami, S., Shima, T., Akayama, S., Matsuda, T., ... Namba-Hamano, T. (2020). LC3 lipidation is essential for TFEB activation during the lysosomal damage response to kidney injury. *Nature Cell Biology, 22*(10), 1252–1263.

Nielsen, I., Clemmensen, K. K. B., Fogde, D. L., Dietrich, T. N., Giacobini, J. D., Bilgin, M., ... Maeda, K. (2024). Cationic amphiphilic drugs induce accumulation of cytolytic lysoglycerophospholipids in the lysosomes of cancer cells and block their recycling into common membrane glycerophospholipids. *Molecular Biology of the Cell, 35*(3), ar25.

Nishimura, S., Linares, J. F., L'Hermitte, A., Duran, A., Cid-Diaz, T., Martinez-Ordonez, A., ... Heikenwalder, M. (2024). Opposing regulation of the STING pathway in hepatic stellate cells by NBR1 and p62 determines the progression of hepatocellular carcinoma. *Molecular Cell, 84*(23), 4660–4676.e10.

Nixon, R. A., & Rubinsztein, D. C. (2024). Mechanisms of autophagy-lysosome dysfunction in neurodegenerative diseases. *Nature Reviews. Molecular Cell Biology, 25*(11), 926–946.

Noguchi, T., Sekiguchi, Y., Shimada, T., Suzuki, W., Yokosawa, T., Itoh, T., ... Hirata, Y. (2023). LLPS of SQSTM1/p62 and NBR1 as outcomes of lysosomal stress response limits cancer cell metastasis. *Proceedings of the National Academy of Sciences of the United States of America, 120*(43), e2311282120.

Ogura, M., Kaminishi, T., Shima, T., Torigata, M., Bekku, N., Tabata, K., ... Hamasaki, M. (2023). Microautophagy regulated by STK38 and GABARAPs is essential to repair lysosomes and prevent aging. *EMBO Reports, 24*(12), e57300.

Papadopoulos, C., Kirchner, P., Bug, M., Grum, D., Koerver, L., Schulze, N., ... Arhzaouy, K. (2017). VCP/p97 cooperates with YOD1, UBXD1 and PLAA to drive clearance of ruptured lysosomes by autophagy. *The EMBO Journal, 36*(2), 135–150.

Park, J. W., Lee, E. J., Moon, E., Kim, H. L., Kim, I. B., Hodzic, D., ... Kim, J. W. (2023). Orthodenticle homeobox 2 is transported to lysosomes by nuclear budding vesicles. *Nature Communications, 14*(1), 1111.

Park, N. Y., Jo, D. S., Yang, J. Y., Bae, J. E., Kim, J. B., Kim, Y. H., ... Yoshimori, T. (2025). Activation of lysophagy by a TBK1-SCF(FBXO3)-TMEM192-TAX1BP1 axis in response to lysosomal damage. *Nature Communications, 16*(1), 1109.

Peng, K., Zhao, G., Zhao, H., Noda, N. N., & Zhang, H. (2025). The autophagy protein ATG-9 regulates lysosome function and integrity. *The Journal of Cell Biology, 224*(6), e202411092.

Peng, W., Wong, Y. C., & Krainc, D. (2020). Mitochondria-lysosome contacts regulate mitochondrial Ca(2+) dynamics via lysosomal TRPML1. *Proceedings of the National Academy of Sciences of the United States of America, 117*(32), 19266–19275.

Pierre, L., Juszczak, F., Delmotte, V., Decarnoncle, M., Ledoux, B., Bultot, L., ... Arnould, T. (2024). AMPK protects proximal tubular epithelial cells from lysosomal dysfunction and dedifferentiation induced by lipotoxicity. *Autophagy, 21*(4), 860–880.

Pu, J. (2022). Targeting the lysosome: Mechanisms and treatments for nonalcoholic fatty liver disease. *Journal of Cellular Biochemistry, 123*(10), 1624–1633.

Radulovic, M., Schink, K. O., Wenzel, E. M., Nähse, V., Bongiovanni, A., Lafont, F., & Stenmark, H. (2018). ESCRT-mediated lysosome repair precedes lysophagy and promotes cell survival. *The EMBO Journal, 37*(21), e99753.

Rajamäki, K., Lappalainen, J., Oörni, K., Välimäki, E., Matikainen, S., Kovanen, P. T., & Eklund, K. K. (2010). Cholesterol crystals activate the NLRP3 inflammasome in human macrophages: A novel link between cholesterol metabolism and inflammation. *PLoS One, 5*(7), e11765.

Raynor, J. L., Collins, N., Shi, H., Guy, C., Saravia, J., Ah Lim, S., ... Sun, Y. (2024). CRISPR screens unveil nutrient-dependent lysosomal and mitochondrial nodes impacting intestinal tissue-resident memory CD8(+) T cell formation. *Immunity, 57*(11), 2597–2614.e13.

Rebsamen, M., Pochini, L., Stasyk, T., de Araújo, M. E., Galluccio, M., Kandasamy, R. K., ... Bruckner, M. (2015). SLC38A9 is a component of the lysosomal amino acid sensing machinery that controls mTORC1. *Nature, 519*(7544), 477–481.

Reddy, A., Caler, E. V., & Andrews, N. W. (2001). Plasma membrane repair is mediated by Ca(2+)-regulated exocytosis of lysosomes. *Cell, 106*(2), 157–169.

Roh, K., Noh, J., Kim, Y., Jang, Y., Kim, J., Choi, H., ... Kim, M. S. (2023). Lysosomal control of senescence and inflammation through cholesterol partitioning. *Nature Metabolism, 5*(3), 398–413.

Rovira, M., Sereda, R., Pladevall-Morera, D., Ramponi, V., Marin, I., Maus, M., ... Muñoz, J. (2022). The lysosomal proteome of senescent cells contributes to the senescence secretome. *Aging Cell, 21*(10), e13707.

Sancak, Y., Bar-Peled, L., Zoncu, R., Markhard, A. L., Nada, S., & Sabatini, D. M. (2010). Ragulator-Rag complex targets mTORC1 to the lysosomal surface and is necessary for its activation by amino acids. *Cell, 141*(2), 290–303.

Sardiello, M., Palmieri, M., di Ronza, A., Medina, D. L., Valenza, M., Gennarino, V. A., ... Polishchuk, R. S. (2009). A gene network regulating lysosomal biogenesis and function. *Science (New York, N. Y.), 325*(5939), 473–477.

Savini, M., Folick, A., Lee, Y. T., Jin, F., Cuevas, A., Tillman, M. C., ... Hu, P. W. (2022). Lysosome lipid signalling from the periphery to neurons regulates longevity. *Nature Cell Biology, 24*(6), 906–916.

Saxton, R. A., & Sabatini, D. M. (2017). mTOR signaling in growth, metabolism, and disease. *Cell, 168*(6), 960–976.

Scheffer, L. L., Sreetama, S. C., Sharma, N., Medikayala, S., Brown, K. J., Defour, A., & Jaiswal, J. K. (2014). Mechanism of Ca^{2+}-triggered ESCRT assembly and regulation of cell membrane repair. *Nature Metabolism, 5*, 5646.

Schröder, L. F., Peng, W., Gao, G., Wong, Y. C., Schwake, M., & Krainc, D. (2024). VPS13C regulates phospho-Rab10-mediated lysosomal function in human dopaminergic neurons. *The Journal of Cell Biology, 223*(5), e202304042.

Shammas, H., Kloster Fog, C., Klein, P., Koustrup, A., Pedersen, M. T., Bie, A. S., ... Guenther, S. (2025). Mechanistic insights into arimoclomol mediated effects on lysosomal function in Niemann-pick type C disease. *Molecular Genetics and Metabolism, 145*(1), 109103.

Shukla, S., Larsen, K. P., Ou, C., Rose, K., & Hurley, J. H. (2022). In vitro reconstitution of calcium-dependent recruitment of the human ESCRT machinery in lysosomal membrane repair. *Proceedings of the National Academy of Sciences of the United States of America, 119*(35), e2205590119.

Siow, W. X., Kabiri, Y., Tang, R., Chao, Y. K., Plesch, E., Eberhagen, C., ... Grimm, C. (2022). Lysosomal TRPML1 regulates mitochondrial function in hepatocellular carcinoma cells. *Journal of Cell Science, 135*(6), jcs259455.

Skowyra, M. L., Schlesinger, P. H., Naismith, T. V., & Hanson, P. I. (2018). Triggered recruitment of ESCRT machinery promotes endolysosomal repair. *Science (New York, N. Y.), 360*(6384), eaar5078.

Spampanato, C., Feeney, E., Li, L., Cardone, M., Lim, J. A., Annunziata, F., ... Parenti, G. (2013). Transcription factor EB (TFEB) is a new therapeutic target for pompe disease. *EMBO Molecular Medicine, 5*(5), 691–706.

Sun, J., Liu, Y., Hao, X., Lin, W., Su, W., Chiang, E., ... Bi, X. (2022). LAMTOR1 inhibition of TRPML1-dependent lysosomal calcium release regulates dendritic lysosome trafficking and hippocampal neuronal function. *The EMBO Journal, 41*(5), e108119.

Sun, Y., Wang, X., Yang, X., Wang, L., Ding, J., Wang, C. C., ... Wang, X. (2023). V-ATPase recruitment to ER exit sites switches COPII-mediated transport to lysosomal degradation. *Developmental Cell, 58*(23), 2761–2775.e2765.

Tan, J. X., & Finkel, T. (2022). A phosphoinositide signalling pathway mediates rapid lysosomal repair. *Nature, 609*(7928), 815–821.

Tan, J. X., & Finkel, T. (2023). Lysosomes in senescence and aging. *EMBO Reports, 24*(11), e57265.

Taylor, S. K. B., Hartman, J. H., & Gupta, B. P. (2024). The neurotrophic factor MANF regulates autophagy and lysosome function to promote proteostasis in Caenorhabditis elegans. *Proceedings of the National Academy of Sciences of the United States of America, 121*(43), e2403906121.

Teranishi, H., Tabata, K., Saeki, M., Umemoto, T., Hatta, T., Otomo, T., ... Hamasaki, M. (2022). Identification of CUL4A-DDB1-WDFY1 as an E3 ubiquitin ligase complex involved in initiation of lysophagy. *Cell Reports, 40*(11), 111349.

Terman, A., Kurz, T., Gustafsson, B., & Brunk, U. T. (2006). Lysosomal labilization. *IUBMB Life, 58*(9), 531–539.

Toifl, S., Didusch, S., Ehrenreiter, K., Desideri, E., Dorard, C., & Baccarini, M. (2025). RAF1 kinase contributes to autophagic lysosome reformation. *Cell Reports, 44*(4), 115490.

Villalobos, T. V., Ghosh, B., DeLeo, K. R., Alam, S., Ricaurte-Perez, C., Wang, A., ... Das, S. (2023). Tubular lysosome induction couples animal starvation to healthy aging. *Nature Aging, 3*(9), 1091–1106.

Wang, A., Chen, C., Mei, C., Liu, S., Xiang, C., Fang, W., ... Zhang, Q. (2024a). Innate immune sensing of lysosomal dysfunction drives multiple lysosomal storage disorders. *Nature Cell Biology, 26*(2), 219–234.

Wang, F., Gómez-Sintes, R., & Boya, P. (2018). Lysosomal membrane permeabilization and cell death. *Traffic (Copenhagen, Denmark), 19*(12), 918–931.

Wang, H., Yi, X., Qu, D., Wang, X., Wang, H., Zhang, H., ... Li, C. (2025). Tumorous cholesterol biosynthesis curtails anti-tumor immunity by preventing MTOR-TFEB-mediated lysosomal degradation of CD274/PD-L1. *Autophagy,* 1–20.

Wang, R., & Baehrecke, E. H. (2025). Atg9 is a conserved regulator of lysosome repair. *The Journal of Cell Biology, 224*(7), e202504129.

Wang, S., Tsun, Z. Y., Wolfson, R. L., Shen, K., Wyant, G. A., Plovanich, M. E., ... Comb, W. (2015a). Metabolism. Lysosomal amino acid transporter SLC38A9 signals arginine sufficiency to mTORC1. *Science (New York, N. Y.), 347*(6218), 188–194.

Wang, W., Gao, Q., Yang, M., Zhang, X., Yu, L., Lawas, M., ... Marugan, J. (2015b). Up-regulation of lysosomal TRPML1 channels is essential for lysosomal adaptation to nutrient starvation. *Proceedings of the National Academy of Sciences of the United States of America, 112*(11), E1373–E1381.

Wang, X., Geng, J., Rimal, S., Sui, Y., Pan, J., Qin, Z., & Lu, B. (2024b). The p53 target DRAM1 modulates calcium homeostasis and ER stress by promoting contact between lysosomes and the ER through STIM1. *Proceedings of the National Academy of Sciences of the United States of America, 121*(39), e2400531121.

Wróbel, M., Cendrowski, J., Szymańska, E., Grębowicz-Maciukiewicz, M., Budick-Harmelin, N., Macias, M., ... Goryca, K. (2022). ESCRT-I fuels lysosomal degradation to restrict TFEB/TFE3 signaling via the Rag-mTORC1 pathway. *Life Science Alliance, 5*(7), e202101239.

Xing, Y., Wang, M. M., Zhang, F., Xin, T., Wang, X., Chen, R., ... Qian, X. (2025). Lysosomes finely control macrophage inflammatory function via regulating the release of lysosomal Fe(2+) through TRPML1 channel. *Nature Communications, 16*(1), 985.

Xu, Y., & Wan, W. (2024). Lysosomal control of the cGAS-STING signaling. *Trends in Cell Biology, 34*(8), 622–625.

Yim, W. W., Yamamoto, H., & Mizushima, N. (2022). Annexins A1 and A2 are recruited to larger lysosomal injuries independently of ESCRTs to promote repair. *FEBS Letters, 596*(8), 991–1003.

Yin, L., Zhou, Y., Liu, H., & Li, Y. (2023). LYECs: lysosome-enhancing compounds as potential therapeutic approaches for Alzheimer disease. *Autophagy, 19*(6), 1863–1864.

Yoshida, Y., Yasuda, S., Fujita, T., Hamasaki, M., Murakami, A., Kawawaki, J., ... Matsuda, N. (2017). Ubiquitination of exposed glycoproteins by SCF(FBXO27) directs damaged lysosomes for autophagy. *Proceedings of the National Academy of Sciences of the United States of America, 114*(32), 8574–8579.

Yu, L., McPhee, C. K., Zheng, L., Mardones, G. A., Rong, Y., Peng, J., ... Wan, F. (2010). Termination of autophagy and reformation of lysosomes regulated by mTOR. *Nature, 465*(7300), 942–946.

Zeng, J., Acin-Perez, R., Assali, E. A., Martin, A., Brownstein, A. J., Petcherski, A., ... Shum, M. (2023). Restoration of lysosomal acidification rescues autophagy and metabolic dysfunction in non-alcoholic fatty liver disease. *Nature Communications, 14*(1), 2573.

Zeng, W., Li, C., Qu, L., & Cang, C. (2024a). Optogenetic manipulation of lysosomal physiology and autophagic activity. *Autophagy, 20*(11), 2591–2592.

Zeng, W., Li, C., Wu, R., Yang, X., Wang, Q., Lin, B., ... Qu, L. (2024b). Optogenetic manipulation of lysosomal physiology and autophagy-dependent clearance of amyloid beta. *PLoS Biology, 22*(4), e3002591.

Zhang, W., Yang, X., Chen, L., Liu, Y. Y., Venkatarangan, V., Reist, L., ... Li, M. (2021). A conserved ubiquitin- and ESCRT-dependent pathway internalizes human lysosomal membrane proteins for degradation. *PLoS Biology, 19*(7), e3001361.

Zhang, X., Cheng, X., Yu, L., Yang, J., Calvo, R., Patnaik, S., ... Lawas, M., et al. (2016). MCOLN1 is a ROS sensor in lysosomes that regulates autophagy. *Nature Communications,* 7, 12109.

Zhang, Z., Chen, C., Yang, F., Zeng, Y. X., Sun, P., Liu, P., & Li, X. (2022). Itaconate is a lysosomal inducer that promotes antibacterial innate immunity. *Molecular Cell, 82*(15), 2844–2857 e2810.

Zhou, L., Xue, X., Yang, K., Feng, Z., Liu, M., & Pastor-Pareja, J. C. (2023). Convergence of secretory, endosomal, and autophagic routes in trans-Golgi-associated lysosomes. *The Journal of Cell Biology, 222*(1), e202203045.

Zhu, H., Li, Q., Liao, T., Yin, X., Chen, Q., Wang, Z., ... Miao, C. (2021). Metabolomic profiling of single enlarged lysosomes. *Nature Methods, 18*(7), 788–798.

Zhu, L., Jorgensen, J. R., Li, M., Chuang, Y. S., & Emr, S. D. (2017). ESCRTs function directly on the lysosome membrane to downregulate ubiquitinated lysosomal membrane proteins. *Elife, 6,* e26403.

Zi, Z., Zhang, Z., Feng, Q., Kim, C., Wang, X. D., Scherer, P. E., ... Yu, Y. (2022). Quantitative phosphoproteomic analyses identify STK11IP as a lysosome-specific substrate of mTORC1 that regulates lysosomal acidification. *Nature Communications, 13*(1), 1760.

Zoncu, R., Bar-Peled, L., Efeyan, A., Wang, S., Sancak, Y., & Sabatini, D. M. (2011). mTORC1 senses lysosomal amino acids through an inside-out mechanism that requires the vacuolar H(+)-ATPase. *Science (New York, N. Y.), 334*(6056), 678–683.

CHAPTER TWO

Autophagy meets plant virology: From molecular crosstalk to outstanding challenges

Miryam Pérez-Cañamás and Carmen Hernández*
Instituto de Biología Molecular y Celular de Plantas (IBMCP; CSIC-UPV) (Consejo Superior de Investigaciones Científicas-Universidad Politécnica de Valencia), Ciudad Politécnica de la Innovación, Calle Ingeniero Fausto Elio, Valencia, Spain
*Corresponding author. e-mail address: cahernan@ibmcp.upv.es

Contents

Abstract

Autophagy is an evolutionarily conserved degradation and recycling pathway that plays essential roles in plant development, stress adaptation, and immunity. Over the past decade, increasing evidence has revealed that autophagy also shapes the outcome of plant–virus interactions. This review summarizes current understanding of the plant autophagy machinery as well as the experimental approaches used to monitor autophagic flux during viral infection. We then explore the diverse roles of autophagy in the plant–virus interface, highlighting its function as an antiviral mechanism that targets viral components for degradation, and the counterstrategies viruses employ to inhibit or evade this defense. Conversely, some viruses exploit autophagy or its components to facilitate replication, movement, or systemic infection, underscoring its dual nature in plant virology. Recent studies have expanded knowledge of selective autophagy receptors that mediate targeted degradation of viral factors, yet the diversity and specificity of these receptors are likely far from fully

International Review of Cell and Molecular Biology, Volume 406
ISSN 1937-6448, https://doi.org/10.1016/bs.ircmb.2026.02.002

characterized. Finally, we discuss outstanding questions such as the mechanisms underlying virus-induced autophagy, the factors determining its switch between antiviral and proviral roles, its crosstalk with other cellular pathways, and the molecular signatures that guide the selective recognition of viral or host proteins. Overall, autophagy emerges as an important regulatory hub in plant–virus interactions, integrating antiviral defense and viral adaptation in a finely tuned balance.

1. Introduction

Viruses are among the most devastating microorganisms across all kingdoms of life, posing serious threats to human and animal health and causing massive crop losses (Rao & Reddy, 2020). One of the most defining features of viral agents is their limited genetic information, which often falls several orders of magnitude below those of the hosts they rely on to survive. Despite this apparently unbalanced situation, viruses are, as obligatory intracellular parasites, able to subdue the host metabolism for their own benefit, just as if it were the case of David meeting Goliath. To achieve this fruitful result, viruses have evolved a vast array of mechanisms for strategically hijacking and/or subverting host resources to thrive. The host, on its side, deploys diverse countermeasures to prevent or limit the progress of viral infection, which, in turn, are targeted by the virus to avoid being defeated (Medina-Puche & Lozano-Duran, 2019; Ramage & Cherry, 2015). Still, the host may activate counter-counter defense mechanisms to cope with virus trickery. The outcome of this never-ending war may be variable and can go from non-infection (when the virus is incapable to overcome defenses mounted by the host or there is incompatibility with the host factors) to a highly productive infection which, moreover, may cause, sometimes fatal, host disease (Fermin, 2018; García & Pallás, 2015).

Focusing the attention on plants, these organisms have developed a plethora of mechanisms to fight off viruses, some of which are continuously operative as physical barriers (e.g., waxy cuticles, cell walls) or passive resistance (e.g., lack of components required for virus life cycle), whereas others will be induced upon pathogen attack (Marwal & Gaur, 2020). Of special relevance in this context is RNA silencing, an evolutionarily conserved mechanism in eukaryotes that mediates sequence-specific regulation of gene expression through the action of small RNAs (sRNAs), including microRNAs (miRNAs) and small interfering RNAs (siRNAs), which are stabilized by 2′-O-methylation at their 3′ ends catalyzed by the methyltransferase HEN1. In the last two decades, intensive investigations have highlighted the role of RNA silencing in viral clearance, particularly in plants where it has been equated with the immune system of animals (Lopez-Gomollon & Baulcombe, 2022). The antiviral RNA silencing is triggered by double-stranded (ds) RNAs of viral origin (e.g., replication intermediates or self-complementary stretches in viral RNAs) that are processed by RNase III Dicer-like (DCL) enzymes generating primary siRNAs with a size of 21–24 nucleotides. One of the strands of these small duplexes is incorporated into an effector complex known as RISC (RNA-induced silencing complex) that invariably contains an endonuclease of the Argonaute (AGO) family. RISC is then directed by the associated siRNA to a target RNA of complementary sequence, promoting its degradation or, in some cases, blocking its translation. The whole process may be strengthened by host RNA-dependent RNA polymerases (RDRs), which facilitate siRNA-primed generation of additional ssRNA templates from RISC-cleaved or other aberrant RNA products that are subsequently processed into secondary siRNAs by DCLs. Among host RDRs, RDR6 and its associated RNA-binding protein, suppressor of gene silencing 3 (SGS3), have been found to be particularly relevant for antiviral silencing, while among AGO proteins, AGO1 and AGO2 appear to function as the main effectors of the pathway. Notably, RNA silencing operates as a cell- and non-cell-autonomous process, with siRNAs serving as mobile signals that spread silencing from the initial infection site to distal tissues (Yang & Li, 2018). The importance of RNA silencing in the virus-host interplay is highlighted by the observation that essentially all plant viruses encode at least one protein able to counteract this host defense mechanism, the so-called viral suppressors of RNA silencing (VSRs). These proteins, which exhibit high variability at structural level, can block RNA silencing at different steps. Inhibiting dsRNA processing or sequestering sRNAs are common strategies employed by VSRs to prevent RISC assembly and block systemic silencing. In addition, direct or indirect interactions with key protein components of the silencing machinery may also contribute to their suppressive activity (Csorba et al., 2015; Jin et al., 2021).

In addition to the primary role of RNA silencing, other inducible defense mechanisms may contribute to plant antiviral immunity, including effector-triggered immunity (ETI), pathogen-associated molecular pattern (PAMP)-triggered immunity (PTI), non-sense mediated mRNA decay (NMD), resistance mediated by the ubiquitin–proteasome system (UPS), and hormone-regulated defense responses (Garcia-Ruiz, 2019; Lozano-Durán, 2024; Marwal & Gaur, 2020). Besides these mechanisms, in recent years autophagy has emerged as a major player in the arms race between plants and viruses, although our understanding of its role in this context is still limited (Kushwaha et al., 2019; Yang et al., 2020).

In this review, we summarize current knowledge on the dual role of autophagy in plant–virus interactions, highlighting its function both as a key component of plant defense responses and, alternatively, as a facilitator of viral infection. We also discuss potential future research directions, including key unanswered questions and possible biotechnological applications of manipulating the autophagy pathway to enhance crop resistance and overall plant health.

2. Autophagy pathway in plants

Autophagy is a highly conserved catabolic mechanism in eukaryotes through which dysfunctional, dispensable or unwanted cytoplasmic components are delivered to vacuoles (yeast and plants) or lysosomes (animals) for breakdown and eventual recycling of the resulting molecules. This pathway plays a key housekeeping role in sustaining cellular homeostasis under normal physiological conditions. However, in response to stressors such as nutrient deprivation, aging, or microbial invasion, autophagic activity is upregulated. This upregulation facilitates the removal of superfluous or malfunctioning intracellular materials, thereby helping to preserve cellular equilibrium and promoting survival of the organism under adverse conditions (Marshall & Vierstra, 2018; Ortega et al., 2024).

In plants, the occurrence of macro-, micro- and mega-autophagy has been reported, though knowledge on the former (hereafter called autophagy) largely excesses that of the latter ones. Autophagy initiates with the assembly of the pre-autophagosomal structure, known in yeast as the phagophore assembly site (PAS) or in other organisms as the autophagosome formation site. At this location, a cup-shaped isolation membrane (phagophore) forms and expands, ultimately generating a sealed, double-membrane vesicle — the autophagosome — which engulfs cytoplasmic material of endogenous origin (proteins, RNA, or organelles) or exogenous origin (pathogens). In the latter case, the process is known as xenophagy or, more specifically, virophagy when referred to viral agents (Wu et al., 2022). At the end of the process, the autophagosome fuses with the vacuole to release the sequestered cargo for degradation by resident hydrolases (Marshall & Vierstra, 2018). The biogenesis and subsequent fate of autophagosomes thus involve several distinct steps, including initiation, nucleation, elongation, completion, fusion with vacuoles, and breakdown, that can be disrupted pharmacologically or through other approaches (see below).

The study of autophagy in plants has expanded rapidly over the past few years. Although much remains to be uncovered, significant progress has been made, particularly using the model organism *Arabidopsis thaliana*. The extensive membrane rearrangements required for autophagosome formation are orchestrated by a conserved set of proteins encoded by autophagy-related genes (ATGs), many of which have been identified in *A. thaliana* as well as in other plant species, including tobacco and rice (Petersen et al., 2024; Tang & Bassham, 2018; Zhou et al., 2015).

As previously noted, autophagy primarily fulfills housekeeping roles, maintaining cellular homeostasis under basal conditions. In response to developmental cues or environmental stress, autophagy is rapidly activated. These stimuli inhibit the activity of the Target of Rapamycin Complex (TORC; composed of the TOR kinase and its accessory proteins RAPTOR and LST8), which acts a central negative regulator of autophagy, primarily through phosphorylation of ATG13 at inhibitory residues (Fig. 1). Concomitantly, SnRK1 (SNF1-related protein kinase 1), the plant ortholog of AMPK (AMP-activated protein kinase), undergoes activation under energy or stress conditions through its phosphorylation by GRIK kinases. SnRK1 not only antagonizes TORC signaling but also promotes the engagement of downstream autophagy machinery, thereby reinforcing autophagy initiation. TORC inhibition leads to dephosphorylation of inhibitory residues of ATG13, whereas SnRK1 phosphorylates ATG1 as well as activation residues of ATG13, which then associate thereby promoting the assembly of the ATG1 kinase complex (comprising ATG1, ATG13, ATG11, and ATG101), a key regulator of autophagy initiation. Once assembled, this complex phosphorylates downstream effectors to initiate autophagosome formation. Membrane dynamics during autophagy involve the ATG9 complex, composed of the transmembrane protein ATG9, ATG2, and ATG18, which facilitates membrane trafficking and recycling. Simultaneously, the ATG6 (Beclin1 in mammals) complex — comprising ATG6, the class III phosphatidylinositol 3-kinase (PI3K) known as VPS34, and regulatory components such as VPS15 and ATG14 — generates phosphatidylinositol-3-phosphate (PI3P) on nascent autophagosomal membranes. PI3P acts as a docking platform for PI3P-binding

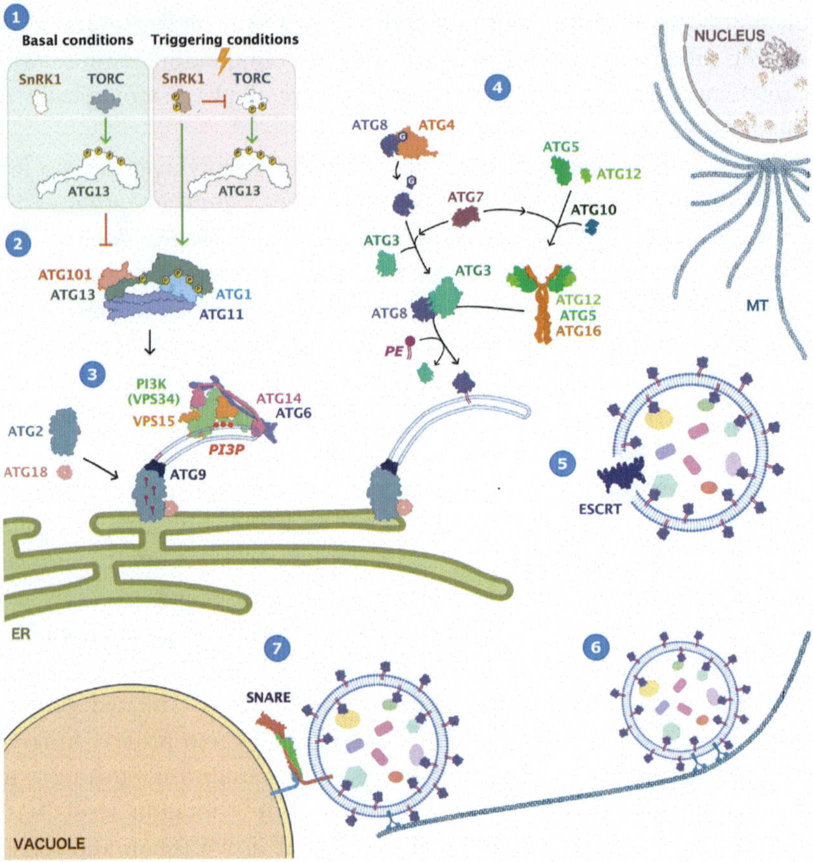

Fig. 1 Schematic representation of plant autophagy. ① Under basal conditions, autophagy is inhibited by TORC (Target of Rapamycin Complex) through hyperphosphorylation of ATG13. Upon developmental cues or environmental stress, activation of SnRK1 inhibits TORC, leading to changes in the phosphorylated state of ATG13 and induction of autophagy. ② ATG13 associates with ATG1 and promotes the assembly of ATG1 kinase complex (ATG1, ATG13, ATG11 and ATG101). ③ This complex phosphorylates downstream effectors to initiate the autophagosome formation. ATG9 complex (ATG9, ATG2 and ATG18) delivers lipids to the expanding membranes, while the ATG6 complex (ATG6, PI3K/VPS34, VPS15 and ATG14) generates phosphatidylinositol 3-phosphate (PI3P) to recruit additional autophagy machinery. ④ Concurrently, ATG8 is processed by ATG4 and then conjugated to phosphatidylethanolamine (PE) through a cascade involving ATG7, ATG3, and ATG12-ATG5-ATG16 complex. ⑤ Following membrane expansion and ATG8-mediated cargo recognition, autophagosomes are closured through the action of ESCRT complexes and ⑥ transported along microtubules to the vacuole. ⑦ Fusion with the vacuolar membrane, mediated by v-SNARE proteins, enables cargo release and subsequent degradation within the vacuolar lumenpeip

proteins, including ATG18s, which contain WD40 repeats and serve as scaffolds to recruit additional autophagy components (Isola & Elazar, 2024; Marshall & Vierstra, 2018).

The ATG12–ATG5–ATG16 complex functions in a ubiquitin-like conjugation system that is essential for autophagosome expansion (Fig. 1). ATG12 is covalently attached to ATG5 through the action of the E1-like enzyme ATG7 and the E2-like enzyme ATG10. The ATG12-ATG5 conjugate then associates non-covalently with ATG16 to form a multimeric complex that facilitates membrane elongation and curvature. In parallel, ATG8 (LC3 in mammals) acts as a ubiquitin-like protein that undergoes covalent conjugation to the lipid phosphatidylethanolamine (PE), a modification crucial for membrane expansion, cargo selection, and autophagosome closure. Similar to ubiquitin, ATG8 is first synthesized as an inactive precursor with an extended C-terminal region that prevents membrane association. This pro-form is processed by the cysteine protease ATG4, which cleaves at a conserved C-terminal glycine residue, yielding a mature ATG8 ready for lipidation. The processed ATG8 is then conjugated to PE on the autophagosomal membrane through a ubiquitin-like cascade involving ATG7 (E1-like) and ATG3 (E2-like), anchoring ATG8 to the membrane and promoting autophagosome maturation and selective cargo recruitment (Wu et al., 2025).

Once fully formed, the autophagosome is transported along microtubules toward the vacuole, guided in part by the endosomal sorting complex required for transport (ESCRT). Fusion with the vacuolar membrane is mediated by vesicle-associated SNARE proteins (v-SNAREs), enabling the release and degradation of sequestered material within the vacuole (Fig. 1). Following fusion, acidification of the vacuole by vacuolar ATPases appears to be essential for efficient autophagosome maturation and cargo degradation (Wu et al., 2025). Regarding membrane sources for autophagosome biogenesis, accumulating evidence points to the endoplasmic reticulum (ER) as a primary contributor, although other organelles may also supply membranes during autophagosome expansion (Staiano & Zappa, 2019; Zhuang et al., 2018).

In contrast to the single-copy ATGs typically found in yeast, most plant species exhibit extensive diversification of several ATG families, reflecting the need to adapt autophagic processes to complex developmental programs and diverse environmental stresses. The ATG8 family represents one of the best-characterized examples of this expansion, comprising up to eight and nine isoforms in the model species *Nicotiana benthamiana* and *A. thaliana*, respectively (Wu et al., 2025). Recent analyses of *A. thaliana* mutants lacking all nine ATG8 genes, combined with complementation assays and proximity labeling proteomics, have shown that individual ATG8 isoforms associate with distinct sets of binding partners and fulfill specialized functional roles, particularly under carbon and nitrogen starvation conditions (Del Chiaro et al., 2025). Beyond ATG8, other components of the autophagy machinery, including the ATG1–ATG13 kinase complex, ATG9, ATG4, the ATG18–ATG2 lipid-transfer module, and ATG6, also occur as multiple isoforms, suggesting regulatory and functional diversification. In contrast, conjugation system components such as ATG5, ATG7, ATG10, ATG12, and ATG16 remain largely single-copy, reflecting their conserved roles in the core autophagy machinery. Collectively, this differential expansion highlights the evolutionary flexibility of the plant autophagy network, enabling precise spatial, developmental, and environmental regulation of degradative and recycling pathways (Marshall & Vierstra, 2018; Wu et al., 2025).

Although autophagy was initially recognized as a primarily bulk catabolic process that remobilizes nutrients and supports energy demands during nutrient deprivation, stress, and developmental transitions, accumulating evidence indicates that it can also selectively target specific cellular components for degradation through the action of cargo receptors. Such receptors recruit specific structures to autophagosomes mainly through interaction with lipidated-ATG8 family proteins anchored in the membrane of the forming autophagosomes. The number of selective autophagy receptors that have been identified and functionally characterized in plants remains relatively low compared to those described in animals or yeast. While some of these receptors are highly conserved, the majority appear to be plant-specific or even specific to certain plant species. Most selective autophagy receptors interact with the membrane-anchored protein ATG8 via ATG8-interacting motifs (AIMs). These AIMs fit the consensus sequence W/F/Y–X–X–L/I/V, mostly of which are present in intrinsically disordered regions to remain accessible to ATG8 docking site. One example is Neighbor of BRCA1 (NBR1), a highly conserved receptor in animals, plants, and several other eukaryotic lineages (Rasmussen et al., 2022), which preferentially—though not exclusively—recognizes ubiquitinated cargoes through its ubiquitin-associated (UBA) domain. Consequently, both the ubiquitinated proteins and NBR1 itself are targeted as selective autophagy substrates and degraded via the autophagy pathway (Zhang & Chen, 2020). Interestingly, a new class of selective autophagy receptors has been identified that interacts with ATG8 through ubiquitin-interacting motif (UIM)-like sequences. These correspond to an eight–amino acid motif with the consensus Ψ–ζ–X–A–Ψ–X–X–S, where Ψ represents a small hydrophobic residue and ζ denotes a polar or charged residue (Rogov et al., 2023).

3. Approaches for monitoring autophagy in plant–virus interactions

Reliable assessment of autophagy remains a major challenge in elucidating its role in plant–virus interactions. Here, we provide a concise overview of the principal methods employed to investigate autophagy, accompanied by one or two illustrative references for each approach. For detailed methodological guidance, we refer readers to comprehensive reviews (Bassham, 2015; Klionsky et al., 2016; Mitou et al., 2009; Qi et al., 2023).

Autophagy in plant cells can be studied through both structural and molecular approaches. Transmission electron microscopy (TEM) allows high-resolution visualization of autophagosomes, identifiable by their double-membrane structure and cytoplasm-like electron density, typically ranging from 100 to 1000 nm in diameter. While conventional TEM does not reveal specific proteins, autophagy-related proteins and, particularly, ATG8, a key marker of autophagosomes, can be localized at the ultrastructural level using immunogold labeling with specific antibodies, providing both morphological and molecular insight into autophagy (Luo et al., 2023; Wang et al., 2019).

As an alternative or complement to TEM, ATG8-based assays using fluorescent protein tags are widely used to visualize autophagosomes. Fusion proteins such as mCherry– or GFP–ATG8 are typically observed by laser scanning confocal microscopy, allowing real-time detection of autophagosomes in living plant cells as discrete fluorescent puncta. Importantly, this approach provides a static readout of autophagosome abundance at a given time point, rather than a direct measure of autophagic activity. In plants, multiple ATG8 homologs can serve as autophagic markers and are commonly expressed either stably in transgenic lines or transiently via *Agrobacterium tumefaciens*-mediated infiltration (Agbemafle et al., 2025; Kurotani et al., 2025).

Consistent with this distinction, an increase in ATG8-labeled puncta alone cannot be interpreted as enhanced autophagic activity without complementary flux measurements. Autophagic flux, which represents the dynamic progression of autophagy from cargo sequestration through vacuolar degradation, is the standard measure of autophagy pathway activity. Assessment of autophagic flux in plants typically involves monitoring the turnover of autophagy-related markers such as ATG8 and selective cargo receptors like NBR1. Western blotting is commonly used to detect the levels of ATG8 and its lipidated form, ATG8–PE, whose accumulation or degradation reflects autophagic activity. Similarly, NBR1 serves as a substrate of selective autophagy, and its protein level decreases with increased autophagic flux. To distinguish increased autophagosome formation from impaired degradation, pharmacological inhibitors can be applied to block vacuolar degradation, thereby permitting the accumulation of autophagic bodies that would otherwise be degraded and enabling accurate flux assessment. In plants expressing GFP–ATG8, the cleavage of GFP from ATG8, detected as free GFP by immunoblotting, serves as an additional readout of autophagic flux. Combining these approaches provides a comprehensive evaluation of autophagy dynamics *in planta* (Yuen et al., 2024).

A variety of chemical compounds to modulate autophagy for experimental purposes is available. Inducers such as rapamycin or AZD8055 inhibit the TOR kinase within the TORC, a negative regulator of autophagy as indicated above, thereby promoting autophagosome formation. Conversely, inhibitors are used to block specific stages of the pathway. For example, 3-methyladenine (3-MA) inhibits class III PI3-kinase activity, thereby preventing autophagosome initiation. Concanamycin A, which inhibits vacuolar H^{+}-ATPase, and E64d, a cysteine protease inhibitor, block vacuolar degradation of autophagic bodies. However, like many small-molecule inhibitors, these compounds can have pleiotropic effects on cellular processes beyond autophagy, such as lysosomal function or endocytic trafficking, underscoring the need for complementary genetic or biochemical assays to confirm specific effects on autophagic flux. These inhibitors are commonly applied by spraying or infiltration in plant tissue to assess autophagic flux indirectly, by blocking vacuolar degradation and thereby enabling the accumulation of autophagosomes or autophagic cargo that would otherwise be turned over. Together, chemical inducers and inhibitors provide valuable tools to study the regulation and dynamics of autophagy under various physiological and stress conditions (Whitmarsh-Everiss & Laraia, 2021).

Beyond chemical treatments, genetic approaches targeting key ATGs offer valuable tools for inhibiting autophagy in plants. Virus-induced gene silencing (VIGS) has been widely used to transiently suppress the expression of ATGs, enabling the investigation of autophagy's role in various stress responses and developmental processes (Haxim et al., 2017; Li et al., 2018). In *A. thaliana*, a range of T-DNA insertion mutants for various ATG*s* have been characterized (Young et al., 2019). More recently, the CRISPR/Cas9 genome editing system has emerged as a powerful alternative for generating targeted ATG knockouts across multiple plant species. This technology has expanded the toolkit for dissecting autophagy's functions and regulatory mechanisms in diverse plant systems (Del Chiaro et al., 2025; Lin & Jones, 2022).

4. Autophagy in the context of plant-virus interplay

Although much less explored than in animals, growing evidence indicates that autophagy contributes to antiviral immunity in plants by facilitating the removal of viral components or structures, and in some cases, by degrading host factors essential for viral infection. Conversely, many plant viruses have evolved mechanisms to antagonize, subvert, or even hijack the autophagy machinery to promote their own replication and spread.

Current knowledge in this area is primarily based on investigations involving a relatively small but increasing number of plant viruses. These include representatives from a diverse range of viral groups, encompassing ds- and single-stranded (ss) DNA viruses, as well as positive (+) and negative-sense (-) ssRNA viruses. In the following sections, we summarize key findings on the antiviral roles of autophagy in plants, examine the viral strategies that counteract or exploit autophagy, and highlight the role of plant-specific selective autophagy receptors in mediating these interactions.

4.1 Autophagy as antiviral pathway in plants and virus countermeasures

Distinct studies have highlighted the important role of autophagy in antiviral defense (Table 1; Fig. 2). One of the first pieces of evidence in this area came from work on tobacco mosaic virus (TMV; ssRNA+; genus *Tobamovirus*; family *Virgaviridae*) and *N. benthamiana* plants carrying the N resistance gene (Liu et al., 2005). The N gene encodes a receptor that recognizes the TMV replicase protein, triggering a hypersensitive response (HR) characterized by localized programmed cell death (PCD) to limit viral spread. Using molecular, cellular, and genetic approaches, Liu et al.

Table 1 Autophagy-mediated antiviral immune responses.

<table>
<tr><th></th><th>Virus[a] (host)[b]</th><th>Targeted viral protein[c] (function)[d]</th><th>Autophagy- related protein[c]</th><th>Effect</th><th colspan="2">Virus countermeasures</th><th colspan="2">Reference(s)</th></tr>
<tr><td rowspan="7">DNA viruses</td><td>CaMV
(At)</td><td>P4
(CP)</td><td>NBR1</td><td>P4 and viral particles are targeted by NBR1 autophagy receptor for degradation</td><td colspan="2">Viral P6 (a pathogenicity factor) sequesters CPs, coordinating particle assembly and their storage in cytoplasmic inclusions, thereby preventing their degradation
P6 also activates TOR kinase which inhibits autophagy</td><td colspan="2">Hafrén et al., 2017
Zvereva et al., 2016</td></tr>
<tr><td>CLCuMuV
TYLCCNV, TYLCV
(Nb)</td><td>βC1
(VSR)</td><td>ATG8f</td><td>βC1 interacts with ATG8f, mediating its degradation through autophagy</td><td colspan="2"></td><td colspan="2">Haxim et al., 2017</td></tr>
<tr><td>TYLCCNV
(At, Nb)</td><td>βV1
(pathogenicity factor)</td><td></td><td>βV1 is degraded via an ATG18a-dependent branch of the autophagy pathway</td><td rowspan="3">Viral C2 (a VSR) binds to ATG7, interfering with autophagosome formation and reducing autophagic flux</td><td></td><td rowspan="3">Cao et al., 2023</td><td>Hu et al., 2025</td></tr>
<tr><td>TYLCYnV
(Nb, Sl)</td><td>C1
(Rep)</td><td>ATG8h</td><td>C1, a nuclear protein, is exported to the cytoplasm through its interaction with ATG8h and subsequently degraded via autophagy</td><td></td><td>Li et al., 2020b</td></tr>
<tr><td>TYLCV
(Sl)</td><td>CP</td><td></td><td>CP and other viral proteins are susceptible to degradation via autophagy and other cellular pathways</td><td>Viral proteins are assembled into small- to mid-sized aggregates, that shelters viral components from autophagy clearance</td><td>Gorovits et al., 2014
Gorovits et al., 2016</td></tr>
<tr><td>TbLCYnV
(Nb)</td><td></td><td></td><td></td><td colspan="2">Viral C4 (a VSR) interacts with SnRK1 β2 subunit and promotes its degradation via the autophagy pathway</td><td colspan="2">Li et al., 2024</td></tr>
<tr><td rowspan="2">RNA viruses</td><td>TMV
(Nb-resistance gene N)</td><td></td><td></td><td>Autophagy restricts hypersensitive response to infection sites following recognition of the TMV virulence protein p50 by the N gene</td><td colspan="2"></td><td colspan="2">Liu et al., 2005</td></tr>
<tr><td>TuMV
CGMMV,
PepMV</td><td>RDR</td><td>ATG6</td><td>Interaction between RDR and ATG6 suppresses viral polymerase activity and targets RDR to autophagosomes for degradation</td><td colspan="2"></td><td colspan="2">Li et al., 2018</td></tr>
</table>

RNA viruses	TuMV TEV, WMV	HC-Pro (VSR)	NBR1	HC-Pro is degraded via NBR1-dependent selective autophagy	Viral VPg (a VSR), and to a lesser extent 6K2 (a viral replication protein), impair NBR1/HC-Pro interaction	Hafrén et al. 2018
	CMV TAV (At, Nb, Nt)	2b (VSR)		HC-Pro and 2b are recognized by rgs-CaM which promotes their degradation through the autophagy pathway		Nakahara et al., 2012 Jeon et al., 2017
	BSMV (Nb)				Viral γb (a VSR) physically binds to ATG7, thereby disrupting the ATG7–ATG8 interaction and impairing autophagosome assembly	Yang et al., 2018
	RSV (Nb, Os)	P3 (VSR)	P3IP ATG8b	P3 is tethered by the plant-specific autophagy receptor P3IP, which mediates its degradation		Jiang et al., 2021a
		NSvc4 (MP)		NSvc4 is degraded predominantly through autophagy as a consequence of excessive or prolonged NSvc4-induced UPR activation	NSvc4 co-opts host chaperones (J -domain proteins) activated during UPR, to protect itself for degradation	Li et al., 2021
	RSMV TYMaV, SYNV (Nb, Os)	glycoprotein G	SnRK1B ATG6b ATG8a	Viral glycoprotein G physically interacts with and enhances the kinase activity of OsSnRK1B, thereby promoting G protein degradation through an ATG6b-ATG8a branch of autophagy		Huang et al., 2024
	TSWV (At, Nb)				Viral NSs (a VSR) interacts with ATG6 and impairs ATG6-mediated autophagy induction	Zhang et al., 2024
	CWMV (Nb)				Viral CP upregulates and interacts with the host factor GAPC, reinforcing GAPC's negative regulation of autophagy	Niu et al., 2022
	CMV CabLCV, BSCTV (Nb, At)	2b AC2, C2 (VSR)	VISP1	2b, AC2 and C2 proteins are targeted, at late stages of infection, by the selective autophagy receptor VISP1 for degradation		Tong et al., 2023

[a] BSCTV, beet severe curly top virus; BSMV, barley stripe mosaic virus; CabLCV, cabbage leaf curl virus; CaMV; cauliflower mosaic virus; CGMMV, cucumber green mottle mosaic virus; CLCuMuV, cotton leaf curl Multan virus; CMV, cucumber mosaic virus; CWMV, Chinese wheat mosaic virus; PepMV, pepino mosaic virus; RSMV, rice stripe mosaic virus; RSV, rice stripe virus; SYNV, sonchus yellow net virus; TAV, tomato aspermy virus; TbLCYnV, tobacco leaf curl Yunan virus; TEV, tobacco etch virus; TMV, tobacco mosaic virus; TSWV, tomato spotted wilt virus; TuMV, turnip mosaic virus; TYLCCNV, tomato yellow leaf curl China virus; TYLCYnV, tomato yellow leaf curl Yunnan virus; TYLCV, tomato yellow leaf curl virus; TYMaV, tomato yellow mottle-associated virus; WMV, watermelon mosaic virus

[b] At, *Arabidopsis thaliana*; Nb, *Nicotiana benthamiana*; Nt, *Nicotiana tabacum*; Os, *Oryza sativa*; Sl, *Solanum lycopersicum*

[c] When known

[d] CP, coat protein; MP, movement protein; RDR, RNA dependent RNA polymerase; Rep, replication initiator protein; VSR, viral suppressor of RNA silencing

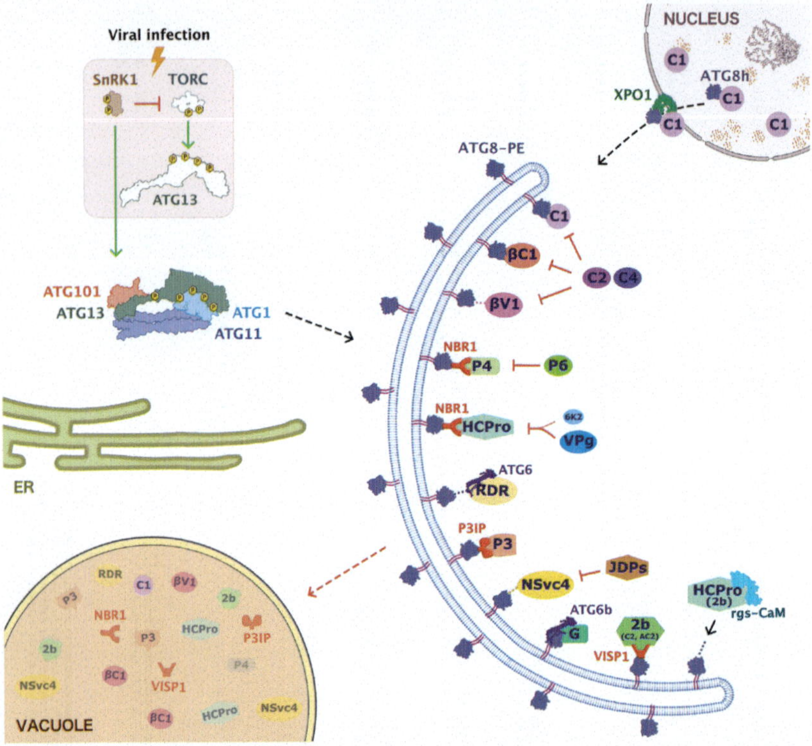

Fig. 2 **The antiviral role of autophagy.** Autophagy is activated during plant virus infections to restrict viral spread. Viral proteins encoded by different viruses are targeted for degradation through this pathway. Some are directly recognized by ATG8 (C1 from tomato yellow leaf curl Yunnan virus, TYLCYnV, a nuclear protein exported by XPO1; βC1 from cotton leaf curl Multan virus, CLCuMuV; protein G from rice stripe mosaic virus, RSMV, a recognition mediated by ATG6b), while others interact with specific receptors that mediate their association with ATG8 (NBR1 in case of P4 from cauliflower mosaic virus, CaMV, and HC-Pro from potyviruses; (Nb)P3IP in case of P3 from rice stripe virus, RSV; VISP1 in case of 2b from cucumber mosaic virus, CMV, AC2 from cabbage leaf curl virus, CabLCV, and C2 from beet severe curly top virus, BSCTV). In some instances, it remains unclear whether viral proteins are directly bound by ATG8 or require autophagy receptors for their recognition (represented by a dashed connecting line between ATG8 and the viral protein βV1 from tomato yellow leaf curl China virus, TYLCCNV, NSvc4 from RSV, RDR from turnip mosaic virus, TuMV, cucumber green mottle mosaic virus, CGMMV, and pepino mosaic virus, PepMV). The specific recognition of HC-Pro and 2b by rgs-CaM from *Solanaceae*, which promotes their autophagic degradation, is also depicted. Viral proteins (P6 from CaMV, C2 and C4 from geminiviruses, VPg and 6K2 from potyviruses) or virus-induced host proteins (JDPs) involved in counterdefense are also depicted.

demonstrated that autophagy is activated both at infection sites and in distant, uninfected cells during HR. Silencing of ATG6 or other core autophagy-related genes (ATG3, ATG7, or PI3K) did not affect HR initiation but led to uncontrolled expansion of PCD into surrounding and distal tissues, resulting in systemic cell death. TMV accumulation was enhanced in the silenced plants, most likely due to the general loss of immune control and to a conceivable direct effect of autophagy on viral components. These findings provided strong evidence that autophagy is not merely a general stress response, but a key regulator of plant immunity.

A subsequent study provided further insight into the mechanisms underlying N gene–mediated resistance to TMV. Plant Bax inhibitor-1 (BI-1), a conserved regulator of cell death, was shown to interact with ATG6 to promote autophagy during this immune response (Xu et al., 2017). Silencing of BI-1 led to a reduction in autophagic activity and an intensified N-mediated PCD response. In contrast, overexpression of BI-1 enhanced autophagy and, notably, triggered autophagy-dependent cell death in a dose-dependent manner. These findings suggested that BI-1 can play both protective and pro-death roles during pathogen infection, depending on its expression level and on the regulation of autophagy under stress conditions.

In addition to studies demonstrating the role of autophagy in the above incompatible plant–RNA virus interaction, several reports have also highlighted its antiviral function in compatible systems involving both DNA and RNA viruses. With regard the former ones, evidence on the contribution of selective autophagy on the defense against cauliflower mosaic virus (CaMV; dsDNA; genus *Caulimovirus*, family *Caulimoviridae*) in the model plant *A. thaliana* was obtained (Hafrén et al., 2017). More specifically, the selective autophagy receptor NBR1 was shown to play a central role in restricting CaMV infection by targeting its viral capsid protein (CP), known as P4, and viral particles for degradation. *A. thaliana* mutants deficient in NBR1 or core autophagy genes (e.g., ATG5, ATG7) showed increased viral accumulation and enhanced disease symptoms, confirming the antiviral function of autophagy. Interestingly, the CaMV multifunctional P6 protein was proposed to exert a protective role against virus autophagic destruction by sequestering CPs, coordinating particle assembly and storage in cytoplasmic inclusions (Hafrén et al., 2017). P6 may also suppress antiviral autophagy through an alternative mechanism, as it was shown to bind and activate TOR kinase, which in turn led to autophagy inhibition and facilitated CaMV infection (Zvereva et al., 2016).

The effect of autophagy in limiting cotton leaf curl Multan virus (CLCuMuV; ssDNA; genus *Begomovirus*, family *Geminiviridae*) plus a betasatellite infection in *N. benthamiana* (Nb) plants was also reported (Haxim et al., 2017). It was observed that such infection triggers autophagy in host plants, and that betasatellite-encoded βC1 protein, a VSR, directly binds the key autophagy protein ATG8, particularly NbATG8f, through an AIM (YDVV) that conforms to the consensus W/F/Y–X–X–L/I/V pattern. This binding takes place in autophagy vesicles and targets βC1 for degradation. Supporting this view, a V32A point mutation in the AIM of βC1, that disrupted its interaction with ATG8, prevented its autophagic clearance. Moreover, CLCuMuV plus the betasatellite with this mutation caused symptoms three days earlier and significantly higher viral DNA levels than wild-type (wt) virus in infected plants, highlighting that autophagic degradation of βC1 dampens infection. Silencing of ATG5 or ATG7 increased susceptibility and viral DNA levels for CLCuMuV and, also, for two other geminiviruses, tomato yellow leaf curl virus (TYLCV) and tomato yellow leaf curl China virus (TYLCCNV), suggesting that the defense role of autophagy is general for members of this viral group. Consistently, autophagy activation through silencing of cytoplamic glyceraldehyde-3-phosphate dehydrogenase (GAPC), a negative regulator of autophagy (Han et al., 2015), delayed symptom appearance and decreased viral loads (Haxim et al., 2017).

Another betasatellite-encoded geminivirus protein, βV1 from TYLCCNV-betasatellite, has been characterized as a pathogenicity factor that is also targeted by host autophagy. In *N. benthamiana*, βV1 localizes to the ER, where it triggers ER stress and activates the bZIP17/28 axis of the unfolded protein response (UPR), a process that facilitates viral infection. Chemical inhibition of autophagy or loss of core autophagy genes (ATG5, ATG7) in *A. thaliana* led to βV1 accumulation, stronger UPR activation, and enhanced viral pathogenicity and these effects were even more pronounced in the *atg18a* mutant. Together, these findings indicated that autophagy restrains virus-induced UPR signaling by targeting βV1 for degradation, with this regulatory event being particularly dependent on the ATG18a-mediated branch of the autophagy pathway (Hu et al., 2025).

Beyond betasatellite-encoded proteins, geminivirus-encoded proteins themselves can also be targeted by autophagy. For example, the replication initiator protein C1 of tomato yellow leaf curl Yunnan virus (TYLCYnV; genus *Begomovirus*; family *Geminiviridae*) has been reported to undergo autophagic degradation. Interestingly, C1 localizes to the nucleus but, after binding ATG8h through an AIM motif (AIFVTL), it is exported via exportin 1 (XPO1), relocalizes to the cytoplasm, and is subsequently degraded. This nuclear export is essential for the autophagic removal of C1. The binding of C1 to ATG8h from both *N. benthamiana* and tomato was proven. Inhibition of autophagy in either species, achieved by pharmacological agents or by silencing of ATGs (ATG8h, ATG5, or ATG7), led to increased C1 accumulation, confirming the role of autophagy in C1 turnover. Additional evidence came from a C1 mutant lacking the AIM: this mutant protein remained confined to the nucleus, escaped autophagic degradation, and, when expressed from the viral genome in place of the wt protein, resulted in higher viral accumulation and more severe disease symptoms in *N. benthamiana* and tomato (Li et al., 2020b). The expression of TYLCYnV C1 alone was sufficient to induce autophagy, although how TYLCYnV or the C1 protein activate autophagy remains unclear.

Other geminiviral proteins may also be vulnerable to host degradation through autophagy and additional pathways, including cytoplasmic and chloroplast proteases as well as the UPS. Studies with TYLCV have shown that the CP and other viral proteins form aggregates that are differentially susceptible to these pathways: autophagy preferentially removes intermediate-sized aggregates, whereas the proteasome mainly degrades soluble proteins and large inclusions (Gorovits et al., 2014). TYLCV circumvents these defenses by assembling its proteins into small- to mid-sized aggregates, a conserved strategy in both plants and whitefly vectors that shelters viral components from proteolysis and promotes persistence (Gorovits et al., 2016).

Recent results suggest that the induction of autophagy by geminivirus infection or by specific geminiviral proteins, such as TYLCYnV C1 (Li et al., 2020b), can be alleviated by other pathogen's protein(s). TYLCYnV C2 has been found to suppress autophagy, weakening plant defense and allowing enhanced viral infection (Cao et al., 2023). C2 specifically binds to the ubiquitin-activating domain of plant ATG7. This competitive interaction blocks ATG7's binding to ATG8, a critical step for autophagosome formation and function, as indicated previously. Consequently, autophagic flux is reduced and viral proteins are not degraded, promoting viral accumulation and symptom progression. This mechanism was observed in both *N. benthamiana* and tomato. Similar ATG7-ATG8 interaction impairment was seen with C2 from three additional geminiviruses (TYLCCNV, TYLCV, and mungbean yellow mosaic India virus), suggesting conservation of the mechanism across plant and virus species. In addition, a different strategy to counteract geminivirus-induced autophagy has been proposed for the C4 protein of tobacco leaf curl Yunan virus (TbLCYnV). This protein has been shown to interact with the TORC negative regulator SnRK1 (subunit β2), promoting its degradation through the autophagy pathway. As a consequence, antiviral autophagy is impaired, enhancing viral accumulation in *N. benthamiana* (Li et al., 2024).

The negative influence of autophagy on compatible plant-RNA virus interactions has been also reported. Turnip mosaic virus (TuMV; ssRNA+; genus *Potyvirus*; family *Potyviridae*) activates autophagy in *N. benthamiana* plants, as indicated by increased ATG8 puncta and autophagosomes (Li et al., 2018). The antiviral role of this virus-triggered autophagy was confirmed in plants silenced for ATGs (ATG3, ATG7, ATG8f) that accumulated more TuMV and exhibited enhanced viral symptoms. Interestingly, ATG6 was found to directly interact with the viral RDR, known as NIb, via its key GDD catalytic motif. This binding suppresses polymerase activity and also leads to NIb being delivered to autophagosome for degradation. Similar interactions between ATG6 and RDRs were observed in RNA viruses belonging to distinct viral groups, including cucumber green mottle mosaic virus (CGMMV; ssRNA+; genus *Tobamovirus*; family *Virgaviridae*) and pepino mosaic virus (PepMV; ssRNA+; genus *Potexvirus*; family *Alphaflexiviridae*), suggesting this is a broad-spectrum antiviral mechanism.

HC-Pro, a multifunctional and essential potyviral protein with VSR activity, is also targeted by autophagy. In tobacco, the calmodulin-like protein rgs (regulator of gene silencing)-CaM functions as a secondary antiviral defense factor by recognizing the potyviral HC-Pro from tobacco etch virus (TEV) as well as structurally unrelated VSRs such as the protein 2b from cucumber mosaic virus (CMV) and tomato aspermy virus (TAV), both ssRNA+ and belonging to genus *Cucumovirus*, family *Bromoviridae*. rgs-CaM interacts with these distinct VSRs and promotes their autophagic degradation, most likely by coupling a Ca^{2+}-dependent signaling pathway to the autophagy machinery (Jeon et al., 2017; Nakahara et al., 2012). Subsequently, it was demonstrated that HC-Pro from TuMV is also degraded via NBR1-dependent selective autophagy in *A. thaliana* (Hafrén et al., 2018). Interestingly, during TuMV infection, HC-Pro avoids degradation, suggesting that the virus manipulates the selective autophagy pathway to maintain HC-Pro stability and thereby sustain RNA silencing suppression. This interference appears to be mediated by other potyviral proteins, mainly VPg and, to a lesser extent, 6K2, through an as yet unresolved mechanism that perturbs NBR1–ATG8 flux (Hafrén et al., 2018). Consistently, HC-Pro of watermelon mosaic virus (WMV) also interacts with NBR1 but resists degradation during infection, indicating that evasion of selective autophagy may represent a conserved counterdefensive strategy among potyviruses.

Barley stripe mosaic virus (BSMV; ssRNA+; genus *Hordeivirus*; family *Virgaviridae*) infection also induces autophagy in *N. benthamiana*. The antiviral role of the pathway was demonstrated by the increased susceptibility of plants in which core ATGs (e.g., ATG7 and ATG8) were silenced (Yang et al., 2018). Further work revealed that BSMV has evolved a counterdefense mechanism mediated by its γb protein, a multifunctional polypeptide also known for its role in RNA silencing suppression. The γb protein suppresses autophagy by physically binding to ATG7, thereby disrupting the ATG7–ATG8 interaction and impairing autophagosome assembly. Consistently, *N. benthamiana* plants infected with a γb-deficient virus or with γb mutants unable to interact with ATG7, displayed enhanced autophagy and reduced viral accumulation, underscoring the critical role of this interaction in BSMV infection success.

Autophagy induction has been observed as well in *N. benthamiana* plants infected with rice stripe virus (RSV; ssRNA+; genus *Tenuivirus*; family *Phenuiviridae*), and its antiviral role was demonstrated by the higher viral titers detected in plants with impaired autophagy following silencing of ATG7 or ATG8. Subsequent work identified RSV VSR, protein P3, as target of autophagy (Jiang et al., 2021a).

RSV infection was also found to be finely regulated by ER stress responses in rice (*Oryza sativa*; Os), with a clear involvement of the autophagy pathway. RSV induces ER stress, triggering the UPR, which modulates the stability of the viral movement protein (MP), NSvc4. A moderate UPR level promotes MP accumulation and enhances virus movement between cells, thereby facilitating infection. In contrast, excessive or prolonged UPR activation leads to MP degradation, predominantly through autophagy and possibly also via ER-associated degradation (ERAD), which limits viral spread. This balance is controlled by UPR-related transcription factors such as OsbZIP50, underscoring the virus's need to precisely regulate host ER stress responses for efficient infection (Li et al., 2021). A later study demonstrated that RSV evades autophagic degradation of NSvc4 by hijacking UPR-induced type I J-domain proteins (JDPs), such as OsDjA5. These chaperones are recruited by NSvc4 to shield it from recognition by the autophagy machinery. Silencing of JDPs enhanced NSvc4 degradation and impaired viral movement, while their overexpression stabilized the protein and promoted infection. This reveals a novel viral counter-defense strategy in which RSV co-opts host chaperones activated during ER stress to protect its MP, thereby fostering cell-to-cell spread and systemic infection.

The infection of rice by a distinct virus, rice stripe mosaic virus (RSMV; ssRNA-; genus *Betacytorhabdovirus*, family *Rhabdoviridae*), is also modulated by autophagy. RSMV infection led to autophagy induction and the antiviral role of the pathway was supported by the increased accumulation of the virus in a rice mutant defective in core autophagy machinery (*atg7*). Once enhanced, autophagy mediates the degradation of the viral G protein, reducing its cytotoxicity and partially limiting viral accumulation (Huang et al., 2024).

Autophagy shapes the infection process of tomato spotted wilt virus (TSWV; ssRNA-; genus *Orthotospovirus;* family *Tospoviridae*) as well. In both *A. thaliana* and *N. benthamiana*, TSWV infection activates autophagy through its N protein (the CP), which contributes to viral restriction, although the molecular mechanism by which N triggers autophagy is still unclear. Consistent with the antiviral role of autophagy, silencing key ATGs (e.g., ATG5, ATG7) or chemically inhibiting the pathway enhances TSWV replication and symptom severity in *N. benthamiana*. Likewise, TSWV symptom induction and accumulation is enhanced in an *A. thaliana atg5* mutant. The viral NSs protein, with several functions including VSR, counteracts this defense by interacting with ATG6, thus impairing ATG6-mediated autophagy induction in a dose-dependent manner (Zhang et al., 2024).

Similarly, infection of *N. benthamiana* plants by Chinese wheat mosaic virus (CWMV; ssRNA+; genus *Furovirus*; family *Virgaviridae*) induces autophagy as an antiviral defense. To counteract this response, the viral CP upregulates and interacts with the host factor GAPC, thereby reinforcing GAPC's negative regulation of autophagy and creating a more favorable environment for viral infection. Interaction of viral CP with GAPC isoforms of *A. thaliana*, wheat and *N. benthamiana* was demonstrated, indicating conservation of the mechanism across distinct host species. Consistently with these observations, genetic silencing of GAPC enhanced autophagy and restricted CWMV accumulation in *N. benthamiana*, whereas GAPC overexpression suppressed autophagy and promoted viral infection (Niu et al., 2022).

4.2 Autophagy as proviral pathway in plants

As an indication of the complex and diverse interplays between viruses and their hosts, viruses are not only targets of autophagy but can also manipulate this pathway to favor their infection (Table 2; Fig. 3). Both DNA and

Table 2 Autophagy manipulation by plant viruses.

	Virus[a] (host)[b]	Viral protein[c] (function)[d]	Targeted host factor[e]	Effect	Reference(s)
DNA viruses	TYLCCNV (Nb)	βC1 (VSR)	SGS3	βC1 induces the expression of and physically interacts with NbCaM. This interaction enhances the stability and accumulation of NbCaM, while promoting the degradation of SGS3 through the autophagy pathway	Li et al., 2017
RNA viruses	Poleroviruses Enamoviruses (At, Nb)	P0 (VSR)	AGO1	P0 protein hijacks a specific ER-derived autophagy pathway to mediate the degradation of membrane-bound AGO1 through the plant-specific autophagy receptors ATI1 and ATI2	Derrien et al., 2012 Li et al., 2019 Michaeli et al., 2019
	TuMV (At, Nb)	HC-Pro (VSR)	AGO1	HC-Pro impairs the activity of the methyltransferase HEN1, thereby preventing the 2′-O-methylation of miRNAs and their proper loading into AGO1. As a result, AGO1 becomes unstable and selectively degraded via autophagy	Pan et al., 2025
	TuMV PVY, TEV, PPV (At)	VPg (VSR)	SGS3	VPg directly interacts with SGS3 and promotes its degradation through both the ubiquitin–proteasome system and the autophagy pathway	Cheng and Wang, 2017
	CGMMV (Nb)		SMG7	CGMMV infection induces autophagy, leading to the degradation of SMG7 through its selective recognition by ATG8i	Chen et al., 2024

(continued)

Table 2 Autophagy manipulation by plant viruses. (*cont'd*)

RNA viruses	RSV (Nb, Os)	NSvc4 (MP)	REM1	NSvc4 interacts with the remorin protein REM1, inhibiting its S-acylation and causing its mislocalization to cytosol, which in turn promotes its autophagic degradation	Fu et al., 2018
		CP	GAPCs	CP interacts with GAPCs, negative regulators of autophagy, which relieves their inhibitory effect and enhances autophagic activity, most likely promoting degradation of host antiviral proteins	Zhao et al., 2023
	BaMV TuMV, TBSV (At, Nb)			BaMV, TuMV and TBSV hijack of autophagic membranes or components for viral replication, which not only favours virus multiplication but may divert autophagy proteins away from antiviral degradation	Huang et al., 2019 Li et al., 2020a Nagy et al., 2025
	CMV (At, Nb)		SGS3/RDR6	CMV infection induces expression of the selective autophagy receptor VISP1 which, at early stages of infection, promotes the autophagic degradation of the cytoplasmic SGS3/RDR6 bodies	Tong et al., 2023

[a] BaMV, bamboo mosaic virus; CGMMV, cucumber green mottle mosaic virus; CMV, cucumber mosaic virus; PPV, plum pox virus; PVY, potato virus Y; RSV, rice stripe virus; TBSV, tomato bushy stunt virus; TEV, tobacco etch virus; TuMV, turnip mosaic virus; TYLCCNV, tomato yellow leaf curl China virus

[b] At, *Arabidopsis thaliana*; Nb, *Nicotiana benthamiana*; Os, *Oryza sativa*

[c] When known

[d] CP, coat protein; MP, movement protein; VSR, viral suppressor of RNA silencing

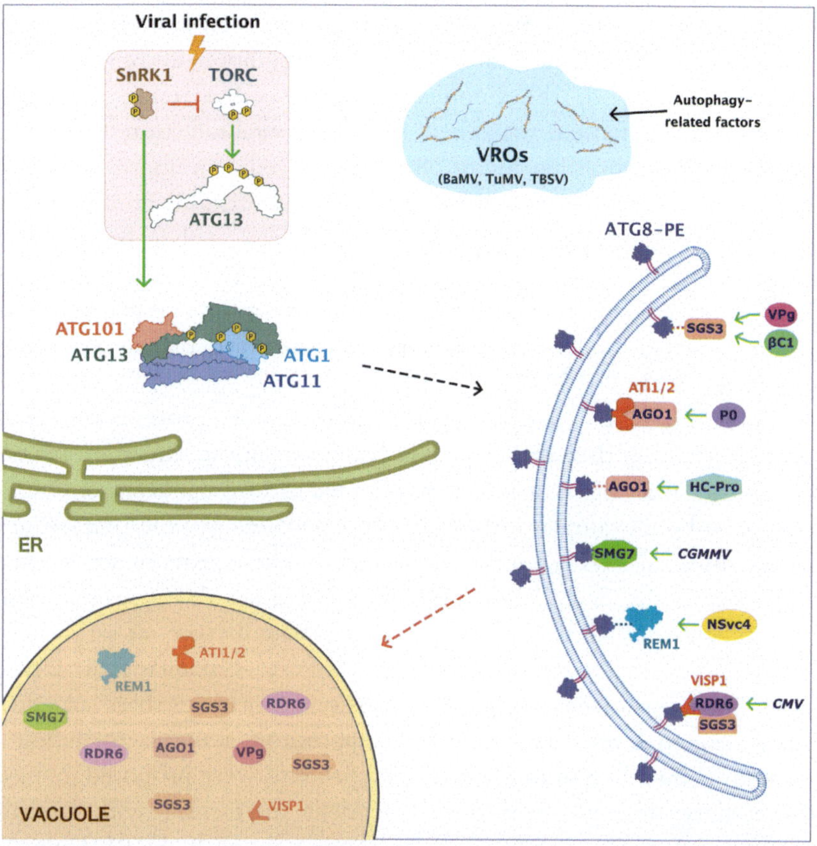

Fig. 3 Manipulation of autophagy by plant viruses. Plant viruses have evolved different strategies to exploit autophagy for their own benefit. Viral proteins can redirect autophagy to degrade antiviral host factors, including SGS3 (VPg from potyviruses; βC1 from tomato yellow leaf curl China virus, TYLCCNV), AGO1 (HC-Pro from turnip mosaic virus, TuMV; P0 from poleroviruses), REM1 (NScv4 from rice stripe virus, RSV) or SMG7 (cucumber green mottle mosaic virus, CGMMV, infection). In several cases, this degradation is mediated by specific receptors such as ATI1/2 (involved in AGO1 degradation by P0 from polerovirus) and VISP1 (mediating RDR6/SGS3 degradation by cucumber mosaic virus, CMV, infection). Some viruses co-opt autophagy components to facilitate the formation of viral replication organelles (VROs).

RNA viruses have been reported to display such capabilities. Among the former, the TYLCCNV betasatellite-encoded βC1 has been shown to induce the expression of, and physically interact with, the *N. benthamiana* calmodulin-like protein (NbCaM), a plant suppressor of RNA silencing.

This interaction enhances the stability and accumulation of NbCaM, while promoting the degradation of SGS3, a key component of the RNA silencing pathway, via the autophagy pathway. As a result, the RNA silencing–based defense is weakened, facilitating viral infection. Consistent with this mechanism, silencing of NbCaM reduced geminivirus infection, whereas its overexpression enhanced viral accumulation, confirming its proviral role. Moreover, inhibition of autophagy, either by chemical treatments or by silencing core components (ATG6, PI3K/VPS34, or VPS15), prevented SGS3 degradation and negatively impacted virus infection, further underscoring the proviral function of the autophagy pathway (Li et al., 2017).

Concerning RNA viruses, several of their encoded proteins have been reported hijacking the autophagy pathway to promote infection, mainly through the targeting of host factors involved in antiviral defense. One of the most striking examples is the P0 protein encoded by poleroviruses and enamoviruses (ssRNA+; family *Solemoviridae*). These proteins act as VSRs by promoting the degradation of AGO1 and other AGO proteins, the main effectors of RNA silencing (e.g., Baumberger et al., 2007; Bortolamiol et al., 2007; Csorba et al., 2010; Pérez-Cañamás et al., 2022). Despite their considerable sequence diversity, P0 proteins share structural features essential for their silencing suppression activity, including a consensus F-box–like motif. Initially, P0 proteins were proposed to exert their activity through an F-box–dependent interaction with S-phase kinase-related protein 1 (SKP1), a core component of the SKP1/Cullin1/F-box (SCF) family of E3 ubiquitin ligases, which mediate the ubiquitination of regulatory and signaling proteins, marking them for proteasomal degradation. However, subsequent studies revealed that SKP1 interaction is not conserved among P0 proteins and, more importantly, that these proteins promote AGO degradation via autophagy rather than through the proteasome (Derrien et al., 2012; Li et al., 2019). Further evidence from the P0 protein of turnip yellows virus (TuYV; genus *Polerovirus*) demonstrated that, in *A. thaliana*, it hijacks a specific ER-derived autophagy pathway to degrade membrane-bound AGO1 in an ATG5- and ATG7-dependent manner (Michaeli et al., 2019).

A distinct mechanism of autophagy-dependent AGO1 clearance has recently been proposed for the potyviral silencing suppressor HC-Pro, particularly from TuMV. Studies performed in *A. thaliana* indicated that, beyond sequestering sRNAs, HC-Pro suppresses RNA silencing by directly inhibiting the methyltransferase activity of HEN1 through its conserved FRNK motif. This inhibition blocks proper 2′-O-methylation of miRNAs, causing them to accumulate in an unmethylated form and fail to load into AGO1. As a result, AGO1 becomes unstable and is selectively degraded via autophagy. HC-Pro facilitates this process by assembling cytoplasmic suppression bodies that recruit both HEN1 and AGO1, thereby preventing their nuclear localization and facilitating autophagic turnover. Though the strength of this mechanism varies among HC-Pro variants, such mechanism adds a novel layer to viral suppression strategies that links interference with sRNA biogenesis to the autophagy-mediated degradation of a core RISC component (Pan et al., 2025).

Another remarkable example of a potyviral protein co-opting autophagy to neutralize antiviral RNA silencing is provided by VPg, which, in addition to its multiple roles in the viral cycle, also functions as VSR. The VPg of TuMV directly interacts with SGS3, a partner of RDR6 in secondary sRNA biogenesis, and promotes its degradation via both the UPS and autophagy. Similar SGS3-targeting activity has been reported for VPg proteins from potato virus Y (PVY), TEV, and plum pox virus (PPV), indicating that this silencing suppressor mechanism is broadly conserved across the *Potyvirus* genus (Cheng & Wang, 2016).

Viruses often manipulate autophagy to weaken antiviral RNA silencing, as mentioned above, but they can also interfere with other defense pathways such as NMD. A recent study demonstrated that NMD restricts CGMMV infection in *N. benthamiana* through the actions of SMG7 and UPF3 (a scaffold protein and a core NMD component, respectively), which target viral RNAs containing premature termination codons. However, CGMMV infection induces autophagy, which drives SMG7 degradation through its selective recognition by ATG8i via an AIM within its sequence, thereby weakening NMD-mediated defense. This creates a negative feedback loop in which autophagy compromises the antiviral activity of NMD (Chen et al., 2024).

Host factors belonging to other functional categories/pathways are also targeted by RNA virus-encoded proteins via autophagy. Remorins are a plant-specific family of membrane-associated proteins that play key roles in cell signaling, pathogen biotic stress response, and membrane organization (Jarsch & Ott, 2011). The MP NSvc4 of RSV was reported to interact with the C-terminus of a remorin protein from *N. benthamiana* and rice (NbREM1/OsREM1.4), inhibiting its S-acylation. Therefore, anchoring of remorins to plasma membrane is reduced and these proteins mislocalize to cytosol which enhances their autophagic degradation. The diminished presence of NbREM1/OsREM1.4 on the plasma membrane seems to increase the size exclusion limit of plasmodesmata (the intercellular channels through which viruses move cell to cell), thereby facilitating virus infection (Fu et al., 2018).

A distinct protein of RSV is also able to manipulate autophagy to facilitate viral infection. The RSV CP has been reported to directly interact with GAPCs from *N. benthamiana* and rice, and this interaction relieves the inhibitory role of these autophagy repressors, resulting in enhanced autophagic activity. This enhancement seems to create a more favorable environment for RSV infection, most likely by promoting degradation of antiviral proteins that would otherwise restrict infection. Consistently, silencing of GAPCs increased both autophagy activity and RSV accumulation in *N. benthamiana* (Zhao et al., 2023).

On its side, bamboo mosaic virus (BaMV; ssRNA+; genus *Potexvirus*; family *Alphaflexiviridae*) seems to subvert autophagy machinery to promote its own replication. BaMV infection has been reported to upregulate ATGs in *N. benthamiana*, leading to autophagosome proliferation. Drug inhibition of autophagy or silencing of key ATGs (such as ATG7 and ATG8) led to a reduction in BaMV RNA accumulation and CP levels, indicating that an intact autophagy pathway facilitates viral infection. Remarkably, BaMV replicase protein was found to colocalize with autophagosome markers, suggesting that autophagic membranes are repurposed as virus replication compartments (Huang et al., 2019).

Despite the fact that some TuMV proteins are targeted by autophagy, as indicated in the previous section, TuMV also appears to exploit this pathway to facilitate infection. Specifically, the virus induces ER stress, which activates the UPR through key transcription factors such as bZIP60. This, in turn, enhances the expression of NBR1-dependent selective autophagy. Disruption of either the UPR or autophagy pathways significantly reduced viral infection in *A. thaliana* (Li et al., 2020a). The viral protein 6K2, which is associated with ER membrane remodeling and the formation of viral replication complexes, was identified as a trigger of both UPR and autophagy induction. These findings highlight a strategy by which TuMV co-opts host ER stress and autophagy responses to create a more favorable environment for replication. Supporting this idea, NIb, the viral RDR, was found to colocalize with autophagy-related structures, suggesting that TuMV may recruit NIb into modified autophagic compartments for replication rather than for degradation.

A similar proviral role of autophagy-related proteins has been considerably studied for tombusviruses, exemplified by tomato bushy stunt virus (TBSV; ssRNA+; genus *Tombusvirus*; family *Tombusviridae*) which broadly exploits host autophagy machinery to support viral replication. Rather than being degraded, viral components co-opt autophagy factors to build and sustain viral replication organelles (VROs). TBSV repurposes phagophore membranes, recruits ATG2 to supply lipids, engages VPS34 to generate PI3P, and uses ATG11 to stabilize membrane contact sites that channel lipids into the VROs. At the same time, the virus forms condensates that divert autophagy proteins away from antiviral degradation. These findings reveal a sophisticated, multilayered viral strategy centered on lipid dynamics, highlighting autophagy as a resource that viruses can rewire for their own benefit (Nagy et al., 2025).

4.3 Plant-specific autophagy receptors in the plant-virus arms race

Besides the involvement of the highly conserved selective autophagy receptor NBR1 in several plant–virus interactions, as indicated in the previous sections, a small set of additional autophagy receptors have been identified as key players in the molecular arms race between plants and viruses. These include specialized cargo receptors and ATG8-interacting proteins that target viral components for degradation or, conversely, are manipulated by viruses to subvert host autophagy pathways.

The *N. benthamiana* P3-interacting protein (NbP3IP) represents a plant-specific selective autophagy receptor involved in antiviral defense. NbP3IP specifically binds the RSV silencing suppressor P3 and recruits it to the autophagy machinery for degradation (Jiang et al., 2021a). Overexpression and knockdown analyses demonstrated that NbP3IP promotes P3 turnover in an autophagy-dependent manner. NbP3IP overexpression enhanced resistance to RSV, correlating with reduced viral accumulation and attenuation of P3-mediated silencing suppression. Consistent with its function as a cargo receptor, NbP3IP interacts with NbATG8f through an AIM of the canonical YxxL/I type. The rice homolog OsP3IP similarly associates with P3 and OsATG8b, mediating P3 degradation and suggesting functional conservation among monocots. Notably, NbP3IP overexpression also activates autophagy, whereas its suppression diminishes autophagic flux, indicating that NbP3IP may additionally act as a regulatory adaptor that positively regulates autophagy beyond its cargo-receptor role (Jiang et al., 2021b).

Other autophagy receptors are co-opted by viruses to counter host antiviral defense as occurs with *Arabidopsis*-specific ATG8-interacting proteins ATI1 and ATI2. These proteins preferentially interact with ATG8f/e isoforms and were first characterized as cargo receptors mediating carbon-starvation–induced autophagy through ER- and plastid-derived vesicles (Honig et al., 2012; Michaeli et al., 2014). They were later shown to function in ER-phagy, assisting the selective degradation of ER components under stress conditions (Wu et al., 2021). This pathway has been shown to be hijacked by the viral F-box protein P0 from TuYV which, as mentioned before, triggers an ER-derived autophagic process to degrade the RNA-silencing effector AGO1 (Michaeli et al., 2019). ATI1 and ATI2 associate with both P0 and membrane-bound AGO1 on the ER, as demonstrated by co-immunoprecipitation and bimolecular fluorescence complementation (BiFC) assays. Through their AIMs**,** that fit the consensus W/F/Y-X-X-L/I/V, ATI1 and ATI2 connect the P0–AGO1 complex to the autophagy machinery, enabling the formation of ER-derived vesicles (ATI bodies) that are subsequently delivered to the vacuole for degradation. Loss of ATI1/ATI2 function or inhibition of autophagy stabilizes AGO1 despite P0 expression, confirming their essential role in this pathway. Thus, ATI1 and ATI2 act as selective autophagy receptors that bridge the viral effector P0 and its host target AGO1 at the ER interface to the autophagic machinery, facilitating viral suppression of RNA silencing.

Recent studies have also identified an *Arabidopsis*-encoded virus-inducible small peptide, VISP1 (Virus-Induced Small Peptide 1), that acts as selective autophagy receptor and contains an UIM (IISALTPS) rather than an AIM, and an arginine/lysine rich motif (ARM) involved in cargo recognition (Tong et al., 2021). Infection by viruses such as CMV has been shown to induce expression of VISP1 in *A. thaliana*. This peptide negatively regulates RNA silencing by promoting the autophagic degradation of the cytoplasmic SGS3/RDR6 bodies, which are essential for secondary siRNA amplification. This degradation suppresses RNA silencing early in infection, creating a window for viral accumulation Interestingly, at later stages, VISP1 switches function by targeting the CMV VSR, protein 2b, for autophagic degradation, contributing to symptom recovery. Despite their sequence divergence, VISP1 is able to target other VSRs such as those encoded by the two geminiviruses, cabbage leaf curl virus (CabLCV-encoded AC2) and beet severe curly top virus (BSCTV-encoded C2). These diverse VSRs are likely recognized via structural features that render them accessible to VISP1 ARM domain (Tong et al., 2023).

Finally, it is worth noting that some viral proteins contain AIMs that mediate direct interaction with ATG8 proteins (e.g., CLCuMuV βC1, TYLCYnV C1) (Haxim et al., 2017; Li et al., 2020b), while others interact with additional autophagy components, such as the association reported between certain viral RDRs and ATG6 (Li et al., 2018). These interactions render viral proteins particularly susceptible to autophagic degradation even in the absence of specific cargo receptors. Such multifaceted interactions highlight the diverse strategies by which autophagy modulates plant-virus interplay.

5. Current challenges and future biotechnological directions

5.1 Open questions

Despite substantial progress in elucidating the role of autophagy in plant–virus interactions, numerous fundamental questions remain unresolved. Among these, the following are particularly noteworthy:

How viral infections induce autophagy?

As noted in previous sections, most, if not all, plant viral infections appear to enhance autophagy, although the underlying mechanisms remain poorly understood, even when the viral component(s) that triggers the pathway has been identified. This issue warrants further investigation, as both transcriptional and post-transcriptional regulation are likely to be involved. Nevertheless, a few mechanistic insights have been reported, notably in infections by CLCuMuV/betasatellite, RSMV and RSV.

In the case of CLCuMuV, the betasatellite-encoded βC1 protein was shown to interact with plant GAPC, which normally binds ATG3 to suppress autophagy (Han et al., 2015). By disrupting the GAPC–ATG3 interaction, βC1 effectively induces autophagy in *N. benthamiana* (Ismayil et al., 2020). A βC1 triple mutant ($\beta C1^{3A}$: I45A, Y48A, I53A) exhibited reduced binding to GAPCs, failed to disrupt the GAPC–ATG3 interaction, and was unable to trigger autophagy, confirming the specific role of βC1 in this process. Plants infected with $\beta C1^{3A}$-carrying virus developed more severe symptoms and accumulated higher viral DNA levels, alongside with reduced autophagy, compared to infections with wt βC1, highlighting the defensive role of βC1-induced autophagy.

The G protein of RSMV, in addition to being targeted by autophagy as previously mentioned, induces autophagy in rice by interacting with and stabilizing OsSnRK1B. This stabilization enhances OsSnRK1B kinase activity, enabling it to phosphorylate an atypical target OsATG6b. OsATG6b, which also interacts directly with the G protein, mediates its autophagic degradation through ATG8a (Huang et al., 2024). Furthermore, glycoproteins from two other rhabdoviruses, tomato yellow mottle-associated virus (TYMaV) and sonchus yellow net virus (SYNV), were shown to interact with SnRK1 or ATG6 orthologs in *N. benthamiana*, suggesting that this autophagy-induction mechanism may be conserved among related viruses.

By contrast, RSV seems to exploit an entirely different route. An RSV-derived siRNA was found to silence the host translation initiation factor eIF4A in *N. benthamiana*, reducing its expression. eIF4A interacts directly with ATG5, and this interaction negatively regulates antiviral autophagy. Thus, silencing eIF4A enhances autophagy and restricts RSV infection, while eIF4A overexpression suppresses autophagy and promotes viral accumulation. In this way, RSV-derived siRNA indirectly promotes autophagy by downregulating eIF4A, thereby relieving its inhibitory effect on ATG5 (Zhang et al., 2021).

Signaling cascades engaged in crosstalk with other immune responses may also contribute to autophagy induction during viral infections (see below). Future research should aim to further dissect both common and virus-specific inducers of autophagy across different plant–virus systems, as elucidating these mechanisms could offer novel tools to fine-tune autophagy for the development of innovative antiviral strategies.

How does autophagy shift between antiviral and proviral roles within the same pathosystem?

Distinct plant–virus pathosystems have revealed that autophagy can exert both antiviral and proviral effects, underscoring its context-dependent nature. The underlying causes of this functional duality within a single host–virus interaction are likely multifactorial.

In the *A. thaliana*-CMV system, for instance, a finely tuned regulatory mechanism has been proposed in which the autophagy receptor VISP1 dynamically balances proviral and antiviral activities by selectively redirecting autophagic degradation toward distinct substrates throughout the course of infection (Tong et al., 2021, 2023).

Similar dual roles have been reported in the *A. thaliana*-CaMV and rice-RSMV systems. In the former, NBR1-mediated selective autophagy targets the viral CP to restrict infection; nevertheless, autophagy-deficient mutants (*atg5*, *atg7*) display accelerated senescence and tissue collapse, whereas *nbr1* mutants retain superior tissue integrity during infection. These findings indicate that while selective autophagy constrains viral accumulation, bulk or NBR1-independent autophagy contributes to host tissue maintenance, potentially prolonging the infection period and enhancing aphid-mediated transmission (Hafrén et al., 2017).

In the rice-RSMV system, autophagic degradation of the viral G protein initially limits viral accumulation but subsequently mitigates G-induced cytotoxicity. This attenuation of cellular damage allows host survival and, paradoxically, supports long-term viral endurance (Huang et al., 2024).

The proviral and antiviral roles of autophagy in other pathosystems, such as *N. benthamiana*/rice-RSV or *N. benthamiana*/*A. thaliana*-TuMV, must also be carefully orchestrated. In these systems, several viral proteins are targeted for autophagic degradation, whereas others manipulate the pathway to remove host factors that restrict infection or subvert it to enhance viral replication (summarized in Tables 1 and 2). Furthermore, paralleling that observed in other plant-virus interactions, *A. thaliana* loss-of-function mutants in ATGs (e.g., *atg5*) showed enhanced TuMV accumulation at early time points of infection whereas viral titers declined at later stages (Hafrén et al., 2018). This pattern suggests that autophagy-mediated maintenance of cellular homeostasis ultimately supports viral persistence by supporting a viable environment for replication.

Collectively, these observations indicate that the balance between antiviral and proviral autophagy is not static but dynamically regulated by multiple interconnected factors, including the temporal modulation of autophagic processes, the nature of the molecular targets involved, the host genetic background, and the cellular/tissue context of the interaction, while not ruling out viral strain–specific effects. Elucidating how these variables collectively shape autophagic outcomes will be crucial for unraveling the complex regulatory networks that govern plant–virus interactions.

How does crosstalk between autophagy and other cellular pathways influence plant–virus interactions?

As highlighted in several of the aforementioned studies, autophagy does not act in isolation during plant–virus interactions but instead frequently engages in extensive crosstalk with other cellular pathways that collectively influence the course of infection. Not surprisingly, autophagy intersects with RNA silencing, a major antiviral mechanism in plants: in some cases, VSRs are selectively targeted as autophagy cargo, while in others, VSRs hijack the autophagy machinery to degrade key components of the RNA silencing pathway (summarized in Tables 1 and 2).

As also illustrated in other previously cited reports (Hu et al., 2025; Li et al., 2020a, 2021), connections with ER stress and its linked UPR have also been found. Interestingly, recent studies are shedding new light on the dual proviral and antiviral roles of the UPR and ER-associated degradation machinery (Adhikari et al., 2025). However, our understanding of how this response modulates plant viral infections at the molecular and genetic level and, moreover, how it is integrated with other pathways, such as autophagy, remains in its infancy.

On the other hand, though connections between plant hormones and autophagy have been considerably documented (Liao et al., 2022), they are relatively unexplored in the context of plant virus-interactions. Nevertheless, some reports showed evidence of such link in this context. For instance, during CMV infection, salicylic acid (SA) activates autophagy to degrade protein 2b, the CMV VSR, and enhance antiviral defense, while 2b counteracts this by suppressing or redirecting autophagy by a yet not well known mechanism (Shukla et al., 2022). This antagonism shows how CMV exploits hormone–autophagy crosstalk to balance host defense and viral persistence.

A similar interplay has been described for CaMV protein P6 and SA signaling. P6 activates the TOR kinase, as indicated previously, which antagonizes SA-induced autophagy, thereby preventing the degradation of viral components (Hafrén et al., 2017; Zvereva et al., 2016). In addition, P6 inhibits the expression of SA-responsive genes such as PR1 and NPR1 and simultaneously enhances jasmonic acid (JA) signaling, shifting the hormonal balance toward a state more favorable for infection Through this dual suppression of SA signaling and autophagy, P6 effectively promotes viral replication and systemic spread (Love et al., 2012).

Additional evidence on the involvement of hormone–autophagy crosstalk in plant viral infections has been obtained with the potato (*Solanum tuberosum*, St)-potyvirus PVY system. The ethylene (ET)-responsive transcription factor StPti5, which acts as PVY susceptibility factor, is degraded via autophagy in healthy plants but accumulates upon infection; its transcription is co-regulated by ET and SA, tying ET/SA signaling to autophagy control during PVY infection (Coll et al., 2024).

Together, these examples highlight the complexity of hormone–autophagy crosstalk in plant–virus interactions, suggesting that viruses may exploit multiple hormonal pathways to modulate autophagic flux and host immunity. Whether similar regulatory mechanisms extend to other phytohormones, such as abscisic acid, brassinosteroids or auxins, and how these networks integrate spatially and temporally during infection remain important open questions for future research.

NMD has been shown to act on viral RNAs in many instances, restricting infection progression (Popp et al., 2020). As exemplified by the *N. benthamiana*- CGMMV pathosystem mentioned previously, virus-induced autophagy can, in turn, act on NMD components, alleviating antiviral effects of the NMD route (Chen et al., 2024). Given the central role of both NMD and autophagy in RNA and protein quality control, it is conceivable that similar regulatory interactions will emerge in other plant–virus systems, revealing a broader crosstalk between these two surveillance pathways in modulating antiviral immunity.

Autophagy may also influence both PTI and ETI. PTI is triggered when cell-surface pattern recognition receptors (PRRs) detect pathogen- or damage-associated molecular patterns (PAMPs/DAMPs). This immune activation initiates signaling cascades, including reactive oxygen species (ROS) production, MAPK activation, and defense gene expression, that can induce autophagy (e.g., through ROS- or SA–dependent signaling). In turn, autophagy contributes to cellular homeostasis by clearing damaged organelles, removing misfolded proteins, or degrading viral components that accumulate during early infection (Zeng et al., 2019).

ETI, on the other hand, is triggered by intracellular immune receptors (often nucleotide-binding leucine-rich repeat, NB-LRR, proteins) that recognize specific effector proteins or their activities, leading to a stronger defense response typically associated with the HR, a form of PCD. As demonstrated in the seminal study using TMV and *N. benthamiana* plants expressing the N resistance gene (Liu et al., 2005), autophagy has been implicated in moderating ETI-associated HR. It can limit excessive cell death by removing damaged organelles, mitigating ROS bursts, and degrading immune components to prevent runaway signaling, while also potentially targeting viral components. Although direct evidence is lacking, it is conceivable that TMV might indirectly modulate autophagy to fine-tune host defense intensity, since successful infection requires a delicate balance between triggering and evading ETI (Zeng et al., 2019).

Finally, while the UPS and autophagy both play major roles in protein quality control, their crosstalk remains largely uncharacterized in viral infections (Sun et al., 2023). Nevertheless, evidence of such crosstalk has been obtained in some instances in which degradation of either viral or antiviral proteins occurs through both autophagic and proteasomal routes (e.g., CP of TYLCV geminivirus, SGS3/RDR6 degradation promoted by potyviral VPg) (Gorovits et al., 2014; Cheng & Wang, 2016). This dual exploitation of host proteostasis machinery suggests that viruses may fine-tune particular branches of the proteostasis systems to favour infection, while plants attempt to selectively deploy degradation of viral factors or stabilize key host proteins. Elucidating the molecular mechanisms underlying this selective modulation will be crucial to understand how the dynamic interplay between plants and viruses reshapes proteostasis networks, influencing both viral success and host resilience.

Which are the molecular signatures and/or post-translational modifications involved in recognition of viral/host antiviral proteins by autophagy receptors?

Another question that warrants further investigation concerns the structural determinants and binding motifs by which specific autophagy receptors/adaptors discriminate viral (or host antiviral) proteins from other viral/cellular proteins. Even less is known on potential post-translational modifications (PTMs) involved in that recognition. In plants, as in animals, the most established "eat-me" signals are ubiquitin conjugation and phosphorylation of key motifs, which recruit ATG8-interacting proteins. However, for viral proteins, or host antiviral proteins whose autophagic degradation is promoted by viral infection, these labeling mechanisms remain poorly understood. Ubiquitination is the best-characterized PTM that marks cargo for autophagy in eukaryotes. In plants, NBR1 primarily recognizes polyubiquitinated proteins through its C-terminal UBA domain and links them to ATG8 via its AIM motif (Zhou et al., 2013). However, accumulating evidence indicates that NBR1 can also target non-ubiquitinated substrates. For instance, in *A. thaliana*, NBR1 mediates the autophagic degradation of CaMV particles and the P4 CP without detectable ubiquitination, suggesting additional modes of substrate recognition (Hafrén et al., 2017). Other autophagy receptors involved in the plant–virus arms race, such as ATI1/2 and P3IP (Jiang et al., 2021a; Michaeli et al., 2019), lack ubiquitin-binding domains like UBA or UIM, suggesting that they mediate cargo recruitment through ubiquitin-independent mechanisms. In any case, ubiquitin-dependent and -independent recognition of autophagy cargoes likely coexist in plant–virus interactions, although the identity of host E3 ligases responsible for the corresponding ubiquitination events remains unknown. Interestingly, some plant viruses encode deubiquitinating proteases (van Vliet et al., 2024), which could form the basis of viral counterdefense strategies, an appealing yet largely unexplored topic.

On the other hand, phosphorylation may modulate the interaction of autophagy receptors or cargo proteins with ATG8, as observed mainly in yeast and mammalian systems (Gubas & Dikic, 2022), and, later, in plants (Nolan et al., 2017). Whether similar phosphorylation-regulatory mechanisms influencing ATG8-binding affinities operate in plant–virus pathosystems remains to be determined.

Other PTMs, such as SUMOylation, acetylation, and S-nitrosylation, can also affect protein stability, conformation, and subcellular localization. However, their potential roles in regulating cargo selection or receptor activation during plant autophagy, either under stress conditions or during development, are still largely unknown.

5.2 Biotechnological perspectives

Effective crop management aimed at reducing the impact of plant pathogens remains a major challenge in agriculture. This is especially true for viruses, whose intimate dependence on host metabolism renders them largely unresponsive to conventional chemical treatments, which are also unsuitable as sustainable control measures.

Natural plant defense mechanisms, such as autophagy, may offer promising strategies for controlling viral infections. Indeed, this pathway could be harnessed in different ways to enhance plant virus resistance. One potential strategy might be focused on strengthening the pathway. Several approaches can be envisioned to reinforce autophagy-mediated antiviral defense:

(i) Genetic enhancement of autophagic flux: Overexpression of key ATGs, such as ATG5, ATG7, or ATG8, can boost the formation of autophagosomes and promote degradation of viral components. Constitutive or inducible upregulation of such genes may increase basal autophagic activity or enable a rapid response upon infection (Li et al., 2018; Wang et al., 2022; Zhang et al., 2024).

(ii) Elicitor-based activation: Application of biostimulants or signaling molecules known to modulate autophagy, such as SA or JA, or TORC inhibitors like rapamycin analog, could transiently enhance antiviral autophagy in infected tissues (Wang et al., 2022).

(iii) Stress- or pathogen-inducible systems: Employing promoters responsive to viral infection or SA signaling to drive ATG expression specifically under stress conditions could minimize potential trade-offs between enhanced defense and plant growth or yield penalties (Zhou et al., 2013).

In addition to manipulating autophagy intensity, targeting the specificity of autophagy receptors or adaptors offers another appealing direction. Engineering or repurposing selective autophagy receptors could improve recognition and clearance of viral components. Strategies might include design of synthetic receptors that fuse viral protein–binding domain (or even nanobodies recognizing viral epitopes) to AIMs or UIMs, modification of existing receptors to enhance their affinity for viral effectors, or overexpression of receptors known to participate in antiviral defense (Hafrén et al., 2017; Rasmussen et al., 2022; Tong et al., 2023). Such synthetic or optimized receptors could provide tailored antiviral autophagy responses while preserving normal autophagic turnover.

Collectively, these strategies position autophagy engineering as a promising frontier for the development of next-generation antiviral crops with enhanced and durable resistance; however, most remain conceptual and await experimental validation in translational contexts.

6. Final remarks

Autophagy has emerged as a central regulator of plant–virus interactions, orchestrating a delicate balance between antiviral defense and viral subversion of host machinery. Despite significant progress in identifying viral effectors and host autophagy components, the spatiotemporal dynamics of autophagy during infection, its intricate crosstalk with other immune pathways, the full repertoire and selectivity of autophagy receptors and adaptors, and the multilayered regulatory mechanisms that fine-tune this process remain only partially understood. Future advances will benefit from integrating classical molecular genetics with systems-level approaches, including multi-omics integration, high-resolution imaging, and computational modeling to uncover emergent properties of autophagy regulation from subcellular to organismal levels.

From a biotechnological perspective, current knowledge already points to tangible applications. One promising strategy is the engineering of selective autophagy receptors with enhanced affinity for conserved viral proteins, enabling their rapid targeting to the vacuole and strengthening basal antiviral immunity. A complementary approach involves genome editing of host factors that viruses exploit to redirect autophagy, for example by modifying specific ATG–virus interaction interfaces to prevent degradation of key defense regulators while preserving essential autophagic functions. Such targeted interventions illustrate how mechanistic insights can be translated into next-generation resistance traits.

Collectively, these efforts will not only deepen our conceptual understanding of autophagy in plant antiviral immunity but also pave the way for the rational engineering of crops with durable, autophagy-mediated resistance to viral pathogens.

Acknowledgments

The authors apologize to colleagues whose relevant contributions may not have been cited due to oversight or space constraints and express their appreciation for the collective efforts driving progress in this rapidly evolving field. Funding in the C.H. lab was provided by the Generalitat Valenciana through grant PROMETEO CIPROM/2022/21 and by the Ministerio de Ciencia e Innovación /Agencia Estatal de Investigación (MICINN/AEI; Spain) through grant TED2021–132160B-I00 within the framework of NextGenerationEU/PRTR. M.P.-C. was the recipient of contracts from the Generalitat Valenciana and MICINN.

References

Adhikari, B., Verchot, J., Brandizzi, F., & Ko, D. K. (2025). ER stress and viral defense: Advances and future perspectives on plant unfolded protein response in pathogenesis. *The Journal of Biological Chemistry, 301*, 108354. https://doi.org/10.1016/j.jbc.2025.108354.

Agbemafle, W., Jayasinghe, V., & Bassham, D. C. (2025). Analysis of autophagy under abiotic stress in arabidopsis seedlings expressing the GFP-ATG8 autophagosome marker. *Current Protocols, 5*, e70166. https://doi.org/10.1002/cpz1.70166.

Bassham, D. C. (2015). Methods for analysis of autophagy in plants. *Methods (San Diego, Calif.), 75*, 181–188. https://doi.org/10.1016/j.ymeth.2014.09.003.

Baumberger, N., Tsai, C.-H., Lie, M., Havecker, E., & Baulcombe, D. C. (2007). The polerovirus silencing suppressor P0 targets ARGONAUTE proteins for degradation. *Current Biology, 17*, 1609–1614. https://doi.org/10.1016/j.cub.2007.08.039.

Bortolamiol, D., Pazhouhandeh, M., Marrocco, K., Genschik, P., & Ziegler-Graff, V. (2007). The polerovirus F box protein P0 targets ARGONAUTE1 to suppress RNA silencing. *Current Biology: CB, 17*, 1615–1621. https://doi.org/10.1016/j.cub.2007.07.061.

Cao, B., Ge, L., Zhang, M., Li, F., & Zhou, X. (2023). Geminiviral C2 proteins inhibit active autophagy to facilitate virus infection by impairing the interaction of ATG7 and ATG8. *Journal of Integrative Plant Biology, 65*, 1328–1343. https://doi.org/10.1111/jipb.13452.

Chen, Y., Jia, M., Ge, L., Li, Z., He, H., Zhou, X., & Li, F. (2024). A negative feedback loop compromises NMD-mediated virus restriction by the autophagy pathway in plants. *Advanced Science (Weinh), 11*, e2400978. https://doi.org/10.1002/advs.202400978.

Cheng, X., & Wang, A. (2016). The potyvirus silencing suppressor protein VPg mediates degradation of SGS3 via ubiquitination and autophagy pathways. *Journal of Virology, 91*, e01478-16. https://doi.org/10.1128/JVI.01478-16.

Coll, A., Lukan, T., Stare, K., Zagorščak, M., Mahkovec Povalej, T., Baebler, Š., ... Gruden, K. (2024). The StPti5 ethylene response factor acts as a susceptibility factor by negatively regulating the potato immune response to pathogens. *The New Phytologist, 244*, 202–218. https://doi.org/10.1111/nph.20004.

Csorba, T., Kontra, L., & Burgyán, J. (2015). Viral silencing suppressors: Tools forged to fine-tune host-pathogen coexistence. *Virology, 479–480*, 85–103. https://doi.org/10.1016/j.virol.2015.02.028.

Csorba, T., Lózsa, R., Hutvágner, G., & Burgyán, J. (2010). Polerovirus protein P0 prevents the assembly of small RNA-containing RISC complexes and leads to degradation of ARGONAUTE1. *The Plant Journal: for Cell and Molecular Biology, 62*, 463–472. https://doi.org/10.1111/j.1365-313X.2010.04163.x.

Del Chiaro, A., Grujic, N., Zhao, J., Papareddy, R. K., Gao, P., Ma, J., ... Dagdas, Y. (2025). An A. thaliana mutant lacking all nine ATG8 isoforms provides genetic evidence for functional specialization of ATG8 in plants. *Journal of Cell Science, 138*, jcs263803. https://doi.org/10.1242/jcs.263803.

Derrien, B., Baumberger, N., Schepetilnikov, M., Viotti, C., De Cillia, J., Ziegler-Graff, V., ... Genschik, P. (2012). Degradation of the antiviral component ARGONAUTE1 by the autophagy pathway. *Proceedings of the National Academy of Sciences of the United States of America, 109*, 15942–15946. https://doi.org/10.1073/pnas.1209487109.

Fermin, G. (2018). Host range, host–virus interactions, and virus transmission. *Viruses*, 101–134. https://doi.org/10.1016/B978-0-12-811257-1.00005-X.

Fu, S., Xu, Y., Li, C., Li, Y., Wu, J., & Zhou, X. (2018). Rice stripe virus interferes with S-acylation of remorin and induces its autophagic degradation to facilitate virus infection. *Molecular Plant, 11*, 269–287. https://doi.org/10.1016/j.molp.2017.11.011.

García, J. A., & Pallás, V. (2015). Viral factors involved in plant pathogenesis. *Current Opinion in Virology, 11*, 21–30. https://doi.org/10.1016/j.coviro.2015.01.001.

Garcia-Ruiz, H. (2019). Host factors against plant viruses. *Molecular Plant Pathology, 20*, 1588–1601. https://doi.org/10.1111/mpp.12851.

Gorovits, R., Fridman, L., Kolot, M., Rotem, O., Ghanim, M., Shriki, O., & Czosnek, H. (2016). Tomato yellow leaf curl virus confronts host degradation by sheltering in small/midsized protein aggregates. *Virus Research, 213*, 304–313. https://doi.org/10.1016/j.virusres.2015.11.020.

Gorovits, R., Moshe, A., Ghanim, M., & Czosnek, H. (2014). Degradation mechanisms of the Tomato yellow leaf curl virus coat protein following inoculation of tomato plants by the whitefly Bemisia tabaci. *Pest Management Science, 70*, 1632–1639. https://doi.org/10.1002/ps.3737.

Gubas, A., & Dikic, I. (2022). A guide to the regulation of selective autophagy receptors. *The FEBS Journal, 289*, 75–89. https://doi.org/10.1111/febs.15824.

Hafrén, A., Macia, J.-L., Love, A. J., Milner, J. J., Drucker, M., & Hofius, D. (2017). Selective autophagy limits cauliflower mosaic virus infection by NBR1-mediated targeting of viral capsid protein and particles. *Proceedings of the National Academy of Sciences of the United States of America, 114*, E2026–E2035. https://doi.org/10.1073/pnas.1610687114.

Hafrén, A., Üstün, S., Hochmuth, A., Svenning, S., Johansen, T., & Hofius, D. (2018). Turnip mosaic virus counteracts selective autophagy of the viral silencing suppressor HCpro. *Plant Physiology, 176*, 649–662. https://doi.org/10.1104/pp.17.01198.

Han, S., Wang, Y., Zheng, X., Jia, Q., Zhao, J., Bai, F., ... Liu, Y. (2015). Cytoplastic glyceraldehyde-3-phosphate dehydrogenases interact with ATG3 to negatively regulate autophagy and immunity in Nicotiana benthamiana. *The Plant Cell, 27*, 1316–1331. https://doi.org/10.1105/tpc.114.134692.

Haxim, Y., Ismayil, A., Jia, Q., Wang, Y., Zheng, X., Chen, T., ... Liu, Y. (2017). Autophagy functions as an antiviral mechanism against geminiviruses in plants. *eLife, 6*, e23897. https://doi.org/10.7554/eLife.23897.

Honig, A., Avin-Wittenberg, T., Ufaz, S., & Galili, G. (2012). A new type of compartment, defined by plant-specific Atg8-interacting proteins, is induced upon exposure of Arabidopsis plants to carbon starvation. *The Plant Cell, 24*, 288–303. https://doi.org/10.1105/tpc.111.093112.

Hu, T., Li, C., Liu, H., Su, C., Wang, Y., Li, F., & Zhou, X. (2025). Geminivirus βV1 protein activates bZIP17/28-mediated UPR signaling to facilitate viral pathogenicity but its activity is attenuated by autophagic degradation in plants. *Plant Communications, 6*, 101198. https://doi.org/10.1016/j.xplc.2024.101198.

Huang, X., Wang, J., Chen, S., Liu, S., Li, Z., Wang, Z., ... Zhang, T. (2024). Rhabdovirus encoded glycoprotein induces and harnesses host antiviral autophagy for maintaining its compatible infection. *Autophagy, 20*, 275–294. https://doi.org/10.1080/15548627.2023.2252273.

Huang, Y.-P., Huang, Y.-W., Hsiao, Y.-J., Li, S.-C., Hsu, Y.-H., & Tsai, C.-H. (2019). Autophagy is involved in assisting the replication of Bamboo mosaic virus in Nicotiana benthamiana. *Journal of Experimental Botany, 70*, 4657–4670. https://doi.org/10.1093/jxb/erz244.

Ismayil, A., Yang, M., Haxim, Y., Wang, Y., Li, J., Han, L., ... Liu, Y. (2020). Cotton leaf curl Multan virus βC1 protein induces autophagy by disrupting the interaction of autophagy-related protein 3 with glyceraldehyde-3-phosphate dehydrogenases. *The Plant Cell, 32*, 1124–1135. https://doi.org/10.1105/tpc.19.00759.

Isola, D., & Elazar, Z. (2024). Phospholipid supply for autophagosome biogenesis. *Journal of Molecular Biology, 436*, 168691. https://doi.org/10.1016/j.jmb.2024.168691.

Jarsch, I. K., & Ott, T. (2011). Perspectives on remorin proteins, membrane rafts, and their role during plant-microbe interactions. *Molecular Plant-Microbe Interactions: MPMI, 24*, 7–12. https://doi.org/10.1094/MPMI-07-10-0166.

Jeon, E. J., Tadamura, K., Murakami, T., Inaba, J.-I., Kim, B. M., Sato, M., ... Nakahara, K. S. (2017). rgs-CaM detects and counteracts viral RNA silencing suppressors in plant immune priming. *Journal of Virology, 91*, e00761-17. https://doi.org/10.1128/JVI.00761-17.

Jiang, L., Lu, Y., Zheng, X., Yang, X., Chen, Y., Zhang, T., ... Yan, F. (2021a). The plant protein NbP3IP directs degradation of Rice stripe virus p3 silencing suppressor protein to limit virus infection through interaction with the autophagy-related protein NbATG8. *The New Phytologist, 229*, 1036–1051. https://doi.org/10.1111/nph.16917.

Jiang, L., Zheng, X., Liu, Y., Chen, J., Lu, Y., & Yan, F. (2021b). Plant protein P3IP participates in the regulation of autophagy in Nicotiana benthamiana. *Plant Signaling & Behavior, 16*, 1861768. https://doi.org/10.1080/15592324.2020.1861768.

Jin, Y., Zhao, J.-H., & Guo, H.-S. (2021). Recent advances in understanding plant antiviral RNAi and viral suppressors of RNAi. *Current Opinion in Virology, 46*, 65–72. https://doi.org/10.1016/j.coviro.2020.12.001.

Klionsky, D. J., Abdelmohsen, K., Abe, A., Abedin, M. J., Abeliovich, H., Acevedo Arozena, A., ... Aghi, M. K. (2016). Guidelines for the use and interpretation of assays for monitoring autophagy (3rd edition). *Autophagy, 12*, 1–222. https://doi.org/10.1080/15548627.2015.1100356.

Kurotani, K., Shinozaki, D., Okada, K., Tabata, R., Kawakatsu, Y., Sugita, R., ... Notaguchi, M. (2025). Autophagy is induced during plant grafting to promote wound healing. *Nature Communications, 16*, 3483. https://doi.org/10.1038/s41467-025-58519-6.

Kushwaha, N. K., Hafrén, A., & Hofius, D. (2019). Autophagy–virus interplay in plants: From antiviral recognition to proviral manipulation. *Molecular Plant Pathology, 20*, 1211–1216. https://doi.org/10.1111/mpp.12852.

Li, C., Xu, Y., Fu, S., Liu, Y., Li, Z., Zhang, T., ... Zhou, X. (2021). The unfolded protein response plays dual roles in rice stripe virus infection through fine-tuning the movement protein accumulation. *PLoS Pathogens, 17*, e1009370. https://doi.org/10.1371/journal.ppat.1009370.

Li, F., Zhang, C., Li, Y., Wu, G., Hou, X., Zhou, X., & Wang, A. (2018). Beclin1 restricts RNA virus infection in plants through suppression and degradation of the viral polymerase. *Nature Communications, 9*, 1268. https://doi.org/10.1038/s41467-018-03658-2.

Li, F., Zhang, C., Tang, Z., Zhang, L., Dai, Z., Lyu, S., ... Wang, A. (2020a). A plant RNA virus activates selective autophagy in a UPR-dependent manner to promote virus infection. *New Phytologist, 228*, 622–639. https://doi.org/10.1111/nph.16716.

Li, F., Zhang, M., Zhang, C., & Zhou, X. (2020b). Nuclear autophagy degrades a geminivirus nuclear protein to restrict viral infection in solanaceous plants. *The New Phytologist, 225*, 1746–1761. https://doi.org/10.1111/nph.16268.

Li, F., Zhao, N., Li, Z., Xu, X., Wang, Y., Yang, X., ... Zhou, X. (2017). A calmodulin-like protein suppresses RNA silencing and promotes geminivirus infection by degrading SGS3 via the autophagy pathway in Nicotiana benthamiana. *PLoS Pathogens, 13*, e1006213. https://doi.org/10.1371/journal.ppat.1006213.

Li, X., Zhao, M., Yang, W., Zhou, X., & Xie, Y. (2024). The C4 protein of TbLCYnV promotes SnRK1 β2 degradation via the autophagy pathway to enhance viral infection in N. benthamiana. *Viruses, 16*, 234. https://doi.org/10.3390/v16020234.

Li, Y., Sun, Q., Zhao, T., Xiang, H., Zhang, X., Wu, Z., ... Han, C. (2019). Interaction between Brassica yellows virus silencing suppressor P0 and plant SKP1 facilitates stability of P0 in vivo against degradation by proteasome and autophagy pathways. *The New Phytologist, 222*, 1458–1473. https://doi.org/10.1111/nph.15702.

Liao, C.-Y., Wang, P., Yin, Y., & Bassham, D. C. (2022). Interactions between autophagy and phytohormone signaling pathways in plants. *FEBS Letters, 596*, 2198–2214. https://doi.org/10.1002/1873-3468.14355.

Lin, Y., & Jones, M. L. (2022). CRISPR/Cas9-mediated editing of autophagy gene 6 in petunia decreases flower longevity, seed yield, and phosphorus remobilization by accelerating ethylene production and senescence-related gene expression. *Frontiers in Plant Science, 13*. https://doi.org/10.3389/fpls.2022.840218.

Liu, Y., Schiff, M., Czymmek, K., Tallóczy, Z., Levine, B., & Dinesh-Kumar, S. P. (2005). Autophagy regulates programmed cell death during the plant innate immune response. *Cell, 121*, 567–577. https://doi.org/10.1016/j.cell.2005.03.007.

Lopez-Gomollon, S., & Baulcombe, D. C. (2022). Roles of RNA silencing in viral and non-viral plant immunity and in the crosstalk between disease resistance systems. *Nature Reviews. Molecular Cell Biology, 23*, 645–662. https://doi.org/10.1038/s41580-022-00496-5.

Love, A. J., Geri, C., Laird, J., Carr, C., Yun, B.-W., Loake, G. J., ... Milner, J. J. (2012). Cauliflower mosaic virus protein P6 inhibits signaling responses to salicylic acid and regulates innate immunity. *PLoS One, 7*, e47535. https://doi.org/10.1371/journal.pone.0047535.

Lozano-Durán, R. (2024). Viral recognition and evasion in plants. *Annual Review of Plant Biology, 75*, 655–677. https://doi.org/10.1146/annurev-arplant-060223-030224.

Luo, M., Law, K. C., He, Y., Chung, K. K., Po, M. K., Feng, L., ... Jiang, L. (2023). Arabidopsis AUTOPHAGY-RELATED2 is essential for ATG18a and ATG9 trafficking during autophagosome closure. *Plant Physiology, 193*, 304–321. https://doi.org/10.1093/plphys/kiad287.

Marshall, R. S., & Vierstra, R. D. (2018). Autophagy: the master of bulk and selective recycling. *Annual Review of Plant Biology, 69*, 173–208. https://doi.org/10.1146/annurev-arplant-042817-040606.

Marwal, A., & Gaur, R. K. (2020). Host plant strategies to combat against viruses effector proteins. *Current Genomics, 21*, 401–410. https://doi.org/10.2174/1389202921999200712135131.

Medina-Puche, L., & Lozano-Duran, R. (2019). Tailoring the cell: a glimpse of how plant viruses manipulate their hosts. *Current Opinion in Plant Biology, Cell biology, 52*, 164–173. https://doi.org/10.1016/j.pbi.2019.09.007.

Michaeli, S., Clavel, M., Lechner, E., Viotti, C., Wu, J., Dubois, M., ... Genschik, P. (2019). The viral F-box protein P0 induces an ER-derived autophagy degradation pathway for the clearance of membrane-bound AGO1. *Proceedings of the National Academy of Sciences of the United States of America, 116*, 22872–22883. https://doi.org/10.1073/pnas.1912222116.

Michaeli, S., Honig, A., Levanony, H., Peled-Zehavi, H., & Galili, G. (2014). Arabidopsis ATG8-INTERACTING PROTEIN1 is involved in autophagy-dependent vesicular trafficking of plastid proteins to the vacuole. *The Plant Cell, 26*, 4084–4101. https://doi.org/10.1105/tpc.114.129999.

Mitou, G., Budak, H., & Gozuacik, D. (2009). Techniques to study autophagy in plants. *International Journal of Plant Genomics, 2009*, 451357. https://doi.org/10.1155/2009/451357.

Nagy, P. D., Pogany, J., & Kang, Y. (2025). Novel exploitation of autophagy by tombusviruses. *Virology, 603*, 110363. https://doi.org/10.1016/j.virol.2024.110363.

Nakahara, K. S., Masuta, C., Yamada, S., Shimura, H., Kashihara, Y., Wada, T. S., ... Uyeda, I. (2012). Tobacco calmodulin-like protein provides secondary defense by binding to and directing degradation of virus RNA silencing suppressors. *Proceedings of the National Academy of Sciences of the United States of America, 109*, 10113–10118. https://doi.org/10.1073/pnas.1201628109.

Niu, E., Ye, C., Zhao, W., Kondo, H., Wu, Y., Chen, J., ... Sun, L. (2022). Coat protein of Chinese wheat mosaic virus upregulates and interacts with cytosolic glyceraldehyde-3-phosphate dehydrogenase, a negative regulator of plant autophagy, to promote virus infection. *Journal of Integrative Plant Biology, 64*, 1631–1645. https://doi.org/10.1111/jipb.13313.

Nolan, T. M., Brennan, B., Yang, M., Chen, J., Zhang, M., Li, Z., ... Yin, Y. (2017). Selective autophagy of BES1 mediated by DSK2 balances plant growth and survival. *Developmental Cell, 41*, 33–46.e7. https://doi.org/10.1016/j.devcel.2017.03.013.

Ortega, M. A., Fraile-Martinez, O., de Leon-Oliva, D., Boaru, D. L., Lopez-Gonzalez, L., García-Montero, C., ... Alvarez-Mon, M. (2024). Autophagy in its (proper) context: Molecular basis, biological relevance, pharmacological modulation, and lifestyle medicine. *International Journal of Biological Sciences, 20*, 2532–2554. https://doi.org/10.7150/ijbs.95122.

Pan, Z. J., Wei, W. L., Tran, P. A., Fang, R. Y., Pham, T. H., Bowman, J. L., ... Lin, S. S. (2025). HC-Pro inhibits HEN1 methyltransferase activity, leading to autophagic degradation of AGO1. *Nature Communications, 16*, 2503. https://doi.org/10.1038/s41467-025-56320-z.

Pérez-Cañamás, M., Bustos, M., Puértolas, V., Castelló, Y., Peiró, S., & Hernández, C. (2022). Assessment of the RNA silencing suppressor activity of protein P0 of Pepper vein yellows virus 5: Uncovering natural variability, relevant motifs and underlying mechanism. *Biology (Basel), 11*, 1801. https://doi.org/10.3390/biology11121801.

Petersen, M., Avin-Wittenberg, T., Bassham, D. C., Dagdas, Y., Fan, C., Fernie, A. R., ... Hofius, D. (2024). Autophagy in plants. *Autophagy Reports, 3*, 2395731. https://doi.org/10.1080/27694127.2024.2395731.

Popp, M. W.-L., Cho, H., & Maquat, L. E. (2020). Viral subversion of nonsense-mediated mRNA decay. *RNA (New York, N. Y.), 26*, 1509–1518. https://doi.org/10.1261/rna.076687.120.

Qi, H., Wang, Y., Bao, Y., Bassham, D. C., Chen, L., Chen, Q.-F., ... Xiao, S. (2023). Studying plant autophagy: Challenges and recommended methodologies. *Advanced Biotechnology, 1*, 2. https://doi.org/10.1007/s44307-023-00002-8.

Ramage, H., & Cherry, S. (2015). Virus-host interactions: From unbiased genetic screens to function. *Annual Review of Virology, 2*, 497–524. https://doi.org/10.1146/annurev-virology-100114-055238.

Rao, G. P., & Reddy, M. G. (2020). *Overview of yield losses due to plant viruses*. Elsevier 531–562. https://doi.org/10.1016/B978-0-12-818654-1.00038-4.

Rasmussen, N. L., Kournoutis, A., Lamark, T., & Johansen, T. (2022). NBR1: The archetypal selective autophagy receptor. *The Journal of Cell Biology, 221*, e202208092. https://doi.org/10.1083/jcb.202208092.

Rogov, V. V., Nezis, I. P., Tsapras, P., Zhang, H., Dagdas, Y., Noda, N. N., ... Johansen, T. (2023). Atg8 family proteins, LIR/AIM motifs and other interaction modes. *Autophagy Reports, 2*, 2188523, 27694127.2023.2188523. https://doi.org/10.1080/27694127.2023.2188523.

Shukla, A., Hoffmann, G., Kushwaha, N. K., López-González, S., Hofius, D., & Hafrén, A. (2022). Salicylic acid and the viral virulence factor 2b regulate the divergent roles of autophagy during cucumber mosaic virus infection. *Autophagy, 18*, 1450–1462. https://doi.org/10.1080/15548627.2021.1987674.

Staiano, L., & Zappa, F. (2019). Hijacking intracellular membranes to feed autophagosomal growth. *FEBS Letters, 593*, 3120–3134. https://doi.org/10.1002/1873-3468.13637.

Sun, H., Jing, X., Wang, C., Wang, P., Huang, Z., Sun, B., ... Zhang, C. (2023). The great game between plants and viruses: A focus on protein homeostasis. *International Journal of Molecular Sciences, 24*, 12582. https://doi.org/10.3390/ijms241612582.

Tang, J., & Bassham, D. C. (2018). Autophagy in crop plants: What's new beyond Arabidopsis? *Open Biology, 8*, 180162. https://doi.org/10.1098/rsob.180162.

Tong, X., Liu, S.-Y., Zou, J.-Z., Zhao, J.-J., Zhu, F.-F., Chai, L.-X., ... Wang, X.-B. (2021). A small peptide inhibits siRNA amplification in plants by mediating autophagic degradation of SGS3/RDR6 bodies. *The EMBO Journal, 40*, e108050. https://doi.org/10.15252/embj.2021108050.

Tong, X., Zhao, J.-J., Feng, Y.-L., Zou, J.-Z., Ye, J., Liu, J., ... Wang, X.-B. (2023). A selective autophagy receptor VISP1 induces symptom recovery by targeting viral silencing suppressors. *Nature Communications, 14*, 3852. https://doi.org/10.1038/s41467-023-39426-0.

van Vliet, V. J. E., De Silva, A., Mark, B. L., & Kikkert, M. (2024). Viral deubiquitinating proteases and the promising strategies of their inhibition. *Virus Research, 344*, 199368. https://doi.org/10.1016/j.virusres.2024.199368.

Wang, J., Miao, S., Liu, Y., & Wang, Y. (2022). Linking autophagy to potential agronomic trait improvement in crops. *International Journal of Molecular Sciences, 23*, 4793. https://doi.org/10.3390/ijms23094793.

Wang, P., Pleskot, R., Zang, J., Winkler, J., Wang, J., Yperman, K., ... Hussey, P. J. (2019). Plant AtEH/Pan1 proteins drive autophagosome formation at ER-PM contact sites with actin and endocytic machinery. *Nature Communications, 10*, 5132. https://doi.org/10.1038/s41467-019-12782-6.

Whitmarsh-Everiss, T., & Laraia, L. (2021). Small molecule probes for targeting autophagy. *Nature Chemical Biology, 17*, 653–664. https://doi.org/10.1038/s41589-021-00768-9.

Wu, J., Michaeli, S., Picchianti, L., Dagdas, Y., Galili, G., & Peled-Zehavi, H. (2021). ATI1 (ATG8-interacting protein 1) and ATI2 define a plant starvation-induced reticulophagy pathway and serve as MSBP1/MAPR5 cargo receptors. *Autophagy, 17*, 3375–3388. https://doi.org/10.1080/15548627.2021.1872886.

Wu, W., Luo, X., & Ren, M. (2022). Clearance or hijack: Universal interplay mechanisms between viruses and host autophagy from plants to animals. *Frontiers in Cellular and Infection Microbiology, 11*, 786348. https://doi.org/10.3389/fcimb.2021.786348.

Wu, Y., Xu, R., & Zhuang, X. (2025). Multifaceted roles of the ATG8 protein family in plant autophagy: From autophagosome biogenesis to cargo recognition. *Journal of Molecular Biology, 437*, 168981. https://doi.org/10.1016/j.jmb.2025.168981.

Xu, G., Wang, S., Han, S., Xie, K., Wang, Y., Li, J., & Liu, Y. (2017). Plant bax inhibitor-1 interacts with ATG6 to regulate autophagy and programmed cell death. *Autophagy, 13*, 1161–1175. https://doi.org/10.1080/15548627.2017.1320633.

Yang, M., Ismayil, A., & Liu, Y. (2020). Autophagy in plant-virus interactions. *Annual Review of Virology, 7*, 403–419. https://doi.org/10.1146/annurev-virology-010220-054709.

Yang, M., Zhang, Y., Xie, X., Yue, N., Li, J., Wang, X.-B., ... Li, D. (2018). Barley stripe mosaic virus γb protein subverts autophagy to promote viral infection by disrupting the ATG7-ATG8 interaction. *The Plant Cell, 30*, 1582–1595. https://doi.org/10.1105/tpc.18.00122.

Yang, Z., & Li, Y. (2018). Dissection of RNAi-based antiviral immunity in plants. *Current Opinion in Virology, 32*, 88–99. https://doi.org/10.1016/j.coviro.2018.08.003.

Young, P. G., Passalacqua, M. J., Chappell, K., Llinas, R. J., & Bartel, B. (2019). A facile forward-genetic screen for Arabidopsis autophagy mutants reveals twenty-one loss-of-function mutations disrupting six ATG genes. *Autophagy, 15*, 941–959. https://doi.org/10.1080/15548627.2019.1569915.

Yuen, E. L. H., Leary, A. Y., Clavel, M., Tumtas, Y., Mohseni, A., Zhao, J., ... Bozkurt, T. O. (2024). A RabGAP negatively regulates plant autophagy and immune trafficking. *Current Biology: CB, 34*, 2049–2065.e6. https://doi.org/10.1016/j.cub.2024.04.002.

Zeng, H.-Y., Zheng, P., Wang, L.-Y., Bao, H.-N., Sahu, S. K., & Yao, N. (2019). Autophagy in plant immunity. *Advances in Experimental Medicine and Biology, 1209*, 23–41. https://doi.org/10.1007/978-981-15-0606-2_3.

Zhang, X., Hong, H., Yan, J., Yuan, Y., Feng, M., Liu, Q., ... Tao, X. (2024). Autophagy plays an antiviral defence role against tomato spotted wilt orthotospovirus and is counteracted by viral effector NSs. *Molecular Plant Pathology, 25*, e70012. https://doi.org/10.1111/mpp.70012.

Zhang, X., Yin, Y., Su, Y., Jia, Z., Jiang, L., Lu, Y., ... Yan, F. (2021). eIF4A, a target of siRNA derived from rice stripe virus, negatively regulates antiviral autophagy by interacting with ATG5 in Nicotiana benthamiana. *PLoS Pathogens, 17*, e1009963. https://doi.org/10.1371/journal.ppat.1009963.

Zhang, Y., & Chen, Z. (2020). Broad and complex roles of NBR1-mediated selective autophagy in plant stress responses. *Cells, 9*, 2562. https://doi.org/10.3390/cells9122562.

Zhao, W., Wang, L., Li, L., Zhou, T., Yan, F., Zhang, H., ... Sun, L. (2023). Coat protein of rice stripe virus enhances autophagy activity through interaction with cytosolic glyceraldehyde-3-phosphate dehydrogenases, a negative regulator of plant autophagy. *Stress Biology, 3*, 3. https://doi.org/10.1007/s44154-023-00084-3.

Zhou, J., Wang, J., Cheng, Y., Chi, Y.-J., Fan, B., Yu, J.-Q., & Chen, Z. (2013). NBR1-mediated selective autophagy targets insoluble ubiquitinated protein aggregates in plant stress responses. *PLoS Genetics, 9*, e1003196. https://doi.org/10.1371/journal.pgen.1003196.

Zhou, X., Zhao, P., Wang, W., Zou, J., Cheng, T., Peng, X., & Sun, M. (2015). A comprehensive, genome-wide analysis of autophagy-related genes identified in tobacco suggests a central role of autophagy in plant response to various environmental cues. *DNA Research: an International Journal for Rapid Publication of Reports on Genes and Genomes, 22*, 245–257. https://doi.org/10.1093/dnares/dsv012.

Zhuang, X., Chung, K. P., Luo, M., & Jiang, L. (2018). Autophagosome biogenesis and the endoplasmic reticulum: A plant perspective. *Trends in Plant Science, 23*, 677–692. https://doi.org/10.1016/j.tplants.2018.05.002.

Zvereva, A. S., Golyaev, V., Turco, S., Gubaeva, E. G., Rajeswaran, R., Schepetilnikov, M. V., ... Pooggin, M. M. (2016). Viral protein suppresses oxidative burst and salicylic acid-dependent autophagy and facilitates bacterial growth on virus-infected plants. *The New Phytologist, 211*, 1020–1034. https://doi.org/10.1111/nph.13967.

CHAPTER THREE

Mitophagy's impact on obesity-related tissues

Rana El Nahas[a], Sandra Cutanda-Tesouro[b], Meritxell Espino-Guarch[a,*], and Pablo M. Garcia-Roves[b,c,d,*]
[a]Laboratory of Immunoregulation, Research Department, Sidra Medicine, Doha, Qatar
[b]Department of Physiological Sciences, Universitat de Barcelona, Barcelona, Spain
[c]Metabolism and Gene Therapy Group, Diabetes and Metabolism Program, Institut D'Investigació Biomèdica de Bellvitge (IDIBELL), Barcelona, Spain
[d]Instituto de Salud Carlos III, Centro de Investigación Biomédica en Red Fisiopatología de La Obesidad y La Nutrición (CIBEROBN), Madrid, Spain
*Corresponding authors. e-mail address: mespinoguarch@sidra.org; pgarciaroves@ub.edu

Contents

Abstract

Obesity, a global health crisis, results from an energy imbalance, leading to metabolic dysfunction and associated conditions such as type 2 diabetes, cardiovascular disease, and metabolic dysfunction-associated steatotic liver disease. A strong link exists between obesity and mitochondrial dysfunction, characterized by reduced mitochondrial mass, impaired oxidative capacity, and decreased ATP production.

International Review of Cell and Molecular Biology, Volume 406
ISSN 1937-6448, https://doi.org/10.1016/bs.ircmb.2025.12.003

Mitophagy, a specific type of autophagy targeting mitochondria for degradation, plays a complex, tissue-specific role in regulating metabolism and mitigating the harmful effects of obesity. Both insufficient and excessive mitophagy can negatively influence disease progression. In white adipose tissue, mitophagy functions as a quality control mechanism that reduces oxidative stress and helps maintain insulin sensitivity. However, chronic obesity impairs mitophagy due to overactivation of mTORC1 and inhibition of AMPK, leading to inflammation and insulin resistance. In brown adipose tissue, mitophagy is essential for thermogenesis and energy expenditure, but excessive activation in obesity may impair thermogenesis. In the liver, mitophagy is crucial for preserving mitochondrial function and preventing metabolic dysfunction-associated steatotic liver disease. In the heart, mitophagy supports cardiac function, particularly in obesity-related cardiomyopathy. In skeletal muscle, obesity worsens impairments in mitochondrial quality control, disrupting the clearance of damaged mitochondria. Therefore, there is a pressing need for therapies that restore mitophagic balance in a context- and depot-specific manner, as the dysregulation of mitophagy contributes not only to local dysfunction but also to broader metabolic phenotypes.

Abbreviations

AMPK	AMP-activated protein kinase.
ATG	Autophagy-related.
ATP	Adenosine Triphosphate.
BAT	Brown Adipose Tissue.
BMI	Body Mass Index.
BNIP3	A mitophagy receptor BCL2/adenovirus E1B 19 kDa interacting protein 3.
ClpP	A mitochondrial matrix protease involved in proteostasis.
CPF	Chlorpyrifos.
DRP1	Fission-related GTPase.
ER	Endoplasmic Reticulum.
FIS1	Fission protein 1.
FFA	Free Fatty Acids.
FNDC5	Fibronectin type III domain-containing protein 5.
FUNDC1	An OMM protein that responds to hypoxia and cellular energy status and contains a LIR motif.
GSIS	Glucose-stimulated insulin secretion.
HFD	High-Fat Diet.
HIIT	High-Intensity Interval Training.
HO-1	Heme Oxygenase-1.
HSP72	Heat Shock Protein 72.
IMM	Inner Mitochondrial Membrane.
I/R	Ischemia-Reperfusion.
IR	Insulin Resistance.
KO	Knockout.
LC3	Microtubule-associated protein 1 A/1B-light chain 3.
LIR	LC3-Interacting Region.
LMPTP	Liver-Specific Low Molecular Weight Protein Tyrosine Phosphatase.

MASLD	Metabolic dysfunction-associated steatotic liver disease.
MICT	Moderate-Intensity Continuous Training.
MT-CYB	Mitochondrially Encoded Cytochrome B.
MUL1	Mitochondrial E3 ubiquitin ligases.
NAFLD	Non-Alcoholic Fatty Liver Disease.
NIX	A protein closely related to BNIP3.that shares structural.functional.and activation similarities with BNIP3.
NL-SDT	Non-lethal sonodynamic therapy.
OMM	Outer Mitochondrial Membrane.
OPA1	Mitochondrial Dynamin Like GTPase.
OXPHOS	Oxidative Phosphorylation.
PBMCs	Peripheral Blood Mononuclear Cells.
PGC-1α	Peroxisome Proliferator-Activated Receptor Gamma Coactivator 1-Alpha.
PINK1	PTEN-Induced Kinase 1.
PPARs	Peroxisome Proliferator-Activated Receptors.
REDD1	Regulated in Development and DNA Damage Response-1.
RIPK3	Receptor-Interacting Protein Kinase 3.
ROS	Reactive Oxygen Species.
T2D	Type 2 Diabetes.
Trx2	Thioredoxin-2.
ULK1	Unc-51-like kinase 1.
UCP1	Uncoupling protein 1.
VCP	Valosin-containing protein.
WAT	White Adipose Tissue.
WHO	World Health Organization.

1. Introduction

1.1 Obesity: A growing global health challenge in the 21st century

Obesity constitutes a worldwide public health epidemic, being a significant threat to global public health. According to the World Health Organization (WHO), individuals with a body mass index (BMI) equal to or greater than 30 are classified as obese. Since 1990, the number of adults affected by obesity has doubled, although notable regional disparities exist. The highest prevalence is observed in Africa (Northern Africa and South Africa) and the Americas (USA and Mexico) (Obesity, 2022) where rates reach, in some regions, up to 67 %.

Individuals with obesity typically exhibit excessive body weight and a persistent imbalance between caloric intake and energy expenditure. The development of obesity is multifactorial, influenced by genetic, environmental, and social determinants. Modern societal shifts in diet and daily activity patterns play a critical role in its onset. Specifically, food choices have increasingly shifted toward lower-quality, ultra-processed products, as healthy options become less economically accessible. This trend, coupled with a continuous rise in sedentary lifestyles, endangers the overall health of the global population (Bluher, 2019; Gonzalez-Muniesa, Martinez-Gonzalez et al. 2017).

Excess nutrition and/or inactivity over an extended period contribute to the maladaptation of primary organs (such as adipose tissue, liver, skeletal muscle, and pancreas), in combination with genetic and epigenetic factors, lead to insulin resistance (IR) and obesity-related Type 2 Diabetes (T2D) development (Czech, 2017).

1.2 Obesity-induced metabolism dysfunction

In the context of obesity, adipocytes, which are the primary cells in the organism for fat storage, undergo both hypertrophy (increase in cell size) and hyperplasia (increase in cell number) in an attempt to adapt to the stress caused by excessive lipid accumulation, a phenomenon associated with lipotoxicity (Slawik & Vidal-Puig, 2006; White & Stephens, 2010).

Hypertrophied adipocytes loss their ability to secrete adipokines and act as endocrine cells (Gilani, Stoll et al., 2024). In turn, lipotoxicity leads to the induction of a pro-inflammatory state as well as the induction of hypoxia within the WAT, through the ability of circulating free fatty acids (FFA) to induce uncoupling of mitochondrial oxidative metabolism leading to an increase in oxygen consumption and the induction of a hypoxic state (Lee, Kim et al., 2014). The lack of sufficient WAT oxygenation over time leads to a chronic hypoxic state, further exacerbating the inflammatory environment and induces a chronic pro-inflammatory condition leading to adverse metabolic effects (Kusminski, Bickel et al., 2016). Such changes affect not only the metabolic function of WAT but also its extracellular matrix, which becomes increasingly fibrotic (Dahdah, Tercero-Alcazar et al., 2024). When the storage capacity of WAT depots becomes saturated or dysfunctional, excess lipids are redirected and deposited ectopically in critical organs such as the liver, pancreas, skeletal muscle, and cardiac muscle, among others. This ectopic fat accumulation adversely impacts the function of these tissues and contributes to systemic metabolic disturbances and, hence, increasing the incidence of metabolic diseases (Guilherme, Virbasius et al., 2008).

Strongly linked to obesity is mitochondrial dysfunction, characterized by reduced mitochondrial mass, impaired oxidative capacity, and decreased ATP production in disease related organs (Pinti, Fink et al., 2019). An inverse correlation between mitochondrial function and diabetic complications had been observed in obese and T2D patients (Czajka, Ajaz et al., 2015; Hoffman & Brookes, 2009; Khan, Raghuram et al., 2011). Studies have shown that obesity-induced mitochondrial dysfunction in adipose tissue does not fully return to baseline levels following lifestyle interventions, indicating a loss of plasticity (Gonzalez-Franquesa et al., 2022; Morino, Petersen et al.. 2005). Additionally, analysis of mitochondrial transfer has revealed that adipocytes possess the ability to expel mitochondria into the extracellular space in small vesicles, which can then travel through the bloodstream to aid multiorgan responses to nutrient stress (Crewe, Funcke et al., 2021), a process that is significantly disrupted following exposure to a western diet (Borcherding, Jia et al., 2022).

Several genes implicated in obesity are epigenetically regulated, where obesity associated modulation has significant and lasting consequences on pathways effecting hypoxia, inflammatory cytokines, mitochondrial function, dyslipidaemia, glucose homeostasis, hormonal imbalance, and endocrine disruptors (Gonzalez-Franquesa et al., 2022; Hinte, Castellano-Castillo et al., 2024; Mahmoud, 2022). Post-intervention epigenetic signature of adipose tissue, mesothelial populations, adipose progenitor cells, PBMCs, and muscle in formerly obese individuals shows persistent dysregulation indicating an incomplete reversal or a lasting obesogenic imprint driven by functional and metabolic loss of plasticity (Fischer, Irmler et al., 2018; Gonzalez-Franquesa et al., 2022; Hinte, Castellano-Castillo et al., 2024; Jacobsen, Brons et al., 2012; Reinisch, Ghosh et al., 2025).

1.3 Mitochondrial dysregulation in obesity-associated disorders

Responses in primary organs to excess nutrition lead to development of metabolic disease, as increased energy substrate levels including circulating glucose, triglycerides, and fatty acids, trigger IR, dyslipidemia, elevated mitochondria-derived ROS, lipotoxicity, and systemic inflammation (Muoio & Newgard, 2008). These elements collectively act as key determinants of obesity-associated pathophysiology among the most prevalent are T2D, cardiovascular disease, certain types of cancer, and metabolic dysfunction-associated steatotic liver disease (MASLD) (Longo, Zatterale et al., 2019; Slawik & Vidal-Puig, 2006).

Excessive fat accumulation within adipose tissue and/or other organs leads to dysregulation of mitochondrial homeostasis. In such conditions, the dynamic and regulatory mechanisms of mitochondria are compromised by the lipotoxic environment (Ahmed, Sultana et al., 2021; Boone & Lewis, 2024; Mu, Wang et al., 2025; Sethi & Vidal-Puig, 2007; Slawik & Vidal-Puig, 2006). For instance, in MASLD, mitochondria exhibit impairments in the oxidative phosphorylation (OXPHOS) system, resulting in increased oxygen consumption that is uncoupled from ATP synthesis. This dysfunction is accompanied by elevated production of reactive oxygen species (ROS), further exacerbating oxidative stress (Stefan, Yki-Jarvinen et al., 2025). Lipotoxicity also contributes to the pathogenesis of diabetic cardiomyopathy. In this condition, cardiac mitochondria display an increased reliance on fatty acid oxidation, producing metabolic intermediates that promote mitochondrial stress and dysfunction (Qian, Tian et al., 2020).

Mitochondria are highly dynamic organelles capable of modifying their structure, activity, and dynamics in response to nutrient availability. However, in the context of nutrient excess, as seen in obesity, an increase in nutrient supply does not necessarily translate into increased ATP production. The capacity of OXPHOS and ATP synthesis is limited by enzymatic constraints. When these limits are exceeded, mitochondrial oxygen consumption becomes increasingly uncoupled from ATP production, a phenomenon known as "LEAK", which results in elevated production of ROS.

In addition to compromising mitochondrial efficiency, nutrient overload alters mitochondrial dynamics, including fusion, fission, biogenesis, and the turnover of mitochondria and their components. In obesity, mitochondrial fission is upregulated while fusion processes are suppressed, leading to excessive mitochondrial fragmentation and further promoting ROS generation. Autophagy and mitophagy are also inhibited, impairing the organelle's quality control and repair mechanisms under conditions of energetic inefficiency. Moreover, mitochondrial biogenesis is disrupted in the presence of elevated ROS levels (Liesa & Shirihai, 2013; Protasoni & Zeviani, 2021).

Thus, under these conditions of chronic energy excess, such as obesity, increased mitochondrial fission and ROS production hinder mitophagy and mitochondrial renewal, resulting in the accumulation of damage and the dysfunction in the onset of mitochondrial. This dysfunction not only affects ATP production and ROS balance but also impacts other key mitochondrial functions. These include regulation of apoptosis, interorganelle communication, and the modulation of epigenetic marks. The failure of these processes contributes to tissue dysfunction and aging (Monzel, Enriquez et al., 2023).

1.4 Mitophagy

Mitophagy is a specific type of autophagy that targets mitochondria for degradation through distinct mechanisms. During this process, damaged mitochondria are selectively eliminated, allowing for the generation of new, more efficient organelles as the mitochondrial pool is renewed. This mechanism is essential for maintaining cellular energy balance, preventing oxidative damage caused by excessive ROS, and ensuring the proper functioning of apoptotic pathways and cellular quality control systems (Lu, Li et al., 2023). In pathological conditions, decreased mitophagy activity have been observed, potentially leading to the accumulation of dysfunctional mitochondria (Lu, Li et al., 2023).

Fatty acids are activators of the transcription of Peroxisome Proliferator-Activated Receptors (PPARs), thereby enhancing the activity of the fatty acid oxidation machinery, mitochondrial biogenesis, and uncoupling enzymes (Garcia-Roves, Huss et al., 2007). However, in an environment of lipid oversupply mitochondria become progressively dysfunctional due to the high levels of damage caused by ROS, leading to the disruption of key mitochondrial functions such as energy production, biogenesis, dynamic processes, and cellular quality control and homeostatic mechanisms, including mitophagy (Prasun, 2020).

1.4.1 General mechanisms of mitophagy

As mentioned before, mitophagy is a complex process that enables the selective elimination of damaged mitochondria and the renewal of the mitochondrial pool. Mitochondria are marked for degradation through various signaling pathways and subsequently engulfed by double-membrane structures formed by the autophagic machinery. They are then transported to lysosomes for degradation. Multiple regulatory mechanisms and labeling strategies exist to identify these dysfunctional mitochondria. Below, we provide a brief overview of the main general pathways involved in the recognition and subsequent degradation of damaged mitochondria.

1.4.2 PINK1/Parkin-mediated mitophagy

PINK1 (PTEN-induced kinase 1) is a protein synthesized in the cytosol and normally imported into healthy mitochondria. This import relies on the mitochondrial membrane potential and allows PINK1 to reach the inner mitochondrial membrane (IMM), where it is cleaved by proteases. The cleaved fragment is then released into the cytosol and rapidly degraded by the proteasome. This continuous processing prevents PINK1 accumulation and keeps it inactive under normal conditions. Proper mitochondrial import, intact membrane potential, and efficient degradation are all essential to maintaining this inactive state. When any of these regulatory mechanisms fail, PINK1 accumulates on the outer mitochondrial membrane (OMM), where it phosphorylates OMM proteins. At this point, the E3 ubiquitin-ligase Parkin is recruited and activated by PINK1. Parkin ubiquitinates the phosphorylated substrates on the OMM (Eiyama & Okamoto, 2015), marking them for recognition by autophagy adaptor proteins, which interact with autophagy-related (ATG)-like components of the autophagic machinery. Ultimately, these marked mitochondria are engulfed within a double-membrane structure and targeted for lysosomal degradation (Wang, Long et al., 2023; Zimmermann, Madeo et al., 2024) (Fig. 1).

1.4.3 Receptor-mediated mitophagy

Several OMM proteins such as BNIP3, NIX, MUL-1 and FUNDC1, contain LC3-interacting region (LIR) motifs that directly interact with LC3, a canonical protein in autophagy, to initiate mitophagy (Fig. 2).

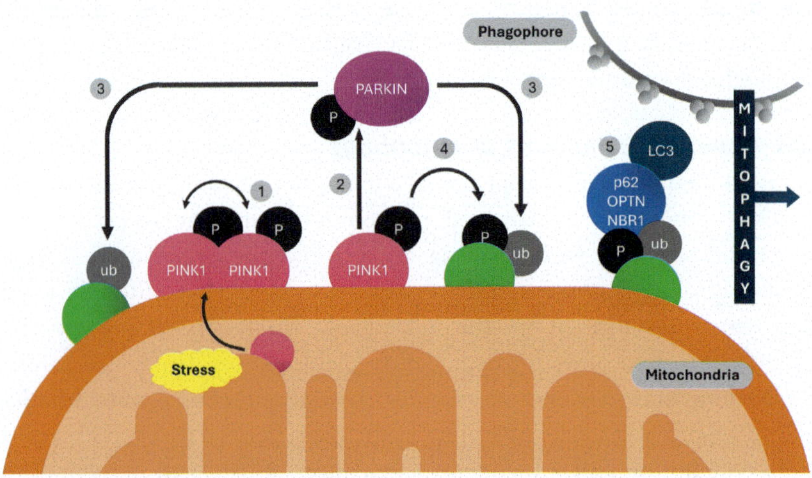

Fig. 1 PINK1/Parkin-mediated mitophagy. 1. In response to stress stimuli, PINK1 is not translocated to the inner mitochondrial membrane (IMM) and instead accumulates at the outer mitochondrial membrane (OMM). PINK1 dimerizes and undergoes autophosphorylation. 2. PINK1 phosphorylates (P) Parkin to promote its E3 ubiquitin ligase activity. 3. Activated Parkin ubiquitinates (ub) various OMM-resident proteins such as Mfn1, Mfn2, and VDAC1. 4. These ubiquitinated proteins are subsequently recognized and further phosphorylated by PINK1. 5. The proteins marked by ubiquitination and phosphorylation are then recognized by adaptor proteins (p62, OPTN, and NBR1), which contain a LC3-interacting region (LIR) domain that enables the recruitment of LC3. Finally, LC3 interacts with ATG-family proteins to form the phagophore around the tagged mitochondrion.

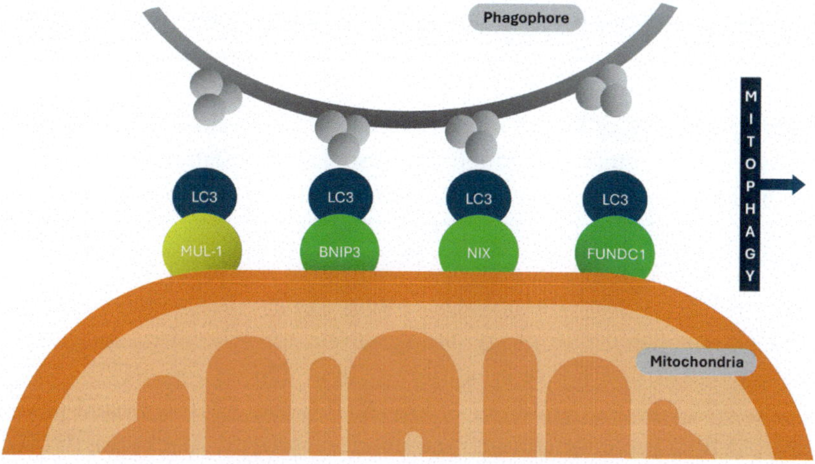

Fig. 2 Receptor-mediated mitophagy. Certain OMM proteins such as MUL1, BNIP3, NIX, and FUNDC1 are activated under stress conditions. These proteins contain an LIR domain that, when in an active state, recruits LC3, thereby allowing the engagement of ATG-family proteins and the initiation of mitophagy.

BNIP3, a mitophagy receptor, is found in the cytosol in its inactive, monomeric state. In response to stress signals, such as hypoxia (Onishi, Yamano et al., 2021) BNIP3 forms a homodimer that inserts into the OMM. Its N-terminal domain, exposed to the cytosol, contains LIR motif, which facilitates the recruitment of the canonical autophagic machinery and the subsequent removal of the targeted mitochondrion.

NIX is a closely related protein that shares structural, functional, and activation similarities with BNIP3. Like BNIP3, NIX also contains a LIR motif and recruits ATG family proteins to mediate mitophagy through its LIR motif. NIX responds to oxidative stress induced by ROS. While both BNIP3 and NIX are capable of independently initiating mitophagy, evidence also indicates that they can cooperate with the PINK1/Parkin pathway. Specifically, BNIP3 contributes to the stabilization and accumulation of PINK1 on the OMM, while NIX is among the OMM proteins ubiquitinated by Parkin (Onishi, Yamano et al., 2021).

FUNDC1 is an OMM protein that responds to hypoxia and cellular energy status and contains a LIR motif. Under normal conditions, this LIR motif is phosphorylated, which prevents autophagy maker (LC3) binding. Upon stress stimuli, the LIR becomes dephosphorylated, allowing interaction with LC3 and the initiation of mitophagy through the ATG proteins. Simultaneously, the interaction between FUNDC1 and OPA1

(IMM fusion protein) is disrupted, thereby inhibiting mitochondrial fusion. At the same time, FUNDC1 can promote mitochondrial fission by activating DRP1 (fission-related GTPase), contributing to mitochondrial quality control and cellular homeostasis (Chen, Liu et al., 2017). FUNDC1 activity is also regulated through proteasomal degradation by the mitochondrial E3 ubiquitin ligase MARCH5, which targets it for turnover and limits its mitophagic function (Zhou, Zhu et al., 2017). Additionally, FUNDC1 function can be negatively regulated through phosphorylation by RIPK3 (Receptor-Interacting Protein Kinase 3), which suppresses FUNDC1-mediated mitophagy and promotes mitochondrial apoptosis in ischemia/reperfusion injury conditions (Zhou, Zhu et al., 2017).

Conversely, under conditions of low ATP levels, AMP-activated protein kinase (AMPK) becomes activated and stimulates Unc-51-like kinase (ULK1), which in turn phosphorylates FUNDC1 (Ser555) to promote mitophagy (Cai, Guo et al., 2022). Interestingly, AMPK can also negatively regulate ULK1 activity under hypoxic conditions (Longo, Bishnu et al., 2024).

MUL1 (Mitochondrial Ubiquitin Ligase 1) is another OMM–anchored E3 ubiquitin ligase that can act as a mitophagy receptor via its LIR-mediated interaction with LC3. MUL1 promotes mitophagy independently of the PINK1/Parkin pathway by interacting with GABARAP and inducing autophagic degradation of damaged mitochondria, particularly under stress conditions. It also regulates mitochondrial dynamics by ubiquitinating mitofusin 2 (MFN2), thereby promoting fission, and by SUMOylating DRP1 to stabilize it. (Calle, Garrido-Moreno et al., 2022).

1.4.4 Alternative mitophagy

In certain contexts, mitophagy can occur independently of the canonical PINK1-Parkin or receptor/LC3-dependent pathways. Autophagosomes may form from Golgi-derived membranes, instead of a cell membrane-lysosome, in a process that bypasses the requirement for Atg5 and Atg7. This mechanism is referred to as *alternative mitophagy* or *ULK1-dependent autophagy*. ULK1 is activated through phosphorylation by AMPK or RIPK3 and subsequently initiates autophagosome formation by activating key proteins such as Beclin1, RIP1, and Rab9. Additionally, activation of Drp1 promotes mitochondrial fission, enabling partial recycling of mitochondrial components. This alternative pathway has been particularly described in cardiac tissue and has been shown to play a protective role in various pathological conditions, including ischemia and obesity-related cardiomyopathy (Tong, Saito et al., 2021) (Fig. 3).

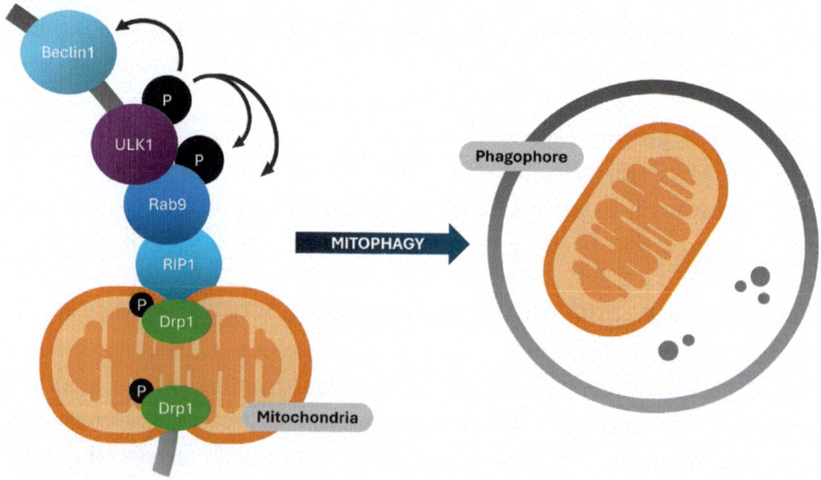

Fig. 3 Alternative mitophagy. Upon phosphorylation, ULK1 becomes activated and is capable of recruiting and activating key autophagy regulators such as Beclin1, Rab9, and RIP1. These molecules cooperate to promote the formation of a phagophore in a manner independent of the canonical pathway. Concurrently, the mitochondrial fission factor Drp1 is also activated to mediate the fragmentation of mitochondria. This fission event facilitates the isolation of dysfunctional mitochondrial fragments, allowing their complete engulfment by the forming autophagosome. The resulting structure enables the targeted degradation of damaged mitochondrial components via the lysosomal pathway.

2. Mitophagy in adipose tissue in the context of obesity

Adipose tissue, comprising white (WAT) and brown (BAT) depots, plays a central role in energy storage, endocrine signaling, and thermogenesis. Mitochondria are critical to these functions, especially in regulating lipid metabolism, oxidative balance, and adaptive responses to energy demand. Maintaining mitochondrial quality through mitophagy is therefore essential for adipocyte function.

In healthy states, mild mitochondrial stress triggers mitohormesis, a protective process that activates key regulators such as PGC-1α, AMPK, and SIRT1/3. These pathways enhance mitochondrial turnover, biogenesis, and antioxidant defenses (Ramanathan, Ali et al., 2022; Rosa-Caldwell, Poole et al., 2022; Su, Guo et al., 2021). Mitophagy is mediated by both ubiquitin-dependent mechanisms, primarily the PINK1/Parkin axis and receptor-mediated autophagy involving proteins like BNIP3,

FUNDC1, and ULK1 (Jun, Tao et al., 2024). These pathways coordinate the removal of damaged mitochondria, supporting energy balance in WAT and thermogenic capacity in BAT.

While the precise contribution of each mitophagy pathway varies by depot and physiological state, both WAT and BAT rely on intact mitochondrial turnover to preserve metabolic health. The next section discusses how these processes become dysregulated under conditions of obesity.

2.1 Mitophagy and mitochondrial dysfunction in adipose tissue

WAT is a central organ in systemic energy regulation, functioning not only as an energy reservoir but also as an endocrine tissue that secretes adipokines and modulates immune and metabolic signals across the body. In obesity, WAT undergoes pathological remodeling characterized by adipocyte hypertrophy, macrophage infiltration, extracellular matrix expansion, and altered insulin signaling, which contribute to systemic IR and chronic low-grade inflammation (Jun, Tao et al., 2024; Liao, Xiong et al., 2023). These alterations are closely associated with mitochondrial dysfunction, which is emerging as a pivotal contributor to obesity-related metabolic disease.

Adipose tissue, consisting of both energy-storing WAT and thermogenic BAT, is particularly sensitive to disruptions in mitochondrial quality control due to its central role in energy regulation. Obesity-driven mitochondrial dysfunction in adipose tissue results from disrupted mitochondrial dynamics (impaired fission and fusion), heightened oxidative stress, and compromised mitochondrial biogenesis (Jun, Tao et al., 2024). Persistent metabolic stress associated with chronic obesity disrupts mitohormetic responses. Sustained oxidative damage and nutrient excess lead to mTORC1 overactivation and AMPK inhibition, ultimately impairing mitophagy and worsening metabolic dysfunction (Lin, Li et al., 2021; Ro, Jang et al., 2019; Taylor & Gottlieb, 2017).

Impaired mitophagy in adipose tissue directly correlates with distinct tissue-specific pathologies observed in obesity. In WAT, defective mitophagy contributes to local inflammation, endoplasmic reticulum stress, and exacerbated IR, as supported by both animal and human studies (Delgado-Angles, Blasco-Roset et al., 2024; Liao, Xiong et al., 2023). Restoration of mitophagy in obese mice has been shown to reduced proinflammatory cytokine expression, improve insulin sensitivity, and mitigated obesity-associated metabolic disturbances. In contrast, in BAT, where thermogenic activity relies on high mitochondrial content and quality, mitophagy dysfunction impairs energy expenditure and heat production. Experimental enhancement of mitophagy in BAT via pharmacologic or genetic approaches has improved mitochondrial function and restored thermogenesis, underscoring its therapeutic potential (Taylor & Gottlieb, 2017; Zhu, Gu et al., 2022).

Given this context, understanding the regulation of mitophagy in adipose tissue is critical. The subsequent sections explore canonical regulatory pathways, including PINK1/Parkin (ubiquitin-dependent) and FUNDC1/ULK1 (autophagy-dependent) axes, as well as emerging modulators of mitophagy. Comparative insights from human, murine, and in vitro models are also discussed to provide a comprehensive understanding of mitophagy's pivotal role in obesity-linked adipose tissue dysfunction.

2.2 Mitophagy mechanisms in brown adipose tissue

BAT is distinguished by its high mitochondrial content and its capacity to generate heat through non-shivering thermogenesis, especially in response to cold or metabolic stress (Neira, Hernandez-Pardos et al., 2024). The maintenance of mitochondrial quality through mitophagy is essential for supporting BAT's thermogenic function and overall energy homeostasis. However, the regulation of mitophagy in BAT is highly context-dependent and becomes increasingly complex under obesity and metabolic stress.

Some studies suggest that enhancing mitophagy improves mitochondrial turnover and thermogenic efficiency. For example, thyroid hormone (T3) promotes mitophagy via mTOR inhibition and SIRT1 activation in non-obese mice, supporting mitochondrial quality control (Yau, Singh et al., 2019). In contrast, findings from obese human and rodent models indicate that suppressed mitophagy may preserve mitochondrial content and sustain thermogenic gene expression, such as UCP1, during cold adaptation (Neira, Hernandez-Pardos et al., 2024).

Thus, mitophagy regulation in BAT is not unidirectional. Its optimal level likely depends on the metabolic state, where both excessive suppression and overactivation may impair thermogenic function. The following sections examine the major pathways that regulate mitophagy in BAT and how they are altered in the context of obesity and metabolic dysfunction.

2.2.1 PINK1/Parkin, ubiquitin-dependent mitophagy in BAT

BAT is essential for non-shivering thermogenesis, a process highly dependent on mitochondrial integrity and function. The PINK1/Parkin pathway is a key regulator of mitophagy in BAT, particularly under conditions of metabolic stress. In a 4-month high-fat diet (HFD) rat model, BAT displayed impaired mitophagic activity, accompanied by reduced expression of FNDC5/Irisin, structural mitochondrial abnormalities, and downregulation of UCP1 and mitochondrial genes such as MT-CYB (Neira, Hernandez-Pardos et al., 2024). FNDC5 has been proposed to modulate mitophagy by regulating PINK1/Parkin-mediated clearance of damaged mitochondria, particularly during cold adaptation (Yau, Singh et al., 2019). In obesity, reduced FNDC5 expression may lead to inadequate mitophagy, impaired coordination with mitochondrial biogenesis, and compromised thermogenic capacity.

Additional evidence comes from PINK1 knockout (KO) mice exposed to HFD, which exhibited mitochondrial dysfunction and increased activation of the NLRP3 inflammasome in BAT (Yau, Singh et al. 2019). This highlights a potential mitochondria–inflammasome axis, in which defective mitophagy contributes to inflammation and thermogenic failure in metabolically stressed adipose tissue.

2.2.2 FUNDC1/Vaspin/ULK1 – autophagy-dependent mitophagy in BAT

Although direct studies of ULK1-dependent pathway in BAT under obesity are limited, indirect evidence suggests that this pathway plays a role in mitochondrial remodeling. In lean mice, thyroid hormone (T3) stimulates mitophagy in BAT and primary brown adipocytes through mTOR inhibition and SIRT1 activation, both of which are upstream modulators of ULK1 activity leading to enhanced mitochondrial turnover and improved thermogenic efficiency (Yau, Singh et al., 2019). While these findings were observed in the absence of obesity, they suggest that ULK1-mediated mitophagy could contribute to BAT remodeling in response to metabolic cues and may be susceptible to disruption in obesity.

Similarly, proteins such as Vaspin and FUNDC1, although primarily studied in other tissues, have been implicated in hypoxia-responsive autophagy and could influence mitophagy in BAT (Wu, Lin et al., 2016). Given the importance of mitochondrial quality for BAT thermogenesis and the responsiveness of ULK1 and SIRT1 to nutrient and hormonal changes, further studies are needed to clarify the role of this axis in BAT under obese conditions.

2.2.3 *Other regulatory influences on mitophagy in BAT*

BNIP3 plays a critical role in regulating mitophagy and mitochondrial dynamics in BAT, particularly in response to metabolic stress. It is naturally enriched in BAT and becomes upregulated during adipogenesis, starvation, and treatment with PPARγ agonists such as rosiglitazone (Choi, Jo et al., 2016). In models of obesity, including diet-induced and db/db mice, BNIP3 expression is significantly reduced in BAT adipocytes, correlating with impaired mitochondrial function and thermogenic decline. Restoration of BNIP3 via PPARγ activation improves mitochondrial integrity and metabolic outcomes (Song, Li et al., 2025). Mechanistically, BNIP3 facilitates mitochondrial fission, a necessary step in mitophagy. Its knockdown leads to mitochondrial elongation, oxidative stress, and reduced glucose uptake, underscoring its role in maintaining mitochondrial health.

This role is conserved across species. In a goat model, 12-hour heat exposure increased the expression of BNIP3, LC3B, and PINK1 in BAT, along with elevated levels of UCP1 and PGC1α, suggesting a conserved role for BNIP3-mediated mitophagy in thermogenic adaptation to environmental stress (Song, Li et al., 2025).

ClpP, a mitochondrial matrix protease involved in proteostasis, has also been implicated in BAT mitophagy regulation. In HFD-fed ClpP KO mice, BAT exhibited reduced mitophagy activity and lower expression of UCP1 and other thermogenic markers (Liesa & Shirihai, 2018). These findings suggest that ClpP may function as a regulatory switch balancing mitophagy and thermogenesis in brown adipose tissue.

2.3 Mitophagy mechanisms in white adipose tissue

While mitophagy in BAT is closely tied to thermogenic regulation, its role in WAT appears to be more focused on maintaining cellular integrity under metabolic stress.

WAT primarily serves as an energy reservoir, storing excess energy as triglycerides. During obesity, WAT undergoes significant remodeling characterized by adipocyte hypertrophy, increased inflammation, and mitochondrial stress. While reduced mitophagy in BAT under obese or thermogenic conditions has been proposed to preserve mitochondrial content and support heat production (Neira, Hernandez-Pardos et al., 2024), the role of mitophagy in WAT appears to follow a different trajectory. In WAT, where mitochondrial density is lower and oxidative stress is heightened during obesity, mitophagy acts primarily as a protective quality control mechanism. However, both insufficient and excessive mitophagy in WAT can lead to adverse metabolic consequences, highlighting the need for tissue-specific and context-dependent interpretation of mitophagy regulation.

2.3.1 PINK1/Parkin – ubiquitin-dependent mitophagy in WAT

Among the key regulators of mitophagy in WAT, the PINK1/Parkin pathway has been most extensively studied, particularly for its dual role in both protecting and impairing metabolic health depending on the level and context of its activation. The PINK1/Parkin pathway plays a critical role in WAT mitophagy by mediating the selective clearance of damaged mitochondria in response to metabolic stress (Ko, Yun et al., 2021; Taylor & Gottlieb, 2017). Elevated Parkin expression in WAT has been associated with excessive mitophagy, which in turn depletes mitochondrial content, impairs beiging remodeling, and contributes to metabolic dysfunction. Elevated expression of p53 in adipocytes has been shown to drive PINK1/Parkin-mediated mitophagy, resulting in reduced mitochondrial biogenesis and diminished recruitment of beige adipocytes. Deletion of p53 in adipocytes reverses this phenotype, restoring mitochondrial integrity and improving insulin sensitivity (Fu, Liu et al., 2019). Similar effects have been observed with Asprosin, a WAT-derived hormone that upregulates PINK1 and Parkin under HFD conditions, promoting mitophagy while suppressing thermogenic gene expression. Asprosin deficiency improves mitochondrial gene expression and enhances energy expenditure (Chen, Yuan et al., 2024).

FNDC5, another key regulator, modulates this pathway by inhibiting PINK1/Parkin activation and promoting PGC-1α–driven mitochondrial biogenesis. Reduced FNDC5 levels in obesity lead to increased mitophagy and mitochondrial depletion in WAT, which can be reversed by cold exposure or FNDC5 restoration (Neira, Hernandez-Pardos et al., 2024). Similarly, deletion of the redox protein thioredoxin-2 (Trx2) enhances ROS generation and NF-κB activation, further driving PINK1/Parkin-mediated mitophagy and contributing to inflammation and IR (He, Huang et al., 2021).

Importantly, suppression of PINK1/Parkin activity can also be metabolically protective. Parkin KO mice exhibit reduced WAT inflammation, preserved mitochondrial content, and enhanced beiging capacity (Delgado-Angles, Blasco-Roset et al., 2024). These observations are further supported by studies using sesamol, a natural compound that inhibits Parkin-mediated mitophagy and simultaneously stimulates mitochondrial biogenesis and thermogenesis via β3-adrenergic signaling (Lin, Chen et al., 2021). Altogether, these findings highlight the importance of balanced PINK1/Parkin signaling in WAT, where both hyperactivation and suppression can have distinct metabolic outcomes depending on context.

2.3.2 FUNDC1/Vaspin/ULK1 – autophagy-dependent mitophagy in WAT

In addition to ubiquitin-dependent pathways, receptor-mediated mitophagy mechanisms such as FUNDC1 and ULK1 also contribute to mitochondrial maintenance in WAT. These pathways are regulated by nutrient-sensing kinases including AMPK and mTORC1. Under normal conditions, AMPK activates ULK1 to initiate autophagy, whereas chronic nutrient excess in obesity leads to mTORC1 overactivation and inhibition of ULK1, impairing mitophagic flux (Cai, Zhao et al., 2019).

Experimental evidence demonstrates that disruption of this axis contributes to mitochondrial dysfunction. Mice overexpressing the ER-phagy receptor FAM134B develops excessive Parkin-independent mitophagy, resulting in mitochondrial depletion, adipose expansion, and systemic IR. These effects can be reversed by autophagy inhibition (Cai, Zhao et al., 2019). FUNDC1 has been shown to function independently of Parkin in WAT, where its deletion under HFD conditions impairs mitophagy, increases inflammation, and worsens metabolic outcomes (Zhu, Gu et al., 2022).

The oxidative stress sensor heme oxygenase-1 (HO-1) also supports mitochondrial quality control in WAT. Loss of HO-1 leads to disrupted mitochondrial fusion, elevated ROS levels, and impaired ULK1-dependent autophagy. These findings position HO-1 as an upstream regulator of mitochondrial integrity during metabolic stress (Singh, Grant et al., 2017). While other regulators such as Vaspin are known to influence autophagy in various tissues, their direct role in WAT mitophagy remains to be clarified.

2.3.3 Other regulatory influences on mitophagy in WAT

Additional factors modulate mitophagy in WAT through both cell-intrinsic and intercellular mechanisms. BNIP3, a mitophagy receptor and transcriptional target of PPARγ, is highly context dependent. In adipocytes, reduced BNIP3 expression during obesity appears to redirect mitophagy through alternative pathways involving FUNDC1, DRP1, and OPA1. These changes are associated with enhanced mitochondrial biogenesis via PGC1α/TFAM and improved insulin sensitivity. Conversely, inhibition of FUNDC1 or core autophagy proteins reverses these benefits (Shi, Yang et al., 2024).

Upstream of BNIP3, the Hippo kinases STK3 and STK4 are upregulated in obese WAT and promote BNIP3 phosphorylation and mitophagy. Inactivation of STK3/STK4 enhances mitochondrial content, stabilizes UCP1, and protects against diet-induced obesity in mice, with similar trends observed in human adipose (Cho, Son et al., 2021).

In macrophages residing within WAT, BNIP3 expression is upregulated via HIF1A in response to hypoxia, contributing to mitophagy and pro-inflammatory polarization. Deletion of BNIP3 in macrophages reduces WAT inflammation and improves metabolic function, highlighting cell-specific roles of mitophagy in adipose remodeling (Kim, Choi et al., 2025).

Intercellular coordination of mitophagy is also impaired in obesity. In the Mito-Fat mouse model, mitochondrial transfer from adipocytes to macrophages is disrupted, resulting in impaired mitophagic communication and worsening metabolic phenotype. This is partly driven by the suppression of the heparan sulfate biosynthetic pathway, which mediates extracellular matrix interactions required for mitochondrial exchange (Sabaratnam, Hansen et al., 2023).

Mitochondrial dynamics further intersect with mitophagy regulation. MARCH5, a mitochondrial E3 ligase, supports fusion and the degradation of damaged proteins. It interacts with PPARγ signaling to promote mitophagy and maintain mitochondrial function in WAT during high-fat feeding (Bond, Moody et al., 2019). Finally, ClpP, a mitochondrial protease previously linked to BAT function, also modulates mitophagy in WAT. In ClpP KO mice, WAT exhibits altered oxidative phosphorylation and depot-specific changes in mitophagy signaling under HFD (Liesa & Shirihai, 2018).

Together, these studies illustrate that mitophagy in WAT is orchestrated by a complex network of nutrient-sensitive, stress-responsive, and intercellular mechanisms. Both excessive and insufficient mitophagy can disrupt adipocyte function, mitochondrial homeostasis, and systemic metabolism in obesity.

2.3.4 Exercise or pharmacological interventions and environmental influences on mitophagy in adipose tissue

A pharmacological approach targeting mitophagy have also shown promise in modulating adipose tissue metabolism. Melatonin supplementation has been found to significantly increase Parkin expression and improve mitochondrial morphology in WAT via the PINK1/Parkin pathway. These effects lead to reduced adipocyte hypertrophy and enhanced insulin sensitivity, reinforcing the importance of antioxidant-induced mitophagy in aging adipose tissue (Goyal, Hoit et al., 2020). Nutritional interventions such as peanut shell extract (PSE), rich in luteolin, have demonstrated potent anti-diabetic effects in db/db mice by restoring mitochondrial proteins (MFN2, DRP1) and mitophagy regulators (PINK1, LC3-II) in epididymal WAT. This mitochondrial remodeling is associated with decreased macrophage infiltration and TNFα expression, indicating both metabolic and anti-inflammatory effects (Deshmukh, Santos et al., 2024).

Beyond these interventions, mitochondrial uncouplers such as BAM15 have emerged as potential modulators of mitophagy. BAM15 increases energy expenditure independently of food intake, restoring energy balance in metabolically stressed conditions. It upregulates MFN2 and strengthens endoplasmic reticulum–mitochondria interactions, improving mitochondrial calcium handling and reducing stress-induced apoptosis (Dantas, Zunica et al., 2022). In BAT, pharmacological interventions targeting mitochondrial dynamics have shown potential in optimizing thermogenic capacity. Thyroid hormone (T3) enhances mitophagy via mTOR inhibition and SIRT1 activation, improving mitochondrial turnover and thermogenic efficiency (Yau, Singh et al., 2019). Similarly, FNDC5 signaling has been implicated in BAT mitochondrial remodeling, with pharmacological modulation of this pathway offering potential therapeutic benefits (Lu, Fujioka et al., 2018).

Exercise interventions, particularly high-intensity interval training (HIIT) and moderate-intensity continuous training (MICT), improved mitophagy-related markers in WAT of obese older adults. HIIT elevated the expression of TFAM, PARKIN, and LC3-II, indicating enhanced mitochondrial biogenesis and turnover. Interestingly, the addition of L-citrulline to HIIT further augmented reductions in fat mass and improved muscle strength, suggesting potential synergy in promoting adipose mitochondrial health (Marcangeli, Youssef et al., 2022; Youssef, Granet et al., 2022). In BAT, exercise-induced mitophagy plays a distinct role in optimizing thermogenic function. Cold exposure combined with endurance training enhances mitochondrial turnover and improves BAT activity, reinforcing its role in systemic energy balance (Pallio & Mannino, 2024). Given BAT's reliance on mitochondrial integrity for thermogenesis, interventions that enhance mitophagy may be particularly beneficial in metabolic disorders characterized by impaired energy expenditure.

Finally, environmental modulation through heat exposure triggered mitophagy in goat brown adipocytes, as shown by elevated LC3-II, PINK1, and Parkin levels, alongside increased UCP1 expression and thermogenic capacity. This suggests that mitophagy is integral to BAT activation and thermogenesis under thermal stress (Song, Li et al., 2025).

3. Mitophagy in liver in the context of obesity

The liver is a principal organ involved in the regulation of systemic metabolism, with significant implications for obesity. Its primary functions include macronutrient metabolism, blood volume regulation, immune system support, endocrine control of growth signaling pathways, lipid and cholesterol homeostasis, and the breakdown of xenobiotic/drug compounds (Trefts, Gannon et al., 2017). Metabolic dysfunction-associated steatotic liver disease (MASLD) is the most prevalent form of liver disease in the absence of excessive alcohol consumption and is strongly associated with obesity (Diehl & Day, 2017). Liver disease manifests in several stages, progressing from liver steatosis to non-alcoholic liver steatohepatitis, and advancing to progressive fibrosis that may lead to cirrhosis, thereby increasing the risk of developing hepatocellular carcinoma (Diehl & Day, 2017). To understand the connection between liver disease and obesity it is relevant to mention that studies focused on animal models of obesity and/or humans living with obesity have demonstrated a significant reduction in mitochondrial respiratory capacity in vWAT (Dewidar, Mastrototaro et al., 2023; Pafili, Kahl et al., 2022; Pedersen, Rygg et al., 2022). This obesity-related metabolic distress and dysfunction in WAT leads to or coexists with inflammation and altered adipokine secretion patterns, which favor an increased lipid flux to the liver (Azzu, Vacca et al., 2020). Thus, patients with different degrees of MASLD and obesity showed mitochondrial dysfunction in vWAT and scWAT (Pedersen, Rygg et al., 2022) or only in vWAT (Pafili, Kahl et al., 2022) implicating the relevance of preventing mitochondrial dysfunction in vWAT, where mitophagy plays a critical role as discussed earlier.

Increased lipid flux to the liver influences the progression of MASLD and impacts mitochondrial functionality. Assessing liver mitochondrial functionality in humans is challenging due to the scarcity of studies, variability in methodologies, and potential contradictions in results. However, it can be concluded that the liver mitochondria exhibit plasticity to overcome steatosis, enhancing respiratory capacity during the initial stages of ectopic lipid accumulation in obese patients. This plasticity progresses to a decline in mitochondrial respiratory capacity as the condition transitions from steatosis to non-alcoholic steatohepatitis (NASH) and cirrhosis (Fromenty & Roden, 2023). The increased mitochondrial content observed in studies where mitochondrial respiratory capacity is reduced indicates a failure in mitochondrial turnover, where mitophagy plays a major role.

3.1 Mitophagy dysregulation in metabolic liver diseases

Mitophagy is crucial for maintaining mitochondrial function in the liver. The inability to effectively remove defective mitochondria during the progression of MASLD directly implicates mitophagy in the progression of the disease, as evidenced by the loss of BNIP3 and a decrease in proteins involved in mitochondrial dynamics, such as OPA1, MFN2, and DRP1 phosphorylation (Moore, Cunningham et al., 2022). This deterioration could be triggered by excessive liver ROS production under these conditions (Longo, Meroni et al., 2021). Thus, HFD-induced obesity disrupts mitophagy by downregulating key pathways such as PINK1/Parkin, BNIP3/NIX, and AMPK-SIRT1, which are vital regulators of mitochondrial clearance (Cioffi, Giacco et al., 2022; Rosa-Caldwell, Poole et al., 2022; Wang, Tao et al., 2021).

3.1.1 PINK1/Parkin – ubiquitin-dependent mitophagy in liver

In an animal model of HFD-induced obesity, the double deletion of AKT2 and AMPKα2 protects against obesity, hepatic steatosis, and liver injury. This protection in the liver is due, at least in part, to the restoration of autophagy and mitophagy mediated by FOXO1 and Parkin, which are essential cellular processes for lipid droplet breakdown (Wang, Tao et al., 2021). Interestingly, Parkin KO mice are safeguarded from diet-induced obesity and hepatosteatosis, exhibiting improved glucose handling (Kim, Stevens et al., 2011). These animals also demonstrate enhanced hepatic insulin sensitivity, potentially mediated by alterations in diacylglycerol levels, ceramide species, and PKCε activity. Based on these results, it appears that Parkin KO plays a protective role in the liver exposed to excessive fat flux (Edmunds, Huckestein et al., 2019). Further studies suggest that altered intestinal lipid metabolism, rather than direct hepatic changes, contributes to the protection observed in Parkin KO mice from diet-induced obesity and metabolic dysfunction (Costa, Huckestein et al., 2016). However, liver-specific Parkin KO showed a very different phenotype. Those mice had an increases susceptibility to diet-induced hepatic steatosis and IR, presenting impaired mitochondrial respiration, increased liver fat accumulation, and greater IR. Considering the information obtained from the liver-specific Parkin KO, Parkin-mediated mitophagy is crucial for maintaining mitochondrial integrity and protecting against obesity-associated liver diseases (Edmunds, Xie et al., 2020).

Furthermore, studies in Wistar rats on a 14-week HFD demonstrate that reduced expression of PINK1, Parkin, and LC3-II correlates with mitochondrial dysfunction and metabolic disturbances, underscoring the impact of impaired mitophagy in liver metabolic disorders (Cioffi, Giacco et al., 2022). Similar results were previously obtained in mice exposed to a HFD for 18 weeks, hepatic lipid accumulation and a disruption of PINK1/Parkin and BNIP3 mediated mitophagy (Rosa-Caldwell, Poole et al., 2022). Of interest, after 10 weeks on HFD, a subset of mice were separated into two experimental groups subjected to a weight loss program for an additional 8 weeks, either with diet alone or the combination of diet and exercise. The overall results of the weight loss program indicate that diet and exercise in combination are more efficient in restoring mitophagy than diet alone. However, PINK1/Parkin-dependent mitophagy is not fully restored and is partially compensated by the activation of BNIP3-mediated mitophagy (Rosa-Caldwell, Poole et al., 2022). Additional information comes from studies using chronic heat treatment following HFD, which has proven beneficial in reducing lipid accumulation in the liver. These beneficial effects are related to the induction of the mitochondrial chaperone heat shock protein 72 (HSP72) (Archer, Rogers et al., 2018). A later study from the same research group demonstrated that chronic heat treatment enhanced hepatic mitochondrial quality, mitophagy, and respiratory function during HFD. However, HSP72 was not essential for heat-induced mitochondrial adaptations (Von Schulze et al., 2021).

It has been demonstrated that the ablation of liver-specific low molecular weight protein tyrosine phosphatase (LMPTP) in murine models confers protection against HFD-induced diabetes (Stanford, Aleshin et al., 2017). Moreover, research utilizing equine hepatic progenitor cells treated with palmitate revealed that inhibition of LMPTP modulates the expression of genes associated with autophagy, mitochondrial dynamics, and the unfolded protein response. This includes the expression of Parkin, but not PINK1 (Marycz, Kornicka et al., 2018). These preliminary observations necessitate validation and more rigorous experimental design to delineate the role of LMPTP in regulating hepatic PINK1/Parkin-dependent mitophagy. Previous results from the same research group and experimental model, but employing progenitor stem cells derived from adipose tissue, suggest that this may indeed be the case (Kornicka-Garbowska, Bourebaba et al., 2021).

Macrophage stimulating 1 (Mst1) is an upstream regulator of mitophagy. Of interest, it is the fact that the downregulation of PINK1/Parkin-dependent mitophagy with MASLD evolution is due to Mst1 upregulation. Mst1 phosphorylates AMPK reducing the expression of Parkin (Zhou, Chang et al., 2019).

3.1.2 FUNDC1/Vaspin/ULK1 – autophagy-dependent mitophagy in liver

The AMPK-SIRT1-PGC1α axis is a critical pathway regulating mitophagy and mitochondrial homeostasis. AMPK is a key energy sensor that phosphorylates and activates ULK1, thereby initiating autophagy. AMPK also enhances mitophagy by increasing PGC1α (peroxisome proliferator-activated receptor gamma coactivator 1-alpha), a master regulator of mitochondrial biogenesis, and deacetylating mitochondrial proteins through SIRT1 (sirtuin 1) (Ramanathan, Ali et al., 2022; Su, Guo et al., 2021). However, AMPK activity in the liver is suppressed in obesity and IR, leading to defective ULK1 activation, reduced mitophagy, and impaired mitochondrial dynamics. One study demonstrated that AMPK inactivation in HFD-fed mice inhibited mitophagy, increasing hepatic triglyceride accumulation and oxidative stress, whereas AMPK reactivation restored mitophagy, reducing hepatic lipid accumulation and inflammation (Rosa-Caldwell, Poole et al., 2022). It is important to understand the complex regulation of mitophagy depending on the stimuli and targeted mitochondria (dysfunctional or functional). Functional mitochondria could be targeted during programmed mitophagy via NIX and BNIP3 receptors (Longo, Bishnu et al., 2024). Thus, AMPK directly inhibits NIX-dependent mitophagy through AMPK phosphorylation of ULK1 at Ser556 and Ser694, which facilitates ULK1 sequestration by 14–3-3. At the same time, AMPK increases Parkin phosphorylation and enhances the rate of depolarization-induced mitophagy, independently of ULK1 (Longo, Bishnu et al., 2024).

Rats exposed to diet-induced obesity for 14 weeks presented clear signs of hepatic steatosis, increased ROS production and mtDNA damage, and reduced mitochondrial biogenesis. Under these circumstances, AMPK is activated phosphorylating and activating ULK1

(Ser555), which activates mitophagy by its action in downstream effectors such as AMBRA1 (Cioffi, Giacco et al., 2022). Further insights into the role of ULK1-dependent mitophagy in MASLD come from the modulation of a specific phosphorylation site of the AMPK beta subunit (Ser108). This study found that this AMPK phosphorylation site is critical for mitochondrial biogenesis and mitophagy induction under elevated fatty acid conditions (Desjardins, Smith et al., 2022). The stress-induced protein regulated in development and DNA damage response-1 (REDD1) is implicated in HFD-induced MASLD. REDD1 expression is increased in livers of HFD-fed mice and in patients with obesity and MASLD (Dumas, Ayachi et al., 2020). Thus, REDD1-KO mice are protected against HFD induced MASLD, through a mechanism that implicates ULK1, among others. mTORC1 phosphorylation of ULK at Ser757 prevents its activation inhibiting autophagy. REDD1 deficiency in the livers of REDD1-KO mice showed a decrease in ULK1-Ser757 indicating an improvement in autophagy, which is corroborated by the increased expression of beclin, LC3-II and a decrease in p62 protein content (Dumas, Ayachi et al., 2020).

3.1.3 Other regulatory influences on mitophagy in liver

In the context of MASLD, while PINK1/Parkin levels remain unchanged, BNIP3 expression is markedly reduced, indicating a selective failure of mitophagy that contributes to disease progression (Moore, Cunningham et al., 2022). The BNIP3/NIX pathway is another crucial mitophagy regulator, particularly under hypoxic and metabolic stress conditions. In MASLD, BNIP3 is significantly downregulated, impairing mitophagy and leading to the accumulation of damaged mitochondria, which release ROS and trigger inflammatory cascades such as NF-κB and NLRP3 inflammasome activation. This further exacerbates mitochondrial dysfunction, contributing to fibrosis and hepatocyte apoptosis (Dumas, Ayachi et al., 2020; Zhou, Chang et al., 2019). In human liver samples from obese patients undergoing bariatric surgery, BNIP3-mediated mitophagy was suppressed, leading to defective mitochondrial clearance and increased lipid accumulation despite stable PINK1/Parkin expression (Moore, Cunningham et al., 2022). However, in an animal model of HFD-induced obesity the obesity group that was subsequently assigned to Roux-en-Y gastric bypass showed in the liver an improvement in mitochondrial biogenesis, dynamics and mitophagy, illustrated by an increased expression and protein content of BNIP3. All these changes are along with an improved insulin sensitivity, reduced oxidative stress, and increased mitochondrial respiratory capacity (Sacks, Mulya et al., 2018).

The fission protein 1 (FIS1) involved in mitochondrial dynamics is as well a critical regulator that interacts with TBC1D15 and LC3 to link oxidative stress-induced mitochondrial fragmentation with mitophagy. In MASLD, reduced Fis1 expression impairs mitophagy, increasing mitochondrial ROS and metabolic dysregulation (Liou, Personnaz et al., 2022). In HFD-fed mice, overexpression of Fis1 restored mitophagy, lowered oxidative damage, and improved glucose and lipid homeostasis. This study also identified the integrated stress response (ISR) pathway, where Atf5 activation improves metabolic health through enhanced mitophagy activity (Liou, Personnaz et al., 2022).

MUL1 is a mitochondrial E3 ubiquitin ligase located on the OMM and it has been implicated in mitophagy. MUL1 acts in parallel to PINK1/Parkin-dependent pathway and, in addition can acts upstream of PINK1promoting its stabilization (Wang, Long et al., 2023). Mul1(−/−) mice are protected against HFD-induced obesity showing that MUL1 is a key player in the regulation of lipogenesis and fatty acid oxidation (Cilenti et al., 2024).

A broader analysis of mitophagy dysregulation in MASLD revealed that its impairment contributes to mtDNA mutations, mitochondrial fragmentation, and fibrosis, with reduced expression of DRP1 and OPA1, which are essential for maintaining mitochondrial network integrity (Su, Nie et al., 2019).

Overall, the dysregulation of PINK1/Parkin, BNIP3/NIX, AMPK-SIRT1, and Fis1 pathways plays a significant role in MASLD due to mitochondrial dysfunction in metabolic liver diseases. Impaired mitophagy leads to excessive ROS accumulation, defective mitochondrial biogenesis, and increased hepatic lipid accumulation, promoting fibrosis and metabolic decline. These findings highlight mitophagy failure as an early contributor to MASLD and IR, emphasizing the need for therapeutic strategies targeting mitochondrial quality control mechanisms.

3.1.4 Exercise or pharmacological interventions and environmental influences on mitophagy in liver

Several interventions have been shown to restore mitophagy and improve liver metabolic function. A single treadmill session increased mitophagic flux in liver mitochondria by upregulating p62/SQSTM1 accumulation and enhancing oxidative phosphorylation, leading to improved mitochondrial clearance (McCoin, Franczak et al., 2022). Similar effects were observed with heat therapy, as described before, which increased mitophagy even in HSP72 KO mice, indicating an alternative mechanism for mitochondrial adaptation (Fisher-Wellman, 2021). Another study confirmed that exercise, ketogenic diets, metformin, and liraglutide restored mitophagy by activating SIRT3, AMPK-SIRT1, and mitochondrial biogenesis, counteracting HFD-induced mitophagy suppression (Ramanathan, Ali et al., 2022).

Pharmacological agents such as metformin, mitochondrial uncouplers, and phytochemicals have also demonstrated the ability to restore mitophagy. Metformin upregulated Parkin-mediated mitophagy by counteracting cytosolic p53 accumulation, restoring mitochondrial function in ob/ob mice (Song, Lee et al. 2016). Similarly, sorafenib and mitochondrial uncouplers (CRMP) improved hepatic mitophagy and fatty acid oxidation in NAFLD, suggesting potential therapeutic benefits (Georgiev, Granata et al., 2022). Other compounds, such as quercetin, melatonin, and resveratrol, enhanced mitophagy and mitochondrial function while reducing oxidative stress and lipid accumulation in MASLD models (Su, Guo et al., 2021).

Environmental factors, such as toxic exposures, have also been implicated in mitophagy dysregulation. Chlorpyrifos (CPF), an organophosphate pesticide, has been shown to inhibit mitophagy, suppress AMPK activation, and worsen MASLD and IR (Wang, Tao et al., 2021). Cadmium is a widely spread contaminant that facilitates the progression of MASLD. Thus, rats on a HFD regime previously exposed to cadmium showed more severe MASLD, an inhibition of FUNDC1-mediated mitophagy, increased hepatic lipid accumulation and excessive mitochondrial ROS production (Lian, Li et al., 2024).

Overall, mitophagy is impaired in obesity, MASLD, and IR, primarily due to downregulation of PINK1/Parkin and BNIP3 pathways, leading to mitochondrial dysfunction. However, interventions such as exercise, heat therapy, genetic modifications, and pharmacological treatments can restore mitophagy and improve metabolic liver function. While metformin, phytochemicals, and mitochondrial uncouplers show promise in reversing mitophagy impairment, excessive mitophagy activation can be detrimental, emphasizing the need for precise regulation. Future research should refine therapeutic approaches that balance mitophagy activation while minimizing potential metabolic risks.

4. Mitochondrial function in skeletal and cardiac muscle

In the context of obesity, skeletal muscle function and metabolic homeostasis are significantly impacted, with the quality and function of mitochondria playing a critical, though sometimes debated role. Skeletal muscle relies heavily on mitochondrial oxidative metabolism for energy transduction (Pileggi, Parmar et al., 2021). Obesity is associated with skeletal muscle dysfunction and the accumulation of intramuscular lipids, and mitochondrial dysfunction, encompassing defects in biogenesis, fusion, fission, and mitophagy, which has been linked to obesity-induced IR (Pileggi, Parmar et al., 2021). While some findings suggest that mitochondrial dysfunction is intrinsic to the development of IR caused by high-fat feeding, other results challenge this notion (Reynaud, Wang et al., 2023). Generally, sustained mitochondrial dysfunction in obesity is thought to be a key contributor to IR, reflected by lower mitochondrial content and impaired function observed in many human studies (Pileggi, Parmar et al., 2021). This includes decreased mitochondrial gene and protein expression, lower activity of citrate synthase (a marker of mitochondrial content), and altered mitochondrial morphology and surface area. However, studies investigating mitochondrial content in obese participants have yielded conflicting results, highlighting the complexity influenced by factors such as age, body composition, duration of obesity, physical activity, and genetics.

Mitochondrial dynamics, involving the balance between fusion and fission, are also altered in obesity and influence muscle mass and performance. Mitochondrial networking is decreased in the skeletal muscle of individuals with obesity (Liesa & Shirihai, 2013, Romanello & Sandri, 2021). HFD feeding in mice has been shown to impair mitochondrial networking, accompanied by blunted respiratory function, decreased expression of fusion proteins like MFN2 and MFN1, and enhanced expression of fission proteins like DRP1 and FIS1(Pileggi, Parmar et al. 2021). Studies in leptin-deficient mice also show increased DRP1 and FIS1 expression (Pileggi, Parmar et al., 2021). Conversely, inhibiting DRP1 in myotubes has been shown to protect against fatty acid-induced mitochondrial fragmentation and IR (Pileggi, Parmar et al., 2021). Investigations into mitochondrial dynamics in human obesity have been scarce and conflicting. In obese mice with cancer cachexia, electron microscopy revealed stepwise reductions in intermyofibrillar and subsarcolemmal mitochondrial size, along with an accumulation of damaged and dysfunctional mitochondria (Cardaci, VanderVeen et al., 2024). This model also showed dysregulated mitochondrial dynamics, specifically a reduction in fission proteins FIS1 and DRP1, which contributed to the accumulation of damaged mitochondria. These changes coincided with increased pAMPK and suggest that obesity exacerbates functional deficits by impairing the removal of damaged mitochondria.

Regarding mitochondrial activity and function, impaired oxygen respiration has been observed in obese cachectic muscle (Ehrlicher, Stierwalt et al., 2021). State 3 (maximal ADP-dependent) coupled respiration was reduced in both obese and obese cancer cachexia muscle. While some studies show lower mitochondrial function in obese muscle, these effects are sometimes absent when normalized to mitochondrial content (Pileggi, Parmar et al., 2021). Short-term high-fat feeding in mice initially lowered lipid-supported oxidative phosphorylation and increased electron leak after 3 days, although this normalized by 7 days (Ehrlicher, Stierwalt et al., 2021). Long-term HFD can increase mitochondrial lipid oxidation capacity, and excess lipid availability generally increases mitochondrial fatty acid oxidative capacity in muscle (Cardaci, VanderVeen et al., 2024). Mitochondrial respiration is decreased in the skeletal muscle of individuals with T2D, a condition frequently associated with obesity (Cardaci, VanderVeen et al., 2024). Long-term rates of mitochondrial protein synthesis have been reported to be increased in mouse skeletal muscle with high-fat feeding (Cardaci, VanderVeen et al., 2024). Proinflammatory NFkB signaling can promote mitochondrial dysfunction in skeletal muscle in response to cellular fuel overloading (Ehrlicher, Stierwalt et al., 2021). Obesity also exacerbates age-related mitochondrial stress by increasing substrate flow, oxidative stress, inflammation, and ER stress (Dantas, Zunica et al. 2022). While mitochondrial dysfunction is a hallmark of obese skeletal muscle, the exact role of mitophagy is debated.

4.1 Mitophagy in skeletal and cardiac muscle in the context of obesity

Alterations in mitophagy signaling may contribute to the decreases in mitochondrial content and function during obesity (Pileggi, Parmar et al., 2021). Obesity is suggested to exacerbate impairments to mitochondrial quality control, specifically disrupting the removal of damaged mitochondria in the muscle (Cardaci, VanderVeen et al., 2024; Huang, Zhang et al., 2023; Reynaud, Wang et al., 2023). Oppositely other findings indicate that obesity-related stressors like nutrient overload and inflammation can trigger upstream mitophagy activation, potentially mediated by lipid-induced oxidative stress (ROS production) (Ehrlicher, Stierwalt et al., 2021; Nisr, Shah et al., 2019). However, downstream degradation may be impaired. For instance, palmitate treatment in myotubes showed sustained signals to activate mitophagy but impaired autophagosome degradation (Ehrlicher, Stierwalt et al., 2021). In the obese cancer cachexia model, an accumulation of p62 suggested impaired autophagy and clearance of damaged mitochondria, and impaired removal of damaged mitochondria through mitophagic dysregulation contributed to unique myopathic signatures (Cardaci, VanderVeen et al., 2024; Ehrlicher, Stierwalt et al., 2021). Furthermore, the discovery that inhibiting muscle mitophagy can protect against obesity and IR through muscle-adipose tissue communication complicates a simple interpretation of mitophagy as solely beneficial or impaired in this context (Dantas, Zunica et al., 2022; Fu, Xu et al., 2018).

Mitochondrial uncoupling has also been shown to attenuate sarcopenic obesity by enhancing skeletal muscle mitophagy and quality control (Dantas, Zunica et al., 2022; Liesa & Shirihai, 2013; Livshits & Kalinkovich, 2019). Overall, obese skeletal muscle appears metabolically impaired and may struggle to respond properly to obesity, with dysregulated mitochondrial turnover and impaired mitophagy being key aspects.

Additionally, we need to account of metabolic differences between cardiac and skeletal muscle demands. Even though both tissues rely on mitochondria activity their response to obesity and metabolic stress is different. Skeletal muscle is highlighted as a key organ for glucose homeostasis and the major site of postprandial glucose uptake, playing a significant role in the development of obese-related IR. In the other hand, cardiac muscle dysfunction, such as heart failure or hypertrophy is shown to be induced by HFD and associated with impaired myocardial energetics (Yoshii, McMillen et al., 2024).

In the subsequent sections, we present how mitophagy is regulated in cardiac and skeletal muscle under obesity conditions by the two canonical proteolytic pathways (Ubiquitin- and LC3-mediated) alongside emerging modulators and highlight comparative insights from human, mouse, and in vitro studies.

4.2 Mitophagy in cardiac muscle

Mitochondrial quality control, particularly through mitophagy, plays an indispensable role in maintaining cardiac function. This is especially evident in pathological states such as obesity-associated cardiomyopathy, where mitochondrial dysfunction is a hallmark feature (Dantas, Zunica et al., 2022). Studies in mice have demonstrated that chronic HFD exposure progressively enhances mitophagy in the heart, with maximal activation occurring between 20 and 24 weeks (Tong, Saito et al., 2021). However, this compensatory increase may become maladaptive over time, as defective mitophagy has been observed in the hearts of obese animal models (Huang, Zhang et al., 2023).

4.2.1 PINK1/Parkin – Ubiquitin-dependent mitophagy in cardiac muscle

In the context of obesity, cardiac Parkin protein levels are typically reduced. While initial HFD exposure may activate upstream mitophagy signals, this does not always translate into effective downstream responses once cardiac dysfunction is established (Dhanabalan, Mzezewa et al., 2020; Huang, Zhang et al., 2023; Thomas, Marek-Iannucci et al., 2019). Ischemia-reperfusion (I/R) studies reveal that mitophagy responses, driven by the PINK1/Parkin axis, are altered both in control and obese rats fed a high-fat, high-sucrose diet. PINK1/Parkin expression is upregulated during ischemia and downregulated during reperfusion, indicating stress-specific modulation (Dhanabalan, Mzezewa et al., 2020; Yu, Dong et al., 2021). Therapeutic strategies aimed at enhancing fatty acid oxidation or pharmacologically activating mitophagy (e.g., with Urolithin A) have been shown to engage the Parkin-mediated pathway. Furthermore, genetic ablation of PINK1 exacerbates susceptibility to ischemia-reperfusion injury, reinforcing its cardioprotective role under metabolic and oxidative stress (Huang, Zhang et al., 2023; Siddall, Yellon et al., 2013).

4.2.2 FUNDC1/Vaspin/ULK1 – Autophagy-dependent mitophagy in cardiac muscle

Under sustained metabolic stress from chronic HFD, the heart initiates mitophagy via the ULK1-Rab9 pathway, a process that appears to intensify with continued exposure and acts as a defense mechanism against obesity-associated cardiomyopathy, preventing hypertrophy and dysfunction (Tong, Saito et al., 2021). Suppression of this pathway aggravates cardiac dysfunction, while its enhancement offers protection. However, this adaptive response may fail in certain disease models. For instance, in Heart Failure with Preserved Ejection Fraction (HFpEF) combined with obesity, mitophagy is notably impaired, attributed in part to defective fatty acid oxidation (Yoshii, McMillen et al., 2024). Restoration of FAO or pharmacological activation of mitophagy using agents like TAT-Beclin 1 significantly mitigates HFpEF symptoms. Similarly, in an Angiotensin II-induced model of atrial myocyte dysfunction, Vaspin therapy improves mitochondrial function by accelerating mitophagy through a ULK1/FUNDC1-mediated mechanism (Zhu, Gu et al., 2022).

4.2.3 *Other regulatory influences on mitophagy in skeletal muscle*

In a rodent model of LPS-induced endotoxemia, cardiac FTO, a key m6A RNA demethylase, is downregulated, correlating with impaired mitophagy and worsened myocardial function. Restoration of FTO expression attenuates cardiac injury by upregulating BNIP3 via an m6A- and YTHDF2-dependent mechanism, suggesting that RNA methylation dynamics are intimately linked with mitophagy regulation in sepsis-related cardiomyopathy (Qi, Zhang et al., 2024).

4.3 Mitophagy in skeletal muscle

4.3.1 *PINK1/Parkin – Ubiquitin-dependent mitophagy in skeletal muscle*

Though the PINK1/Parkin pathway is a canonical mitophagy regulator, its effectiveness in skeletal muscle under HFD-induced obesity appears to be context-dependent. In mice on a standard diet, Parkin overexpression improves mitochondrial respiration, elevates mitochondrial content, and increases muscle mass. However, in HFD-fed mice, while muscle mass is still augmented, it did not significantly impact mitochondrial function (Reynaud, Wang et al., 2023). Another study using BAM15, a mitochondrial uncoupler, observed elevated LC3-II/LC3-I ratios and increased PINK1 levels in isolated mitochondria from sarcopenic obese mice. Notably, Parkin levels remained unchanged, implying that PINK1 accumulation alone might suffice for mitophagy initiation through downstream ubiquitination and autophagosome recruitment, independent of total Parkin expression (Dantas, Zunica et al., 2022).

4.3.2 *FUNDC1/Vaspin/ULK1 – Autophagy-dependent mitophagy in cardiac muscle*

Selective deletion of FUNDC1 in skeletal muscle results in reduced LC3-II and p62 in mitochondrial fractions, indicative of a mitophagy deficit (Dantas, Zunica et al., 2022). This deficiency impairs mitochondrial energetics and stresses the muscle. Surprisingly, FUNDC1-deficient mice show resistance to HFD-induced obesity and IR. This paradox is explained by a compensatory increase in fibroblast growth factor 21 (FGF21), which boosts thermogenesis and lipid oxidation in adipose tissue. These findings underscore the critical role of muscle-derived mitophagy in orchestrating systemic metabolic homeostasis via muscle-adipose communication.

4.3.3 *Other regulatory influences on mitophagy in skeletal muscle*

In cachectic mouse models bearing Lewis lung carcinoma (LLC), mitophagy markers such as LC3b and BNIP3 are elevated in both lean and obese animals. However, only the obese cachectic mice exhibit signs of defective mitophagy, with elevated p62 levels indicating impaired autophagic flux and disturbed mitochondrial turnover. In contrast, lean cachectic mice exhibit decreased, suggesting a more functional mitophagy response (Cardaci, VanderVeen et al., 2024).

The outer mitochondrial membrane-localized kinase, mitoAMPK, plays a pivotal role in sensing energetic stress and mediating mitophagy during exercise or pharmacological interventions such as metformin administration (Drake, Wilson et al., 2021). Although its specific regulation in obesity has not been fully elucidated, mitoAMPK remains a promising candidate for linking energy status to mitophagy regulation in skeletal muscle.

Additionally, the VCP interactor Washc4 is essential for autophagy-mediated proteostasis in striated muscle. Genetic disruption of VCP or Washc4 leads to impaired autophagy and myopathy, emphasizing the importance of mitochondrial quality control machinery in muscle health. Importantly, these defects are not limited to Washc4 alone, various disruptions in mitophagy regulators can compromise muscle proteostasis, resulting in myopathy or muscle wasting (Romanello & Sandri, 2021).

4.3.4 *Exercise or pharmacological interventions and environmental influences on mitophagy in skeletal muscle*

Exercise is a key regulator of mitophagy in skeletal muscle, supporting mitochondrial health and metabolic balance. Resistance training activates mitochondrial fission and mitophagy, evidenced by increased DRP1Ser616 phosphorylation following a single exercise bout (Diaz-Castro, Tunon-Suarez et al., 2024). Regular exercise promotes mitochondrial proteostasis (Braga, Crisol et al., 2021) and improves glucose tolerance, benefits that are lost when mitophagy is impaired, as seen in ULK1 knockout mice (Chen, Ma et al., 2023).

Different forms of exercise engage mitophagy through distinct signaling pathways. Resistance exercise primarily works through AMPK activation to prevent muscle atrophy, while aerobic training enhances the mitochondrial localization of PINK1 and Parkin (Drake & Yan, 2017). Supporting this, endurance-trained (ET) individuals show higher mitophagy markers compared to sedentary (SED) subjects, without parallel changes in general autophagy. Interestingly, these differences remain unaffected by high-fat meal (HFM) consumption (Lei, Gan et al., 2024). Additionally, the muscle-derived factor FGF21 contributes to mitochondrial clearance under stress, with its protective effects relying on BNIP3; inhibiting BNIP3 reduces FGF21-dependent protection against muscle atrophy (Oost, Kustermann et al., 2019).

Pharmacological agents also modulate mitophagy and muscle integrity. Apigenin reduces oxidative stress and prevents excessive mitophagy in age-related atrophy (Dantas, Zunica et al., 2022), while Urolithin A and tomatidine enhance muscle function via mitophagy in aging models (Lei, Gan et al., 2024). Compounds like isoquercitrin, celecoxib, and salidroside reduce inflammation and preserve muscle in denervation models (Lei, Gan et al., 2024). BAM15, a mitochondrial uncoupler, represents a distinct pharmacological approach. Its administration increases energy expenditure independently of food intake, thereby restoring energy balance in metabolically stressed conditions (Dantas, Zunica et al., 2022). Additionally, BAM15 appears to upregulate MFN2 and strengthen endoplasmic reticulum–mitochondria interactions. By enhancing mitochondrial calcium handling and reducing stress-induced apoptosis, BAM15 fosters adaptive remodeling of muscle cells (Dantas, Zunica et al., 2022).

5. Obese-related mitophagy in other tissues

5.1 Pancreatic beta cells

Mitophagy plays a crucial role in maintaining pancreatic β-cell function, particularly under metabolic stress. RCAN1 (Regulator of Calcineurin 1), a protein involved in inflammation, calcium signaling, and apoptosis, has been shown to impair mitophagy in β-cells under high nutrient conditions. Overexpression of RCAN1 leads to glucose intolerance, lipid imbalance, and reduced glucose-stimulated insulin secretion (GSIS), while promoting an NLRP3-dependent inflammatory response (Zhang, Li et al., 2025). Mechanistically, RCAN1 suppresses mitophagy by inducing DNA methylation of *Beclin1*, leading to Miro1-dependent mitophagy deficiency and β-cell apoptosis. These effects are mediated through the AKT/mTOR pathway, *Beclin1* methylation, and activation of the NLRP3 inflammasome (Zhang, Li et al., 2025).

In a separate study, terazosin, a repurposed drug, was found to improve β-cell mitophagy and function in a mouse model of non-alcoholic fatty pancreatic disease (NAFPD) induced by a HFD (Zhang, Li et al., 2025). Terazosin selectively activated GPR119, increasing insulin secretion in MIN6 cells. In NAFPD mice, it reduced hyperglycemia, obesity, and β-cell dysfunction by inhibiting the MST1-Foxo3a pathway and restoring mitophagy. These benefits were not observed in GPR119-deficient mice, confirming the pathway's essential role (Zhang, Li et al., 2025).

5.2 Bone marrow derived macrophages

Simulating obesity, macrophages were exposed to high glucose and palmitic acid to investigate the effects of non-lethal sonodynamic therapy (NL-SDT) (Wang, Shen et al., 2024). NL-SDT increased *DRP1* expression through mitochondrial ROS, promoting mitophagy in macrophages. This mitophagy limited activation of the NLRP3 inflammasome and reduced IL-1β secretion. In co-culture experiments, NL-SDT-treated macrophages reduced apoptosis and improved glucose-stimulated insulin secretion in beta cells. In db/db mice, NL-SDT lowered blood glucose and increased insulin levels, suggesting that targeting macrophages with NL-SDT can enhance islet function in T2D (Wang, Shen et al., 2024).

5.3 Final summary

Obesity, a global health crisis, results from an imbalance between energy intake and expenditure, leading to metabolic dysfunction and associated conditions such T2D, cardiovascular disease, and NAFLD. A strong link exists between obesity and mitochondrial dysfunction, characterized by reduced mitochondrial mass, impaired oxidative capacity, and decreased ATP production. Mitophagy plays a complex, tissue-specific role in regulating metabolism and mitigating the harmful effects of obesity (Fig. 4). Both insufficient and excessive mitophagy can negatively influence disease progression.

In white adipose tissue (WAT), mitophagy functions as a quality control mechanism that reduces oxidative stress and helps maintain insulin sensitivity. However, chronic obesity impairs mitophagy due to overactivation of mTORC1 and inhibition of AMPK, leading to inflammation and IR. Additionally, excessive mitophagy may reduce mitochondrial content, impairing the browning of WAT. In BAT, mitophagy is essential for

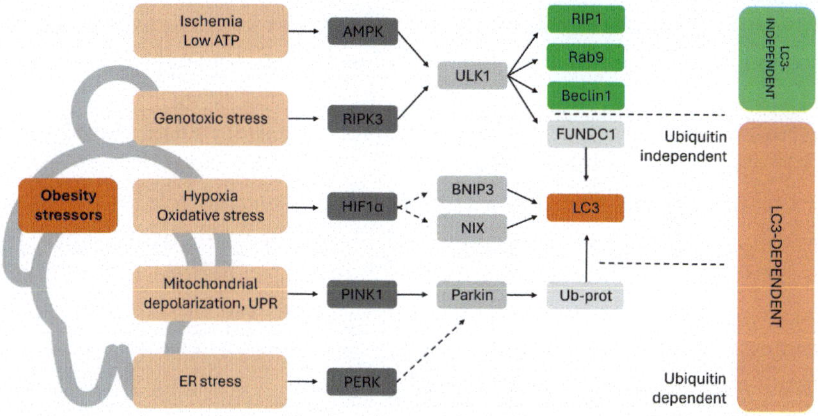

Fig. 4 Summary of stressors and pathways involved in mitophagy and its upstream elements.

thermogenesis and energy expenditure but meanwhile enhanced mitophagy can improve thermogenic efficiency in lean states, excessive activation in obesity may impair thermogenesis.

In the liver, mitophagy is essential for preserving mitochondrial function and preventing metabolic dysfunction-associated steatotic liver disease (MASLD). In MASLD, mitophagy is disrupted, resulting in the accumulation of ROS and hepatic lipid accumulation. Key pathways involved in mitochondrial clearance such as PINK1/Parkin, BNIP3/NIX, and AMPK-SIRT1 are downregulated.

In the heart, mitophagy supports cardiac function, particularly in obesity-related cardiomyopathy. While early exposure to a high-fat diet may initially activate mitophagy, defective mitophagy is commonly observed in obese models. The ULK1-Rab9 pathway is triggered as a compensatory mechanism, although this response may fail in some disease contexts. In skeletal muscle, obesity worsens impairments in mitochondrial quality control, disrupting the clearance of damaged mitochondria. Dysregulated mitochondrial turnover and impaired mitophagy derived from different pathways are central features of obese skeletal muscle.

Exercise, including high-intensity interval training (HIIT) and moderate-intensity training, enhances mitophagy-related markers in both WAT and brown adipose tissue (BAT), improving mitochondrial turnover and thermogenic capacity. Pharmacological agents such as melatonin, BAM15, and peanut shell extract (PSE) support mitochondrial integrity and reduce adipose tissue inflammation. Drugs like metformin, mitochondrial uncouplers, and various phytochemicals restore mitophagy and improve liver metabolic function. Other compounds like urolithin A and tomatidine have been shown to enhance muscle function through mitophagy. Additionally, environmental factors, such as heat exposure, can also stimulate mitophagy in BAT, promoting thermogenesis.

Overall, mitophagy plays a complex, tissue-specific role in metabolic regulation and in counteracting the detrimental effects of obesity. There is a pressing need for therapies that restore mitophagic balance in a context- and depot-specific manner because the dysregulation of mitophagy contributes not only to local dysfunction but also to broader metabolic phenotypes. Future research should focus on identifying reliable biomarkers and developing targeted interventions to modulate mitophagy and improve metabolic outcomes in obesity-related conditions.

References

Ahmed, B., Sultana, R., & Greene, M. W. (2021). Adipose tissue and insulin resistance in obese. *Biomedicine & Pharmacotherapy = Biomedecine & Pharmacotherapie, 137*, 111315.

Archer, A. E., Rogers, R. S., Von Schulze, A. T., Wheatley, J. L., Morris, E. M., McCoin, C. S., ... Geiger, P. C. (2018). Heat shock protein 72 regulates hepatic lipid accumulation. *American Journal of Physiology-Regulatory, Integrative and Comparative Physiology, 315*(4), R696–R707.

Azzu, V., Vacca, M., Virtue, S., Allison, M., & Vidal-Puig, A. (2020). Adipose tissue-liver cross talk in the control of whole-body metabolism: Implications in nonalcoholic fatty liver disease. *Gastroenterology, 158*(7), 1899–1912.

Bluher, M. (2019). Obesity: Global epidemiology and pathogenesis. *Nature Reviews Endocrinology, 15*(5), 288–298.

Bond, S. T., Moody, S. C., Liu, Y., Civelek, M., Villanueva, C. J., Gregorevic, P., ... Drew, B. G. (2019). The E3 ligase MARCH5 is a PPARgamma target gene that regulates mitochondria and metabolism in adipocytes. *American Journal of Physiology. Endocrinology and Metabolism, 316*(2), E293–E304.

Boone, C., & Lewis, S. C. (2024). Bridging lipid metabolism and mitochondrial genome maintenance. *The Journal of Biological Chemistry, 300*(8), 107498.

Borcherding, N., Jia, W., Giwa, R., Field, R. L., Moley, J. R., Kopecky, B. J., ... Brestoff, J. R. (2022). Dietary lipids inhibit mitochondria transfer to macrophages to divert adipocyte-derived mitochondria into the blood. *Cell Metabolism, 34*(10), 1499–1513 e1498.

Braga, R. R., Crisol, B. M., Bricola, R. S., Sant'ana, M. R., Nakandakari, S., Costa, S. O., ... Ropelle, E. R. (2021). Exercise alters the mitochondrial proteostasis and induces the mitonuclear imbalance and UPR(mt) in the hypothalamus of mice. *Scientific Reports, 11*(1), 3813.

Cai, C., Guo, Z., Chang, X., Li, Z., Wu, F., He, J., ... Tan, Y. (2022). Empagliflozin attenuates cardiac microvascular ischemia/reperfusion through activating the AMPKalpha1/ULK1/FUNDC1/mitophagy pathway. *Redox Biology, 52*, 102288.

Cai, M., Zhao, J., Liu, Q., Wang, X., & Wang, Y. (2019). FAM134B improves preadipocytes differentiation by enhancing mitophagy. *Biochimica et Biophysica Acta. Molecular and Cell Biology of Lipids, 1864*(12), 158508.

Cardaci, T. D., VanderVeen, B. N., Bullard, B. M., McDonald, S. J., Unger, C. A., Enos, R. T., ... Murphy, E. A. (2024). Obesity worsens mitochondrial quality control and does not protect against skeletal muscle wasting in murine cancer cachexia. *Journal of Cachexia, Sarcopenia and Muscle, 15*(1), 124–137.

Chen, S., Yuan, W., Huang, Q., Xiong, X., Wang, C., Zeng, W., ... Huang, Q. (2024). Asprosin contributes to pathogenesis of obesity by adipocyte mitophagy induction to inhibit white adipose browning in mice. *International Journal of Obesity (Lond), 48*(7), 913–922.

Chen, Y. L., Ma, Y. C., Tang, J., Zhang, D., Zhao, Q., Liu, J. J., ... Zou, C. G. (2023). Physical exercise attenuates age-related muscle atrophy and exhibits anti-ageing effects via the adiponectin receptor 1 signalling. *Journal of Cachexia, Sarcopenia and Muscle, 14*(4), 1789–1801.

Chen, Z., Liu, L., Cheng, Q., Li, Y., Wu, H., Zhang, W., ... Chen, Q. (2017). Mitochondrial E3 ligase MARCH5 regulates FUNDC1 to fine-tune hypoxic mitophagy. *EMBO Reports, 18*(3), 495–509.

Cho, Y. K., Son, Y., Saha, A., Kim, D., Choi, C., Kim, M., ... Lee, Y. H. (2021). STK3/STK4 signalling in adipocytes regulates mitophagy and energy expenditure. *Nature Metabolism, 3*(3), 428–441.

Choi, J. W., Jo, A., Kim, M., Park, H. S., Chung, S. S., Kang, S., & Park, K. S. (2016). BNIP3 is essential for mitochondrial bioenergetics during adipocyte remodelling in mice. *Diabetologia, 59*(3), 571–581.

Cilenti, L., Di Gregorio, J., Mahar, R., Liu, F., Ambivero, C. T., Periasamy, M., ... Zervos, A. S. (2024). Inactivation of mitochondrial MUL1 E3 ubiquitin ligase deregulates mitophagy and prevents diet-induced obesity in mice. *Frontiers in Molecular Biosciences, 11*, 1397565.

Cioffi, F., Giacco, A., Petito, G., de Matteis, R., Senese, R., Lombardi, A., ... Silvestri, E. (2022). Altered mitochondrial quality control in rats with metabolic dysfunction-associated fatty liver disease (mafld) induced by high-fat feeding. *Genes (Basel), 13*(2).

Costa, D. K., Huckestein, B. R., Edmunds, L. R., Petersen, M. C., Nasiri, A., Butrico, G. M., ... Jurczak, M. J. (2016). Reduced intestinal lipid absorption and body weight-independent improvements in insulin sensitivity in high-fat diet-fed Park2 knockout mice. *American Journal of Physiology. Endocrinology and Metabolism, 311*(1), E105–E116.

Crewe, C., Funcke, J. B., Li, S., Joffin, N., Gliniak, C. M., Ghaben, A. L., ... Scherer, P. E. (2021). Extracellular vesicle-based interorgan transport of mitochondria from energetically stressed adipocytes. *Cell Metabolism, 33*(9), 1853–1868.e1811.

Czajka, A., Ajaz, S., Gnudi, L., Parsade, C. K., Jones, P., Reid, F., & Malik, A. N. (2015). Altered mitochondrial function, mitochondrial DNA and reduced metabolic flexibility in patients with diabetic nephropathy. *EBioMedicine, 2*(6), 499–512.

Czech, M. P. (2017). Insulin action and resistance in obesity and type 2 diabetes. *Nature Medicine, 23*(7), 804–814.

Dahdah, N., Tercero-Alcazar, C., Malagon, M. M., Garcia-Roves, P. M., & Guzman-Ruiz, R. (2024). Interrelation of adipose tissue macrophages and fibrosis in obesity. *Biochemical Pharmacology, 225*, 116324.

Dantas, W. S., Zunica, E. R. M., Heintz, E. C., Vandanmagsar, B., Floyd, Z. E., Yu, Y., ... Kirwan, J. P. (2022). Mitochondrial uncoupling attenuates sarcopenic obesity by enhancing skeletal muscle mitophagy and quality control. *Journal of Cachexia, Sarcopenia and Muscle, 13*(3), 1821–1836.

Delgado-Angles, A., Blasco-Roset, A., Godoy-Nieto, F. J., Cairo, M., Villarroya, F., Giralt, M., & Villarroya, J. (2024). Parkin depletion prevents the age-related alterations in the FGF21 system and the decline in white adipose tissue thermogenic function in mice. *Journal of Physiology and Biochemistry, 80*(1), 41–51.

Deshmukh, H., Santos, J. M., Bender, M., Dufour, J. M., Lovett, J., & Shen, C. L. (2024). Peanut shell extract improves mitochondrial function in db/db mice via suppression of oxidative stress and inflammation. *Nutrients, 16*(13).

Desjardins, E. M., Smith, B. K., Day, E. A., Ducommun, S., Sanders, M. J., Nederveen, J. P., ... Steinberg, G. R. (2022). The phosphorylation of AMPKbeta1 is critical for increasing autophagy and maintaining mitochondrial homeostasis in response to fatty acids. *Proceedings of the National Academy of Sciences of the United States of America, 119*(48), e2119824119.

Dewidar, B., Mastrototaro, L., Englisch, C., Ress, C., Granata, C., Rohbeck, E., ... Roden, M. (2023). Alterations of hepatic energy metabolism in murine models of obesity, diabetes and fatty liver diseases. *EBioMedicine, 94*, 104714.

Dhanabalan, K., Mzezewa, S., Huisamen, B., & Lochner, A. (2020). Mitochondrial oxidative phosphorylation function and mitophagy in ischaemic/reperfused hearts from control and high-fat diet rats: Effects of long-term melatonin treatment. *Cardiovascular Drugs and Therapy / Sponsored by the International Society of Cardiovascular Pharmacotherapy, 34*(6), 799–811.

Diaz-Castro, F., Tunon-Suarez, M., Rivera, P., Botella, J., Cancino, J., Figueroa, A. M., ... Castro-Sepulveda, M. (2024). A single bout of resistance exercise triggers mitophagy, potentially involving the ejection of mitochondria in human skeletal muscle. *Acta Physiologica (Oxford, England), 240*(9), e14203.

Diehl, A. M., & Day, C. (2017). Cause, pathogenesis, and treatment of nonalcoholic steatohepatitis. *The New England Journal of Medicine, 377*(21), 2063–2072.

Drake, J. C., Wilson, R. J., Laker, R. C., Guan, Y., Spaulding, H. R., Nichenko, A. S., ... Yan, Z. (2021). Mitochondria-localized AMPK responds to local energetics and contributes to exercise and energetic stress-induced mitophagy. *Proceedings of the National Academy of Sciences of the United States of America, 118*(37).

Drake, J. C., & Yan, Z. (2017). Mitophagy in maintaining skeletal muscle mitochondrial proteostasis and metabolic health with ageing. *The Journal of Physiology, 595*(20), 6391–6399.

Dumas, K., Ayachi, C., Gilleron, J., Lacas-Gervais, S., Pastor, F., Favier, F. B., ... Giorgetti-Peraldi, S. (2020). REDD1 deficiency protects against nonalcoholic hepatic steatosis induced by high-fat diet. *The FASEB Journal, 34*(4), 5046–5060.

Edmunds, L. R., Huckestein, B. R., Kahn, M., Zhang, D., Chu, Y., Zhang, Y., ... Jurczak, M. J. (2019). Hepatic insulin sensitivity is improved in high-fat diet-fed Park2 knockout mice in association with increased hepatic AMPK activation and reduced steatosis. *Physiological Reports, 7*(21), e14281.

Edmunds, L. R., Xie, B., Mills, A. M., Huckestein, B. R., Undamatla, R., Murali, A., ... Jurczak, M. J. (2020). Liver-specific Prkn knockout mice are more susceptible to diet-induced hepatic steatosis and insulin resistance. *Molecular Metabolism, 41*, 101051.

Ehrlicher, S. E., Stierwalt, H. D., Newsom, S. A., & Robinson, M. M. (2021). Short-term high-fat feeding does not alter mitochondrial lipid respiratory capacity but triggers mitophagy response in skeletal muscle of mice. *Frontiers in Endocrinology (Lausanne), 12*, 651211.

Eiyama, A., & Okamoto, K. (2015). PINK1/Parkin-mediated mitophagy in mammalian cells. *Current Opinion in Cell Biology, 33*, 95–101.

Fischer, I. P., Irmler, M., Meyer, C. W., Sachs, S. J., Neff, F., Hrabe de Angelis, M., ... Ussar, S. (2018). A history of obesity leaves an inflammatory fingerprint in liver and adipose tissue. *International Journal of Obesity (Lond), 42*(3), 507–517.

Fisher-Wellman, K. H. (2021). Heating up to heal-acute heat exposure increases hepatic mitophagy resulting in hormetic improvements in mitochondrial bioenergetic efficiency. *Function (Oxford), 2*(2), zqab011.

Fromenty, B., & Roden, M. (2023). Mitochondrial alterations in fatty liver diseases. *Journal of Hepatology, 78*(2), 415–429.

Fu, T., Xu, Z., Liu, L., Guo, Q., Wu, H., Liang, X., ... Gan, Z. (2018). Mitophagy directs muscle-adipose crosstalk to alleviate dietary obesity. *Cell Reports, 23*(5), 1357–1372.

Fu, W., Liu, Y., Sun, C., & Yin, H. (2019). Transient p53 inhibition sensitizes aged white adipose tissue for beige adipocyte recruitment by blocking mitophagy. *The FASEB Journal, 33*(1), 844–856.

Garcia-Roves, P., Huss, J. M., Han, D. H., Hancock, C. R., Iglesias-Gutierrez, E., Chen, M., & Holloszy, J. O. (2007). Raising plasma fatty acid concentration induces increased biogenesis of mitochondria in skeletal muscle. *Proceedings of the National Academy of Sciences of the United States of America, 104*(25), 10709–10713.

Georgiev, A., Granata, C., & Roden, M. (2022). The role of mitochondria in the pathophysiology and treatment of common metabolic diseases in humans. *American Journal of Physiology-Cell Physiology, 322*(6), C1248–C1259.

Gilani, A., Stoll, L., Homan, E. A., & Lo, J. C. (2024). Adipose signals regulating distal organ health and disease. *Diabetes, 73*(2), 169–177.

Gonzalez-Franquesa, A., Gama-Perez, P., Kulis, M., Szczepanowska, K., Dahdah, N., Moreno-Gomez, S., ... Garcia-Roves, P. M. (2022). Remission of obesity and insulin resistance is not sufficient to restore mitochondrial homeostasis in visceral adipose tissue. *Redox Biology, 54*, 102353.

Gonzalez-Muniesa, P., Martinez-Gonzalez, M. A., Hu, F. B., Despres, J. P., Matsuzawa, Y., Loos, R. J. F., ... Martinez, J. A. (2017). Obesity. *Nature Reviews Disease Primers, 3*, 17034.

Goyal, N., Hoit, D., DiNitto, J., Elijovich, L., Fiorella, D., Pierot, L., ... Arthur, A. S. (2020). How to WEB: A practical review of methodology for the use of the woven EndoBridge. *Journal of NeuroInterventional Surgery*.

Guilherme, A., Virbasius, J. V., Puri, V., & Czech, M. P. (2008). Adipocyte dysfunctions linking obesity to insulin resistance and type 2 diabetes. *Nature Reviews. Molecular Cell Biology, 9*(5), 367–377.

He, F., Huang, Y., Song, Z., Zhou, H. J., Zhang, H., Perry, R. J., ... Min, W. (2021). Mitophagy-mediated adipose inflammation contributes to type 2 diabetes with hepatic insulin resistance. *The Journal of Experimental Medicine, 218*(3).

Hinte, L. C., Castellano-Castillo, D., Ghosh, A., Melrose, K., Gasser, E., Noe, F., ... von Meyenn, F. (2024). Adipose tissue retains an epigenetic memory of obesity after weight loss. *Nature, 636*(8042), 457–465.

Hoffman, D. L., & Brookes, P. S. (2009). Oxygen sensitivity of mitochondrial reactive oxygen species generation depends on metabolic conditions. *The Journal of Biological Chemistry, 284*(24), 16236–16245.

Huang, J. R., Zhang, M. H., Chen, Y. J., Sun, Y. L., Gao, Z. M., Li, Z. J., ... Wu, X. Q. (2023). Urolithin A ameliorates obesity-induced metabolic cardiomyopathy in mice via mitophagy activation. *Acta Pharmacologica Sinica, 44*(2), 321–331.

Jacobsen, S. C., Brons, C., Bork-Jensen, J., Ribel-Madsen, R., Yang, B., Lara, E., ... Vaag, A. (2012). Effects of short-term high-fat overfeeding on genome-wide DNA methylation in the skeletal muscle of healthy young men. *Diabetologia, 55*(12), 3341–3349.

Jun, L., Tao, Y. X., Geetha, T., & Babu, J. R. (2024). Mitochondrial adaptation in skeletal muscle: Impact of obesity, caloric restriction, and dietary compounds. *Current Nutrition Reports, 13*(3), 500–515.

Khan, S., Raghuram, G. V., Bhargava, A., Pathak, N., Chandra, D. H., Jain, S. K., & Mishra, P. K. (2011). Role and clinical significance of lymphocyte mitochondrial dysfunction in type 2 diabetes mellitus. *Translational Research: the Journal of Laboratory and Clinical Medicine, 158*(6), 344–359.

Kim, K. Y., Stevens, M. V., Akter, M. H., Rusk, S. E., Huang, R. J., Cohen, A., ... Sack, M. N. (2011). Parkin is a lipid-responsive regulator of fat uptake in mice and mutant human cells. *The Journal of Clinical Investigation, 121*(9), 3701–3712.

Kim, S., Choi, C., Son, Y., Lee, J., Joo, S., & Lee, Y. H. (2025). BNIP3-mediated mitophagy in macrophages regulates obesity-induced adipose tissue metaflammation. *Autophagy*, 1–19.

Ko, M. S., Yun, J. Y., Baek, I. J., Jang, J. E., Hwang, J. J., Lee, S. E., ... Lee, K. U. (2021). Mitophagy deficiency increases NLRP3 to induce brown fat dysfunction in mice. *Autophagy, 17*(5), 1205–1221.

Kornicka-Garbowska, K., Bourebaba, L., Rocken, M., & Marycz, K. (2021). Inhibition of protein tyrosine phosphatase improves mitochondrial bioenergetics and dynamics, reduces oxidative stress, and enhances adipogenic differentiation potential in metabolically impaired progenitor stem cells. *Cell Communication and Signaling: CCS, 19*(1), 106.

Kusminski, C. M., Bickel, P. E., & Scherer, P. E. (2016). Targeting adipose tissue in the treatment of obesity-associated diabetes. *Nature Reviews. Drug Discovery, 15*(9), 639–660.

Lee, Y. S., Kim, J. W., Osborne, O., Oh, D. Y., Sasik, R., Schenk, S., ... Olefsky, J. M. (2014). Increased adipocyte O2 consumption triggers HIF-1alpha, causing inflammation and insulin resistance in obesity. *Cell, 157*(6), 1339–1352.

Lei, Y., Gan, M., Qiu, Y., Chen, Q., Wang, X., Liao, T., ... Shen, L. (2024). The role of mitochondrial dynamics and mitophagy in skeletal muscle atrophy: from molecular mechanisms to therapeutic insights. *Cellular & Molecular Biology Letters, 29*(1), 59.

Lian, C. Y., Li, H. J., Xia, W. H., Li, Y., Zhou, X. L., Yang, D. B., ... Wang, L. (2024). Insufficient FUNDC1-dependent mitophagy due to early environmental cadmium exposure triggers mitochondrial redox imbalance to aggravate diet-induced lipotoxicity. *Environmental Pollution (Barking, Essex: 1987), 361*, 124724.

Liao, T., Xiong, L., Wang, X., Yang, S., & Liang, Z. (2023). Mitochondrial disorders as a mechanism for the development of obese Sarcopenia. *Diabetol Metab Syndr, 15*(1), 224.

Liesa, M., & Shirihai, O. S. (2013). Mitochondrial dynamics in the regulation of nutrient utilization and energy expenditure. *Cell Metabolism, 17*(4), 491–506.

Liesa, M., & Shirihai, O. S. (2018). Mitochondrial adaptation in obesity is a ClpPicated business. *EMBO Reports, 19*(6).

Lin, C., Chen, J., Hu, M., Zheng, W., Song, Z., & Qin, H. (2021). Sesamol promotes browning of white adipocytes to ameliorate obesity by inducing mitochondrial biogenesis and inhibition mitophagy via beta3-AR/PKA signaling pathway. *Food & Nutrition Research, 65*.

Lin, Q., Li, S., Jiang, N., Jin, H., Shao, X., Zhu, X., ... Ni, Z. (2021). Inhibiting NLRP3 inflammasome attenuates apoptosis in contrast-induced acute kidney injury through the upregulation of HIF1A and BNIP3-mediated mitophagy. *Autophagy, 17*(10), 2975–2990.

Liou, Y. H., Personnaz, J., Jacobi, D., Knudsen, N. H., Chalom, M. M., Starost, K. A., ... Lee, C. H. (2022). Hepatic Fis1 regulates mitochondrial integrated stress response and improves metabolic homeostasis. *JCI Insight*, 7(4).

Livshits, G., & Kalinkovich, A. (2019). Inflammaging as a common ground for the development and maintenance of sarcopenia, obesity, cardiomyopathy and dysbiosis. *Ageing Research Reviews, 56*, 100980.

Longo, M., Bishnu, A., Risiglione, P., Montava-Garriga, L., Cuenco, J., Sakamoto, K., ... Ganley, I. G. (2024). Opposing roles for AMPK in regulating distinct mitophagy pathways. *Molecular Cell, 84*(22), 4350–4367.e4359.

Longo, M., Meroni, M., Paolini, E., Macchi, C., & Dongiovanni, P. (2021). Mitochondrial dynamics and nonalcoholic fatty liver disease (NAFLD): New perspectives for a fairy-tale ending? *Metabolism: Clinical and Experimental, 117*, 154708.

Longo, M., Zatterale, F., Naderi, J., Parrillo, L., Formisano, P., Raciti, G. A., ... Miele, C. (2019). Adipose tissue dysfunction as determinant of obesity-associated metabolic complications. *International Journal of Molecular Sciences, 20*(9).

Lu, Y., Fujioka, H., Joshi, D., Li, Q., Sangwung, P., Hsieh, P., ... Jain, M. K. (2018). Mitophagy is required for brown adipose tissue mitochondrial homeostasis during cold challenge. *Scientific Reports, 8*(1), 8251.

Lu, Y., Li, Z., Zhang, S., Zhang, T., Liu, Y., & Zhang, L. (2023). Cellular mitophagy: Mechanism, roles in diseases and small molecule pharmacological regulation. *Theranostics, 13*(2), 736–766.

Mahmoud, A. M. (2022). An overview of epigenetics in obesity: The role of lifestyle and therapeutic interventions. *International Journal of Molecular Sciences, 23*(3).

Marcangeli, V., Youssef, L., Dulac, M., Carvalho, L. P., Hajj-Boutros, G., Reynaud, O., ... Gouspillou, G. (2022). Impact of high-intensity interval training with or without l-citrulline on physical performance, skeletal muscle, and adipose tissue in obese older adults. *Journal of Cachexia, Sarcopenia and Muscle, 13*(3), 1526–1540.

Marycz, K., Kornicka, K., Szlapka-Kosarzewska, J., & Weiss, C. (2018). Excessive endoplasmic reticulum stress correlates with impaired mitochondrial dynamics, mitophagy and apoptosis, in liver and adipose tissue, but not in muscles in EMS horses. *International Journal of Molecular Sciences, 19*(1).

McCoin, C. S., Franczak, E., Washburn, M. P., Sardiu, M. E., & Thyfault, J. P. (2022). Acute exercise dynamically modulates the hepatic mitochondrial proteome. *Molecular Omics, 18*(9), 840–852.

Monzel, A. S., Enriquez, J. A., & Picard, M. (2023). Multifaceted mitochondria: Moving mitochondrial science beyond function and dysfunction. *Nature Metabolism, 5*(4), 546–562.

Moore, M. P., Cunningham, R. P., Meers, G. M., Johnson, S. A., Wheeler, A. A., Ganga, R. R., ... Rector, R. S. (2022). Compromised hepatic mitochondrial fatty acid oxidation and reduced markers of mitochondrial turnover in human NAFLD. *Hepatology (Baltimore, Md.), 76*(5), 1452–1465.

Morino, K., Petersen, K. F., Dufour, S., Befroy, D., Frattini, J., Shatzkes, N., ... Shulman, G. I. (2005). Reduced mitochondrial density and increased IRS-1 serine phosphorylation in muscle of insulin-resistant offspring of type 2 diabetic parents. *The Journal of Clinical Investigation, 115*(12), 3587–3593.

Mu, C., Wang, S., Wang, Z., Tan, J., Yin, H., Wang, Y., ... Yang, F. (2025). Mechanisms and therapeutic targets of mitochondria in the progression of metabolic dysfunction-associated steatotic liver disease. *Annals of Hepatology: Official Journal of the Mexican Association of Hepatology, 30*(1), 101774.

Muoio, D. M., & Newgard, C. B. (2008). Mechanisms of disease: Molecular and metabolic mechanisms of insulin resistance and beta-cell failure in type 2 diabetes. *Nature Reviews. Molecular Cell Biology, 9*(3), 193–205.

Neira, G., Hernandez-Pardos, A. W., Becerril, S., Ramirez, B., Valenti, V., Moncada, R., ... Rodriguez, A. (2024). Differential mitochondrial adaptation and FNDC5 production in brown and white adipose tissue in response to cold and obesity. *Obesity (Silver Spring, Md.), 32*(11), 2120–2134.

Nisr, R. B., Shah, D. S., Ganley, I. G., & Hundal, H. S. (2019). Proinflammatory NFkB signalling promotes mitochondrial dysfunction in skeletal muscle in response to cellular fuel overloading. *Cellular and Molecular Life Sciences: CMLS, 76*(24), 4887–4904.

Obesity, W. (2022). The world obesity federation. 2025 from https://data.worldobesity.org/maps/?area=trends&group=M&year=2030.

Onishi, M., Yamano, K., Sato, M., Matsuda, N., & Okamoto, K. (2021). Molecular mechanisms and physiological functions of mitophagy. *The EMBO Journal, 40*(3), e104705.

Oost, L. J., Kustermann, M., Armani, A., Blaauw, B., & Romanello, V. (2019). Fibroblast growth factor 21 controls mitophagy and muscle mass. *Journal of Cachexia, Sarcopenia and Muscle, 10*(3), 630–642.

Pafili, K., Kahl, S., Mastrototaro, L., Strassburger, K., Pesta, D., Herder, C., ... Roden, M. (2022). Mitochondrial respiration is decreased in visceral but not subcutaneous adipose tissue in obese individuals with fatty liver disease. *Journal of Hepatology, 77*(6), 1504–1514.

Pallio, G., & Mannino, F. (2024). New insights into adipose tissue metabolic function and dysfunction. *International Journal of Molecular Sciences, 25*(17) 2nd Edition.

Pedersen, J. S., Rygg, M. O., Chrois, K., Sustarsic, E. G., Gerhart-Hines, Z., Wever Albrechtsen, N. J., ... Dela, F. (2022). Influence of NAFLD and bariatric surgery on hepatic and adipose tissue mitochondrial biogenesis and respiration. *Nature Communications, 13*(1), 2931.

Pileggi, C. A., Parmar, G., & Harper, M. E. (2021). The lifecycle of skeletal muscle mitochondria in obesity. *Obesity Reviews: an Official Journal of the International Association for the Study of Obesity, 22*(5), e13164.

Pinti, M. V., Fink, G. K., Hathaway, Q. A., Durr, A. J., Kunovac, A., & Hollander, J. M. (2019). Mitochondrial dysfunction in type 2 diabetes mellitus: an organ-based analysis. *American Journal of Physiology. Endocrinology and Metabolism, 316*(2), E268–E285.

Prasun, P. (2020). Mitochondrial dysfunction in metabolic syndrome. *Biochimica et Biophysica Acta (BBA) - Molecular Basis of Disease, 1866*(10), 165838.

Protasoni, M., & Zeviani, M. (2021). Mitochondrial structure and bioenergetics in normal and disease conditions. *International Journal of Molecular Sciences, 22*(2).

Qi, P., Zhang, W., Gao, Y., Chen, S., Jiang, M., He, R., ... Tang, R. (2024). N6-methyladenosine demethyltransferase FTO alleviates sepsis by upregulating BNIP3 to induce mitophagy. *Journal of Cellular Physiology, 239*(12), e31448.

Qian, P., Tian, H., Wang, Y., Lu, W., Li, Y., Ma, T., ... Yao, W. (2020). A novel oral glucagon-like peptide 1 receptor agonist protects against diabetic cardiomyopathy via alleviating cardiac lipotoxicity induced mitochondria dysfunction. *Biochemical Pharmacology, 182*, 114209.

Ramanathan, R., Ali, A. H., & Ibdah, J. A. (2022). Mitochondrial dysfunction plays central role in nonalcoholic fatty liver disease. *International Journal of Molecular Sciences, 23*(13).

Reinisch, I., Ghosh, A., Noe, F., Sun, W., Dong, H., Leary, P., ... Wolfrum, C. (2025). Unveiling adipose populations linked to metabolic health in obesity. *Cell Metabolism, 37*(3), 640–655.e644.

Reynaud, O., Wang, J., Ayoub, M. B., Leduc-Gaudet, J. P., Mayaki, D., Dulac, M., ... Gouspillou, G. (2023). The impact of high-fat feeding and parkin overexpression on skeletal muscle mass, mitochondrial respiration, and H(2)O(2) emission. *American Journal of, 324*(2), C366–C376.

Ro, S. H., Jang, Y., Bae, J., Kim, I. M., Schaecher, C., & Shomo, Z. D. (2019). Autophagy in adipocyte browning: Emerging drug target for intervention in obesity. *Frontiers in Physiology, 10*, 22.

Romanello, V., & Sandri, M. (2021). The connection between the dynamic remodeling of the mitochondrial network and the regulation of muscle mass. *Cellular and Molecular Life Sciences: CMLS, 78*(4), 1305–1328.

Rosa-Caldwell, M. E., Poole, K. E., Seija, A., Harris, M. P., Greene, N. P., & Wooten, J. S. (2022). Exercise during weight loss improves hepatic mitophagy. *Sports Medicine and Health Science, 4*(3), 183–189.

Sabaratnam, R., Hansen, D. R., & Svenningsen, P. (2023). White adipose tissue mitochondrial bioenergetics in metabolic diseases. *Reviews in Endocrine & Metabolic Disorders, 24*(6), 1121–1133.

Sacks, J., Mulya, A., Fealy, C. E., Huang, H., Mosinski, J. D., Pagadala, M. R., ... Kirwan, J. P. (2018). Effect of Roux-en-Y gastric bypass on liver mitochondrial dynamics in a rat model of obesity. *Physiological Reports, 6*(4).

Sethi, J. K., & Vidal-Puig, A. J. (2007). Thematic review series: Adipocyte biology. Adipose tissue function and plasticity orchestrate nutritional adaptation. *Journal of Lipid Research, 48*(6), 1253–1262.

Shi, L., Yang, J., Tao, Z., Zheng, L., Bui, T. F., Alonso, R. L., ... Cheng, Z. (2024). Loss of FoxO1 activates an alternate mechanism of mitochondrial quality control for healthy adipose browning. *Clinical Science (London), 138*(6), 371–385.

Siddall, H. K., Yellon, D. M., Ong, S. B., Mukherjee, U. A., Burke, N., Hall, A. R., ... Hausenloy, D. J. (2013). Loss of PINK1 increases the heart's vulnerability to ischemia-reperfusion injury. *PLoS ONE, 8*(4), e62400.

Singh, S. P., Grant, I., Meissner, A., Kappas, A., & Abraham, N. G. (2017). Ablation of adipose-HO-1 expression increases white fat over beige fat through inhibition of mitochondrial fusion and of PGC1alpha in female mice. *Hormone Molecular Biology and Clinical Investigation, 31*(1).

Slawik, M., & Vidal-Puig, A. J. (2006). Lipotoxicity, overnutrition and energy metabolism in aging. *Ageing Research Reviews, 5*(2), 144–164.

Song, Y., Li, D., Su, D., Jiang, T., Li, L., Zhan, S., ... Wang, L. (2025). Short-term heat exposure affects thermogenesis and mitophagy in goat brown adipocytes. *BMC Genomics, 26*(1), 272.

Song, Y. M., Lee, W. K., Lee, Y. H., Kang, E. S., Cha, B. S., & Lee, B. W. (2016). Metformin restores parkin-mediated mitophagy, suppressed by cytosolic p53. *International Journal of Molecular Sciences, 17*(1).

Stanford, S. M., Aleshin, A. E., Zhang, V., Ardecky, R. J., Hedrick, M. P., Zou, J., ... Bottini, N. (2017). Diabetes reversal by inhibition of the low-molecular-weight tyrosine phosphatase. *Nature Chemical Biology, 13*(6), 624–632.

Stefan, N., Yki-Jarvinen, H., & Neuschwander-Tetri, B. A. (2025). Metabolic dysfunction-associated steatotic liver disease: Heterogeneous pathomechanisms and effectiveness of metabolism-based treatment. *Lancet Diabetes & Endocrinology, 13*(2), 134–148.

Su, Z., Guo, Y., Huang, X., Feng, B., Tang, L., Zheng, G., & Zhu, Y. (2021). Phytochemicals: Targeting mitophagy to treat metabolic disorders. *Frontiers in Cell and Developmental Biology, 9*, 686820.

Su, Z., Nie, Y., Huang, X., Zhu, Y., Feng, B., Tang, L., & Zheng, G. (2019). Mitophagy in hepatic insulin resistance: Therapeutic potential and concerns. *Frontiers in Pharmacology, 10*, 1193.

Taylor, D., & Gottlieb, R. A. (2017). Parkin-mediated mitophagy is downregulated in browning of white adipose tissue. *Obesity (Silver Spring, Md.), 25*(4), 704–712.

Thomas, A., Marek-Iannucci, S., Tucker, K. C., Andres, A. M., & Gottlieb, R. A. (2019). Decrease of cardiac parkin protein in obese mice. *Frontiers in Cardiovascular Medicine, 6*, 191.

Tong, M., Saito, T., Zhai, P., Oka, S. I., Mizushima, W., Nakamura, M., ... Sadoshima, J. (2021). Alternative mitophagy protects the heart against obesity-associated cardiomyopathy. *Circulation Research, 129*(12), 1105–1121.

Trefts, E., Gannon, M., & Wasserman, D. H. (2017). The liver. *Current Biology: CB, 27*(21), R1147–R1151.

Von Schulze, A. T., Deng, F., Fuller, K. N. Z., Franczak, E., Miller, J., Allen, J., ... Geiger, P. C. (2021). Heat treatment improves hepatic mitochondrial respiratory efficiency via mitochondrial remodeling. *Function (Oxford), 2*(2), zqab001.

Wang, J., Shen, Y., Chen, H., Guan, J., Li, Z., Liu, X., ... Tian, Z. (2024). Non-lethal sonodynamic therapy inhibits high glucose and palmitate-induced macrophage inflammasome activation through mtROS-DRP1-mitophagy pathway. *The FASEB Journal, 38*(22), e70178.

Wang, S., Long, H., Hou, L., Feng, B., Ma, Z., Wu, Y., ... Zhao, G. (2023). The mitophagy pathway and its implications in human diseases. *Signal Transduction and Targeted Therapy, 8*(1), 304.

Wang, S., Tao, J., Chen, H., Kandadi, M. R., Sun, M., Xu, H., ... Ren, J. (2021). Ablation of Akt2 and AMPKalpha2 rescues high fat diet-induced obesity and hepatic steatosis through Parkin-mediated mitophagy. *Acta Pharmaceutica Sinica B, 11*(11), 3508–3526.

White, U. A., & Stephens, J. M. (2010). Transcriptional factors that promote formation of white adipose tissue. *Molecular and Cellular Endocrinology, 318*(1-2), 10–14.

Wu, W., Lin, C., Wu, K., Jiang, L., Wang, X., Li, W., ... Feng, D. (2016). FUNDC1 regulates mitochondrial dynamics at the ER-mitochondrial contact site under hypoxic conditions. *The EMBO Journal, 35*(13), 1368–1384.

Yau, W. W., Singh, B. K., Lesmana, R., Zhou, J., Sinha, R. A., Wong, K. A., ... Yen, P. M. (2019). Thyroid hormone (T(3)) stimulates brown adipose tissue activation via mitochondrial biogenesis and MTOR-mediated mitophagy. *Autophagy, 15*(1), 131–150.

Yoshii, A., McMillen, T. S., Wang, Y., Zhou, B., Chen, H., Banerjee, D., ... Tian, R. (2024). Blunted cardiac mitophagy in response to metabolic stress contributes to HFpEF. *Circulation Research, 135*(10), 1004–1017.

Youssef, L., Granet, J., Marcangeli, V., Dulac, M., Hajj-Boutros, G., Reynaud, O., ... Aubertin-Leheudre, M. (2022). Clinical and biological adaptations in obese older adults following 12-weeks of high-intensity interval training or moderate-intensity continuous training. *Healthcare (Basel), 10*(7).

Yu, L. M., Dong, X., Xue, X. D., Xu, S., Zhang, X., Xu, Y. L., ... Wang, H. S. (2021). Melatonin attenuates diabetic cardiomyopathy and reduces myocardial vulnerability to ischemia-reperfusion injury by improving mitochondrial quality control: Role of SIRT6. *Journal of Pineal Research, 70*(1), e12698.

Zhang, C., Li, J., Wang, L., Ma, J., Li, X., Wu, Y., ... Yang, Y. (2025). Terazosin, a repurposed GPR119 agonist, ameliorates mitophagy and beta-cell function in NAFPD by inhibiting MST1-Foxo3a signalling pathway. *Cell Proliferation, 58*(3), e13764.

Zhou, H., Zhu, P., Guo, J., Hu, N., Wang, S., Li, D., ... Chen, Y. (2017). Ripk3 induces mitochondrial apoptosis via inhibition of FUNDC1 mitophagy in cardiac IR injury. *Redox Biology, 13*, 498–507.

Zhou, T., Chang, L., Luo, Y., Zhou, Y., & Zhang, J. (2019). Mst1 inhibition attenuates non-alcoholic fatty liver disease via reversing Parkin-related mitophagy. *Redox Biology, 21*, 101120.

Zhu, Y., Gu, Z., Shi, J., Chen, C., Xu, H., & Lu, Q. (2022). Vaspin attenuates atrial abnormalities by promoting ULK1/FUNDC1-mediated mitophagy. *Oxidative Medicine and Cellular Longevity, 2022*, 3187463.

Zimmermann, A., Madeo, F., Diwan, A., Sadoshima, J., Sedej, S., Kroemer, G., & Abdellatif, M. (2024). Metabolic control of mitophagy. *European Journal of Clinical Investigation, 54*(4), e14138.

CHAPTER FOUR

Autophagy in endothelial cells: A gatekeeper of immune cell trafficking and function

Guillermo Marcos-Zapatero[a], Dafne García-Mateos[a], Irene Carriles-Moar[a], and Natalia Reglero-Real[a,b,*]

[a]Departamento de Biología Molecular, Instituto Universitario de Biología Molecular (IUBM) and Centro de Biología Molecular Severo Ochoa (CBM), CSIC-Universidad Autónoma de Madrid, Madrid, Spain
[b]Instituto de Investigación Sanitaria La Princesa, Madrid, Spain
*Corresponding author. e-mail address: nreglero@cbm.csic.es

Contents

Abstract

Endothelial cells (ECs) are increasingly recognised as dynamic regulators of immunity rather than passive vascular barriers. By integrating inflammatory, metabolic, and mechanical cues, ECs actively shape leukocyte trafficking and determine inflammatory outcomes. Autophagy, a conserved lysosomal degradation pathway essential for cellular homeostasis, has emerged as a key determinant of endothelial function in health and disease. Recent work identifies EC autophagy as a context-dependent determinant of leukocyte trafficking, vascular inflammation, antigen presentation, and

International Review of Cell and Molecular Biology, Volume 406
ISSN 1937-6448, https://doi.org/10.1016/bs.ircmb.2026.02.001

endothelial phagocytic capacity. Beyond controlling immune cell recruitment, EC autophagy intersects with xenophagy, efferocytosis, and vascular ageing, highlighting its broad immunomodulatory potential. In this review, we synthesise current evidence defining how autophagy within ECs regulates immune responses in a tissue-specific manner during acute and chronic inflammation. Collectively, these insights position EC autophagy as a central integrator of vascular and immune homeostasis and underscore its potential as a therapeutic target in chronic inflammatory and ageing-associated diseases.

1. Introduction

The vasculature is a dynamic organ system essential not only for nutrient and oxygen delivery, fluid balance, and haemostasis, but also as a central regulator of immune surveillance and inflammatory responses. Endothelial cells (ECs), which line all blood and lymphatic vessels, serve as the interface between circulation and tissues and act as highly specialised immunoregulatory units. Beyond providing a physical barrier, ECs integrate mechanical, metabolic, and inflammatory cues from the tissue microenvironment to orchestrate immune cell trafficking, thereby shaping the balance between protective immunity and pathological inflammation (Nourshargh & Alon, 2014; Pober & Sessa, 2007; Amersfoort et al., 2022; Nourshargh et al., 2010).

Inflammation represents a coordinated host defence mechanism indispensable for survival and repair. However, excessive or unresolved inflammation can result in tissue damage and drive chronic disease (Medzhitov, 2008; Furman et al., 2019; Zhao et al., 2021). As key effector components of the inflammatory response, leukocytes are recruited from the bloodstream to the tissues following different inflammatory cues. This requires their sequential migration across postcapillary venules, involving breaching of ECs, pericytes, and the basement membrane (Nourshargh & Alon, 2014). EC activation is induced by a wide range of proinflammatory mediators, including cytokines (such as IL-6, IL-1β and TNF) and pathogen-associated molecular patterns (e.g. lipopolysaccharide: LPS), which are released by tissue resident cells (e.g. macrophages, dendritic cells and mast cells) or invading microorganisms, respectively. During steady state, ECs also enable selective trafficking of patrolling leukocytes in specific tissues, including the liver and lung, to maintain immune surveillance and homeostasis (Moreno-Cañadas et al., 2021; Ficht & Iannacone, 2020; Ballesteros et al., 2020). Beyond regulating trafficking, subsets of ECs display immune effector functions such as antigen presentation, T cell co-regulation, and phagocytosis (Amersfoort et al., 2022), consistent with their developmental link to immune cells (Lange et al., 2021).

The molecular events regulating leukocyte passage through vascular barriers have been extensively studied in postcapillary venules during inflammation and are described by the leukocyte adhesion cascade: rolling, arrest, crawling, and transendothelial migration (TEM) (Ley et al., 2007). This series of sequential but overlapping steps depends on selectins, chemokines, leukocyte β2-integrins and their counter-receptors in ECs (e.g. ICAM-1 and VCAM-1), and culminates in the reorganisation of EC junctional adhesion molecules (Woodfin et al., 2011; Colom et al., 2015; Girbl et al., 2018). EC junctions comprise tight junctions, which include members of the junctional adhesion molecule (JAM) family and claudins, adherent junctions (AJ) containing VE-cadherin and other molecules such as platelet endothelial cell adhesion molecule-1 (PECAM-1) (Dejana et al., 2009; Reglero-Real et al., 2016; Vestweber, 2015). TEM can occur via paracellular or transcellular routes, the latter particularly relevant at restrictive barriers such as the blood–brain barrier (BBB) (Mapunda et al., 2022). This highlights the need to study the molecular mechanisms regulating leukocyte TEM in different vascular beds, which might differ from the general dogma described above and impact the outcome of immune responses in different organs. Although other vascular and perivascular elements, including pericytes, the venular basement membrane, and resident sentinel cells, contribute to leukocyte guidance beyond the endothelial barrier (Proebstl et al., 2012; Voisin & Nourshargh, 2013; Pober & Tellides, 2012; Voisin et al., 2010; De Filippo et al., 2013), ECs remain key gatekeepers of leukocyte entry and thus central determinants of inflammatory responses (Nourshargh & Alon, 2014).

Over the last decade, autophagy has emerged as a central regulator of both innate and adaptive immunity (Deretic & Levine, 2018). Within leukocytes, autophagy regulates multiple immune cell functions, including cytokine production, differentiation, pathogen clearance and antigen presentation (Deretic, 2021). Consistent with this, genetic links between autophagy and inflammation have been identified across a wide spectrum of human disorders characterised by dysregulated immune responses (Deretic & Levine, 2018; Levine & Kroemer, 2019; Klionsky et al., 2021). At a cellular level, autophagy maintains intracellular homeostasis by delivering cytoplasmic material to lysosomes for degradation. This process encompasses the removal or recycling of protein aggregates, membranes, whole organelles such as mitochondria, and invading microorganisms, as well as the generation of metabolic substrates during nutrient stress (Herzig & Shaw, 2018; Lin & Hardie, 2018; Liu & Sabatini, 2020; Morishita & Mizushima, 2019).

Autophagy comprises several distinct but interconnected pathways. Macroautophagy, the best-characterised form and hereafter referred to as canonical autophagy, involves the sequestration of cytoplasmic material within double-membrane autophagosomes that subsequently fuse with lysosomes. This process is tightly regulated by autophagy-related genes (*Atgs*), including the ATG5–ATG12–ATG16L complex, which is essential for autophagosome initiation and elongation (Morishita & Mizushima, 2019; Feng et al., 2014; Yu et al., 2018). Upon autophagy induction, cytosolic microtubule-associated protein light chain 3B (LC3), a member of the ATG8 family, is lipidated and incorporated into the autophagosome double membrane. LC3-interacting regions then facilitate selective cargo recruitment, enabling delivery of autophagosomal contents to lysosomes for degradation (Morishita & Mizushima, 2019). Within canonical autophagy, both bulk and selective forms operate. Bulk autophagy involves the non-specific sequestration of portions of the cytoplasm, including organelles and macromolecular complexes, for degradation. By contrast, selective autophagy targets specific, potentially harmful cellular components for removal, often mediated by ubiquitination and cargo receptors. A prominent example is mitophagy, the selective degradation of damaged or dysfunctional mitochondria, which is essential for maintaining mitochondrial quality and overall cellular homeostasis (Vargas et al., 2023). In addition to canonical autophagy, microautophagy involves direct lysosomal membrane invagination of cytoplasmic material, while chaperone-mediated autophagy (CMA) selectively translocates individual proteins across the lysosomal membrane for degradation. Collectively, these forms of autophagy contribute to tissue homeostasis and immune regulation (Deretic, 2021; Kaushik et al., 2021; Macian, 2019).

Beyond these classical pathways, a growing number of non-canonical autophagy-related processes have been described. These mechanisms employ components of the canonical autophagy machinery but operate on single-membrane vesicles derived from the endocytic system (Galluzzi & Green, 2019; Galluzzi et al., 2017; Martinez-Martin et al., 2017). The prototypical example is LC3-associated phagocytosis (LAP), together with related pathways such as LC3-associated endocytosis (LANDO), collectively referred to as CASM (Conjugation of ATG8s to Single Membranes) (Sanjuan et al., 2007; Heckmann & Green, 2019; Heckmann et al., 2019; Florey et al., 2011). These

Table 1 Immunomodulatory functions of autophagy pathways in ECs. The table summarises the genetic and therapeutic interventions used to modulate EC autophagy pathways in different tissues and how they impact immune outcomes in physiological and pathological settings. Of note, the term autophagy denotes the totality of canonical (macroautophagy) and non-canonical autophagy processes, as their individual contributions cannot be dissected due to the targeting of core autophagy genes shared between both pathways.

Tissue	Vascular bed	Genotype/intervention	Autophagy status	Immune function	Pathological setting/inflammatory reaction	Autophagy role	References
Lung		*Atg7* –/–	Autophagy deficiency	Increased vascular leakage	Homeostasis	Protective	Patella et al. (2016)
		MKK3 –/–	Mitophagy induction	Reduced neutrophil recruitment and macrophage accumulation	LPS-induced sepsis ALI	Protective	Mannam et al. (2013)
		RAB26–/–siATG5siATG7	Autophagy deficiency	Increased vascular leakage	LPS-induced sepsis ALI	Protective	Dong et al. (2018)
Skin	Veins	*Atg5* –/–	Autophagy deficiency	Increased neutrophil recruitment	IL-1β and LPS-induced dermal inflammation	Protective	Reglero-Real et al. (2021a)
	TECs	*Atg5*–/–	Autophagy deficiency	Improved immunosurveillance and $CD8^+$ T cell infiltration in the tumour microenvironment	Melanoma models	Detrimental	Verhoeven et al. (2023)
Peritoneum	Veins	*Atg5* –/–	Autophagy deficiency	Increased neutrophil, monocyte and eosinophil recruitment	Zymosan and LPS-induced peritonitis	Protective	Reglero-Real et al. (2021a)

(continued)

Table 1 Immunomodulatory functions of autophagy pathways in ECs. The table summarises the genetic and therapeutic interventions used to modulate EC autophagy pathways in different tissues and how they impact immune outcomes in physiological and pathological settings. Of note, the term autophagy denotes the totality of canonical (macroautophagy) and non-canonical autophagy processes, as their individual contributions cannot be dissected due to the targeting of core autophagy genes shared between both pathways. (*cont'd*)

Tissue	Vascular bed	Genotype/ intervention	Autophagy status	Immune function	Pathological setting/ inflammatory reaction	Autophagy role	References
Cremaster muscle	Postcapillary venules	*Atg5 −/−*	Autophagy deficiency	EC adhesion molecule accumulationIncreased neutrophil TEM and recruitmentIncreased tissue damage after IR	LPS-induced endotoxemia and IR	Protective	Reglero-Real et al. (2021a)
		Atg16L1 E230/ E230	Non-canonical autophagy deficiency	EC adhesion molecule distribution.Increased neutrophil recruitment	IR		
Aorta	Arteries	*Atg5 −/−*	Autophagy deficiency	Increased inflammation	Atherosclerosis	Protective	Vion et al. (2017)
		miR-100 mimetics	Autophagy induction	Dampened NF-κB signalling.Reduced macrophage accumulation	Atherosclerosis	Protective	Pankratz et al. (2018)
		Cav-1$^{-/-}$ *Ldlr*$^{-/-}$	Autophagy induction	Decreased macrophage accumulation	Atherosclerosis	Protective	Zhang et al. (2020)

Lymph node	Lymphatics	*Atg5* −/−	Autophagy deficiency	Diminished T-reg proliferation.Increased $CD4^{+}$ Th17 cell egress from LNs and recruitment to inflamed joints	Arthritis	Protective	(Harlé et al. 2021)
			Autophagy deficiency	Decreased $CD8^{+}$ T cell egress from LN	Melanoma	Protective	Houbaert et al. (2024)
Liver	Sinusoids	*Atg5* −/−	Autophagy deficiency	Increased liver inflammation (CCL2, CCL5, IL6 and VCAM-1), and macrophage accumulation	NASH	Protective	Hammoutene et al. (2020)
Kidney		KLF4 overexpression	Autophagy induction	Decreased inflammatory signalling (TNFα and MCP-1)	DKD	Protective	Wang et al. (2022)
		Atg7 −/−	Autophagy deficiency	Increased NLRP3 signalling, and IL-1β driven fibrosis	Glomerulopathy	Protective	Kim et al. (2024)
		Gimap6 −/−	Autophagy deficiency	Microangiopathic glomerulosclerosis	Glomerulopathy	Protective	Yao et al. (2022)
Central Nervous System	BBB	WT		Increased macrophage accumulation and increased inflammation	SCI and EAE	Detrimental	Zhou et al. (2019)

(continued)

Table 1 Immunomodulatory functions of autophagy pathways in ECs. The table summarises the genetic and therapeutic interventions used to modulate EC autophagy pathways in different tissues and how they impact immune outcomes in physiological and pathological settings. Of note, the term autophagy denotes the totality of canonical (macroautophagy) and non-canonical autophagy processes, as their individual contributions cannot be dissected due to the targeting of core autophagy genes shared between both pathways. (*cont'd*)

Tissue	Vascular bed	Genotype/ intervention	Autophagy status	Immune function	Pathological setting/ inflammatory reaction	Autophagy role	References
Bone Marrow		siBeclin1	Autophagy deficiency	Decreased haematopoiesis	*In vitro* co-culture	Protective	Lyu et al. (2020)
	Capillaries	*CD200 −/−*	Autophagy deficiency	Poor BM reconstitution after engraftment	BM transplantation	Protective	Furuhashi et al. (2025)
		Atg7 −/−eNOS −/−	Autophagy deficiency				

ALI: Acute Lung Injury; BBB: Blood Brain Barrier; BM: Bone Marrow; DKD: Diabetic Kidney Disease; EAE: Experimental Autoimmune Encephalomyelitis; EC: Endothelial Cell; IR: Ischemia-Reperfusion Injury; LN: Lymph Nodes; NASH: Nonalcoholic Steatohepatitis; SCI: Spinal Cord Injury; TECs: Tumour-associated Endothelial Cells; TEM: Transendothelial Migration; WT: Wild-Type.

pathways couple endolysosomal trafficking to LC3 lipidation and have important implications for immune cell activation and inflammatory regulation.

Within ECs, autophagy has long been recognised as a regulator of fundamental aspects of endothelial physiology, including survival, redox state, thrombosis, angiogenesis, and metabolism (Verhoeven et al., 2021; Nussenzweig et al., 2015; Sachdev & Lotze, 2017; Schaaf et al., 2019). More recently, EC autophagy has been directly implicated in the control of leukocyte trafficking, notably through effects on junctional organisation and adhesion molecule trafficking (Reglero-Real et al., 2021a). EC autophagy therefore emerges as a critical determinant of how immune cells access tissues. In this review, we focus on the role of EC autophagy in the regulation of immune cell functions, with particular emphasis on how EC autophagy shapes leukocyte entry into inflamed tissues (Table 1). We further discuss emerging evidence that EC autophagy influences a broader repertoire of immune processes, including pathogen clearance via xenophagy, tissue remodelling through phagocytosis and efferocytosis, antigen presentation in adaptive immunity, and the regulation of vascular function during ageing.

2. Regulation of immune cell trafficking by EC autophagy

As discussed above, ECs are central regulators of leukocyte extravasation. Autophagy modulates this function by influencing EC junctional dynamics, vesicular trafficking, and inflammatory signalling pathways. Emerging evidence indicates that these effects are context-dependent, varying across leukocyte subsets and vascular compartments. This positions EC autophagy as a critical determinant of immune cell trafficking and vascular immune responses.

2.1 Neutrophils

Neutrophils are a heterogeneous population of innate immune cells that display diverse functional and transcriptional programmes, which vary across tissues and inflammatory settings (Cerezo-Wallis et al., 2025). Accordingly, neutrophils are rapidly recruited to sites of inflammation within hours, yet also traffic into tissues under steady-state conditions to support immune surveillance (Ballesteros et al., 2020; Shim et al., 2022). Pharmacological studies first implicated autophagy in neutrophil recruitment: autophagy-inducing drugs such as rapamycin and resveratrol limited neutrophil infiltration in LPS-induced acute lung injury (ALI) (Fielhaber et al., 2012; Mameli et al., 2022; Chen et al., 2013; Li et al., 2021), whereas inhibition of autophagy via chloroquine treatment exacerbated it (Zhao et al., 2019). By contrast, administration of other autophagy inhibitors, such as 3-MA, has been related to both increased (Zhao et al., 2019) and reduced lung neutrophil sequestration in response to LPS (Slavin et al., 2018). These conflicting outcomes likely reflect non-specific effects of these drugs, highlighting the need for genetic models. In LPS-induced endotoxemia, genetic induction of EC mitophagy (via *MKK3* ablation) limited polymorphonuclear (PMN) leukocyte recruitment to lungs and improved survival of septic mice (Mannam et al., 2013). This effect was linked to reduced mitochondrial ROS production, NF-κB activation and ICAM-1 expression in ECs during LPS challenge, in line with the protective role of mitophagy in preventing stress-induced EC injury (Wu et al., 2015). Conversely, mitophagy defects in ECs cause cytosolic mtDNA leakage, inflammasome/STING activation, and a pro-adhesive phenotype that exacerbates neutrophil recruitment to the lung and kidneys, worsening sepsis outcomes (Wu et al., 2021; Blazer et al., 2022). Moreover, EC *Atg5* and *Atg7* deletion disrupts barrier integrity and enhances vascular leakage in basal conditions (Patella et al., 2016) and during ALI (Dong et al., 2018), though leukocyte trafficking was not initially investigated in these studies.

Our laboratory has shown directly that microvascular EC autophagy restricts neutrophil TEM during acute inflammation in vivo (Reglero-Real et al., 2021a). Intravital confocal imaging of murine cremaster muscle tissues revealed that *Atg5*-deficient ECs exhibit enhanced paracellular and transcellular neutrophil migration across inflamed venules, causing excessive tissue accumulation and damage (Fig. 1A). Similar results were obtained in skin and peritoneum following acute physiological inflammation. Moreover, local pharmacological activation of EC autophagy rescued this phenotype (Reglero-Real et al., 2021a), positioning EC autophagy as a potential anti-inflammatory target. Mechanistically, autophagy restricted the surface availability of junctional adhesion molecules such as PECAM-1 by degrading receptor-enriched membranous pools, thereby reducing neutrophil breaching of ECs (Reglero-Real et al., 2012; Mamdouh et al., 2003). In line with these results, autophagy stabilised VE-cadherin junctions by degrading activated Src kinase, preventing VE-cadherin phosphorylation, internalisation and strengthening barrier function (Dong et al., 2018).

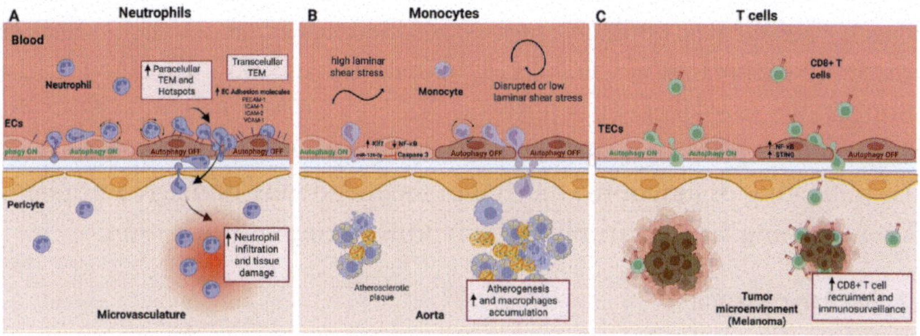

Fig. 1 Impact of blood endothelial cell (EC) autophagy on leukocyte trafficking. The schematic illustrates the role of blood EC autophagy in regulating neutrophil, monocyte / macrophage and T cell trafficking dynamics under inflammatory conditions. (A) Compromised autophagy in microvascular EC leads to increased surface expression of adhesion molecules (PECAM-1, ICAM-1, ICAM-2 and VCAM-1), promotes enhanced neutrophil paracellular and transcellular transendothelial migration (TEM), and the formation of neutrophil TEM hotspots. Collectively, these processes result in increased neutrophil infiltration into inflamed tissues and local tissue damage. (B) In aortic ECs, disturbed or low laminar shear stress (LSS) blocks autophagy, thereby promoting monocyte recruitment and macrophage tissue accumulation, and accelerating atherosclerosis progression. In contrast, high LSS supports EC autophagy, leading to increased expression of the transcription factor Krüppel-like factor (Kfl2), reduced NF-κB signalling, and inhibition of caspase-3 via nuclear miR-126–5p, collectively exerting atheroprotective effects. (C) Compromised autophagy in tumour ECs (TECs) promotes the recruitment of CD8+ T cells into the tumour microenvironment, an effect that correlates with increased NF-κB and STING endothelial signalling, thereby enhancing anti-tumour immune responses.

Apart from remodelling junctions, EC autophagy may also regulate neutrophil TEM through platelet interactions. Neutrophil breaching of ECs requires functional crosstalk with platelets (Sreeramkumar et al., 2014). Deletion of endothelial *Atg5* and *Atg7* has been shown to support the antithrombotic functions of ECs (Verhoeven et al., 2021; Yau et al., 2017), including limiting platelet adhesion to injured vessels (Torisu et al., 2013). These findings raise the possibility that EC autophagy contributes to neutrophil–platelet cooperation during thrombo-inflammatory injury.

Importantly, many functions attributed to EC autophagy involve the core canonical machinery, which also regulates intracellular trafficking beyond canonical degradation. This raises the hypothesis that distinct autophagy-related pathways could control specific aspects of EC biology. Supporting this idea, our group demonstrated that non-canonical autophagy operates in inflamed microvascular ECs in vivo. In ischemia–reperfusion (IR) injury, the WD40 domain of ATG16L1, which is essential for non-canonical autophagy processes (Fletcher et al., 2018), regulates neutrophil extravasation, EC LC3 lipidation, and EC PECAM-1 distribution (Reglero-Real et al., 2021a, 2021b). These findings suggest that non-canonical pathways, such as LAP and LANDO, may modulate vesicular trafficking of junctional molecules independently of macroautophagy, thereby shaping EC-neutrophil interactions during inflammation.

2.2 Monocytes

Monocytes are innate immune cells that are recruited to inflamed tissues later than neutrophils and play a particularly prominent role in chronic inflammatory conditions such as atherosclerosis. During atherogenesis, endothelial activation promotes monocyte adhesion and TEM into the subendothelial space, where monocytes differentiate into macrophages. These macrophages then sustain a local pro-inflammatory environment through cytokine release, ROS production, and lipid-handling programmes, thereby linking monocyte recruitment to macrophage-driven lesion growth and persistence (Prame Kumar et al., 2018).

Autophagy plays a protective role in ECs within the context of atherosclerosis. Blood flow patterns critically influence aortic EC behaviour: regions exposed to laminar shear stress (LSS) are generally atheroprotected, whereas areas of disturbed flow are prone to plaque development, making disturbed flow a key driver of endothelial dysfunction and atherogenesis (Le et al., 2017). Autophagy in ECs is tightly regulated by these hemodynamic forces (Claude-Taupin et al., 2021). Under high LSS, robust autophagy supports EC homeostasis (Vion et al., 2017), while ECs in disturbed flow regions exhibit defective autophagy. Consistently, conditional deletion of autophagy genes in ECs accelerates plaque development (Vion et al., 2017; Torisu et al., 2016; Santovito et al., 2020), highlighting EC autophagy as an anti-atherogenic mechanism. For instance, genetic ablation of *Atg5* in ECs under high LSS enhanced apoptosis, senescence, and inflammation, exacerbating plaque burden in *ApoE*-deficient mice on high-fat diet (Vion et al., 2017). EC autophagy under high LSS also inhibits Caspase-3 via nuclear miR-126–5p, thereby limiting apoptosis and preserving EC survival (Santovito et al., 2020). Furthermore, mitophagy is essential for transcriptional induction of the anti-inflammatory factor Kruppel-like factor 2 (Klf2) in ECs exposed to LSS (Coon et al., 2022). Since shear stress is a critical regulator of vascular inflammation (Humphrey & Schwartz, 2021), the interplay between autophagy and mechanotransduction likely sits at the core of endothelial control over monocyte trafficking and macrophage accumulation into atherosclerotic plaques. One mechanism involves endothelial-enriched microRNA miR-100, which suppresses pro-inflammatory activation in an autophagy-dependent manner (Pankratz et al., 2018). Systemic delivery of miR-100 mimetics reduced plaque size and macrophage accumulation in hypercholesterolemic mice, whereas inhibition of miR-100 enhanced leukocyte adhesion and NF-κB signalling. Similarly, pharmacological autophagy blockade with 3-MA increases EC activation and macrophage accumulation in hypercholesterolemic mice lacking Caveolin-1, a key regulator of atherogenesis (Zhang et al., 2020) (Fig. 1B).

2.3 T cells

T cell trafficking between blood, peripheral tissues, and secondary lymphoid organs is fundamental to adaptive immunity (Fowell & Kim, 2021). Following antigen recognition in lymph nodes (LNs), naïve T cells proliferate and differentiate into effector subsets ($CD4^+$ helper, $CD8^+$ cytotoxic), which exit from LNs to infiltrate inflamed tissues (Hunter et al., 2016; Bauer et al., 2022; Hampton & Chtanova, 2019; Santambrogio et al., 2019). Lymphatic ECs (LECs) critically regulate this step through the production of sphingosine-1-phosphate (S1P), a key lipid mediator of T cell egress (Saxena et al., 2019; Cyster & Schwab, 2012).

Autophagy in LECs has recently emerged as a key modulator of this process. Inflammatory signals induce autophagy in LECs, driving degradation of sphingosine kinase 1 (SphK1) and reducing S1P availability (Harlé et al., 2021). Consequently, autophagy induction in primary LECs suppressed S1P-dependent $CD4^+$ T-cell transmigration in vitro, whereas *Atg5*-deficient LECs impaired the egress of both naïve and antigen-specific $CD8^+$ T cells from tumour-draining LNs in vivo (Houbaert et al., 2024). In a murine model of melanoma, loss of LEC autophagy prevented recruitment of tumour-infiltrating T and natural killer cells and abrogated the response to immunotherapy (Houbaert et al., 2024). By contrast, in arthritis, deletion of LEC *Atg5* promoted efflux of pathogenic $CD4^+$ Th17 cells from LNs, accelerating their accumulation in joints and exacerbating disease progression (Harlé et al., 2021). This duality positions autophagy as a beneficial mechanism in LECs, exerting anti-inflammatory effects in arthritis while simultaneously enhancing T cell immunosurveillance in melanoma. Furthermore, these findings indicate that LEC autophagy differentially regulates the trafficking of distinct lymphocyte subsets (Fig. 2), likely due to varying sensitivities to S1P gradients. Consistent with this,

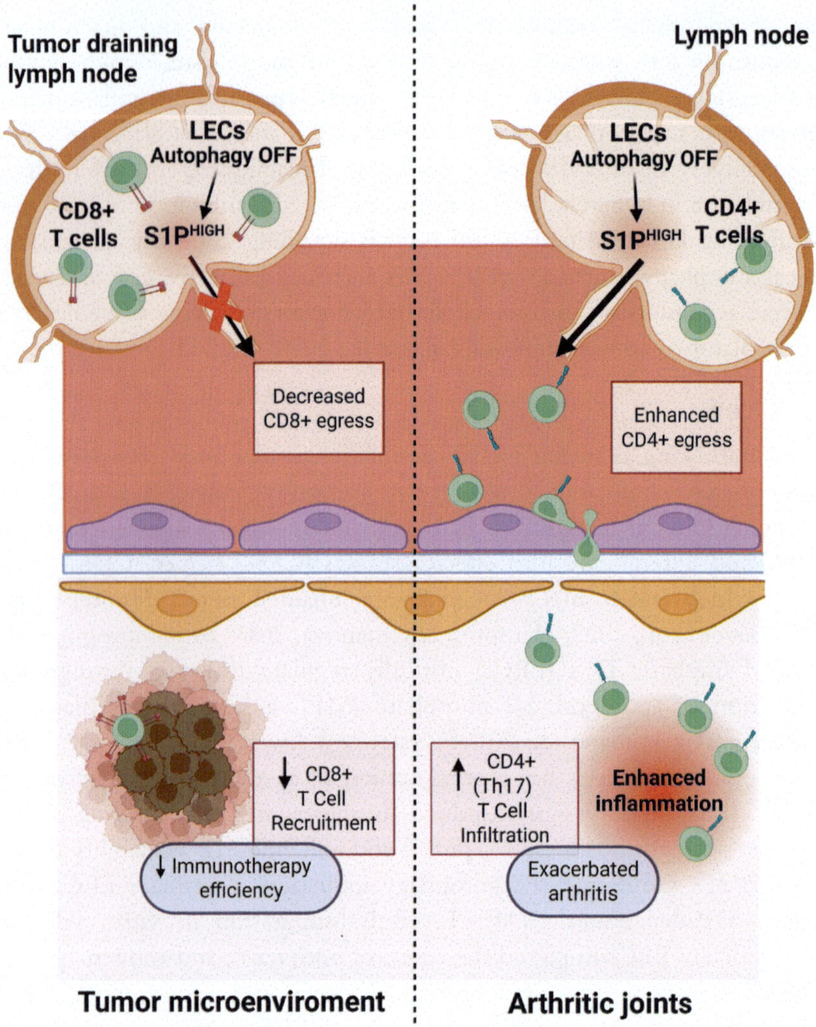

Fig. 2 Impact of lymphatic EC (LEC) autophagy on leukocyte trafficking. Autophagy in LECs plays a key role in regulating sphingosine-1-phosphate (S1P) levels for accurate T cell egress. (Left) Disabled autophagy in LECs of tumour-draining LNs reduces CD8+ T cell egress by disrupting the local S1P gradient, limiting their recruitment into the tumour microenvironment and dampening immunotherapy efficacy. (Right) Loss of LEC autophagy compromises S1P gradient, enhancing CD4+ (Th17) T cell egress into the circulation and promoting their infiltration into inflamed joints, worsening arthritis outcomes.

Th17 cells express higher levels of the type 1 S1P receptor compared with other $CD4^+$ T-cell subsets (Harlé et al., 2021). Whether vascular EC autophagy also influences the dynamics of lymphocyte entry into LNs, either through afferent lymphatics or high endothelial venules, remains less explored to date and represents a significant knowledge gap in the field.

Autophagy within tumour-associated ECs (TECs) further shapes T cell trafficking in cancer. In melanoma, autophagy within TECs is detrimental, limiting $CD8^+$ T cell infiltration and compromising responses to immunotherapy (Fig. 1C). Consistent with this, EC-specific deletion of *Atg5* increased intratumoral $CD8^+$ T cell accumulation and reduced tumour growth. Mechanistically, *Atg5* deletion in ECs correlated with enhanced activation of NF-κB and STING signalling pathways, indicating a shift towards a more inflammatory endothelial phenotype. Supporting the translational relevance of these findings, analysis of human melanoma datasets identified an "Inflammatory High/Autophagy Low" TEC gene signature that correlated with improved responses to immunotherapy (Verhoeven et al., 2023).

Collectively, these studies identify EC autophagy as a key brake on neutrophil trafficking in acute inflammation, a regulator of monocyte recruitment and subsequent macrophage accumulation in chronic vascular disease, and a modulator of T cell egress from LNs and tumour immunity. Together, they underscore how EC autophagy pathways fine-tune leukocyte trafficking across diverse inflammatory contexts, with implications for both tissue injury and therapeutic intervention.

3. Tissue-specific functions of EC autophagy in inflammation and immunity

A growing body of evidence indicates ECs are a heterogeneous population with distinct molecular and functional properties shaped by local metabolic, shear stress, and inflammatory environments (Kalucka et al., 2020; Schupp et al., 2021; Geldhof et al., 2022; Paik et al., 2020). As such, autophagic responses are unlikely to be uniform across vascular beds and instead vary in a tissue- and context-dependent manner. This variation may have important consequences, influencing both the sensitivity of ECs to autophagy modulation and the immune outcomes that follow. As outlined above, EC autophagy controls the trafficking of distinct leukocyte subsets in tissues such as the lung, smooth muscle, aorta, LNs and skin.

Beyond leukocyte recruitment, growing evidence indicates that autophagy in ECs also regulates inflammatory signalling and immune responses in a tissue-specific fashion (Fig. 3).

3.1 Liver

Liver sinusoidal ECs (LSECs) are uniquely exposed to dietary and pathogen-derived antigens. Accordingly, they play central immunoregulatory roles, contributing to tolerance and immunosurveillance through antigen cross-presentation to T cells and scavenger activity (Amersfoort et al., 2022; Poisson et al., 2017). Autophagy supports LSEC survival, phenotype, nitric oxide availability, and antioxidant capacity after acute injury (Guixé-Muntet et al., 2017; Ruart et al., 2019). Although ablation

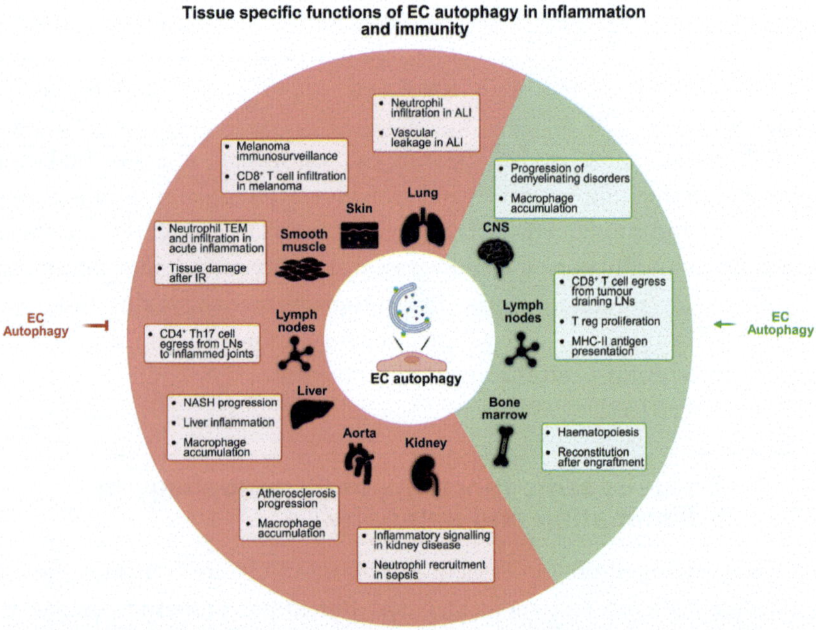

Fig. 3 Tissue-specific functions of EC autophagy in inflammation and immunity. EC autophagy exerts context-dependent effects across organs, influencing immunity and inflammatory outcomes. The left section depicts immune and inflammatory functions that are limited by EC autophagy activity in the lung, skin, smooth muscle, lymph nodes, liver, aorta, and kidney. The right section illustrates immune and inflammatory functions promoted by EC autophagy activity in the central nervous system (CNS), lymph nodes (LNs), and bone marrow. ALI: Acute Lung Injury; EC: Endothelial Cell; IR: Ischemia-Reperfusion Injury; NASH: Nonalcoholic Steatohepatitis; TEM: Transendothelial Migration.

of LSEC autophagy had no impact on the recruitment of inflammatory cells to the liver parenchyma in a model of CCl4-induced acute liver injury (Ruart et al., 2019), autophagy deficient LSECs promoted inflammation during the development of chronic liver diseases such as non-alcoholic steatohepatitis (NASH) (Hammoutene et al., 2020). Following high-fat diet administration, mice deficient in EC *Atg5* exhibited increased liver expression of inflammatory markers, including monocyte- (CCL2) and T cell- (CCL5) recruiting chemokines, adhesion receptors (VCAM-1) and the monocytic marker CD68. The latter was associated with increased liver apoptosis and peri-sinusoidal fibrosis. Together, these results suggest LSEC autophagy plays a protective role during NASH, limiting inflammation and dampening disease pathology. Moreover, LSECs from NASH patients exhibited reduced numbers and size of autophagic vesicles, supporting impaired EC autophagy in human disease (Hammoutene et al., 2020).

3.2 Kidney

Kidney ECs exhibit pronounced phenotypic and metabolic heterogeneity that reflects their anatomical distribution along the nephron and underpins functions such as plasma filtration, solute exchange, and paracrine support of neighbouring cells (Dumas et al., 2021). Glomerular ECs, for example, form fenestrations that facilitate plasma filtration, whereas cortical and medullary ECs adapt to their microenvironment to regulate solute handling and tubular function. In pathological states such as diabetic kidney disease (DKD) and chronic kidney disease (CKD), this specialisation deteriorates, leading to capillary rarefaction, sustained inflammation, fibrosis, and progressive nephron loss (Dumas et al., 2021). Inflammation is a major driver of kidney disease progression, fuelling renal fibrosis and functional decline (Kadatane et al., 2023; Harris & Neilson, 2006), though the precise contribution of each renal cell type to cytokine and chemokine production remains unclear.

In the context of DKD, the transcription factor KLF4, a regulator of vascular homeostasis, is downregulated in both human samples and rodent models, correlating with suppressed autophagy. In high-glucose–treated HUVECs, KLF4 downregulation coincides with impaired autophagy, ROS accumulation, increased TNF-α and MCP-1 secretion, and reduced cell viability. Overexpression of KLF4 rescues autophagy, prevents oxidative stress, and attenuates inflammatory signalling. Although in vivo evidence for this mechanism in ECs is still lacking, these findings suggest EC autophagy may play a protective anti-inflammatory role during DKD (Wang et al., 2022). Evidence also suggests that EC autophagy acts as a protective mechanism by dampening inflammatory responses during septic acute kidney injury (AKI). In a caecal ligation and puncture (CLP) rat model, endothelial injury markers (soluble thrombomodulin, soluble VE-cadherin) were elevated in serum, while kidney EC autophagy was upregulated. Rapamycin-mediated induction of autophagy in this model reduced leukocyte infiltration and glomerular injury, associating increased EC autophagy to reduced septic inflammation (Pang et al., 2023). However, further studies using EC autophagy-deficient models are needed.

Moreover, EC autophagy plays a protective role in preserving kidney architecture and renal function during ageing. EC-specific *Atg7* knockout mice exhibit enhanced kidney ageing-related changes and fibrosis. The latter associates with the activation of the NLRP3 inflammasome and IL-1β-induced c-Myc upregulation, which is necessary for the progression of renal fibrosis (Kim et al., 2024). Supporting these findings, ubiquitous deficiency of GIMAP6, an autophagy-related protein predominantly expressed in lymphocytes and ECs, causes severe glomerulosclerosis in mice. Crucially, bone marrow transplant studies confirmed this phenotype was EC-driven rather than hematopoietic (Yao et al., 2022). Although GIMAP6 mutations in humans manifest with immune dysregulation and vasculitis, the inflammatory consequences of EC-specific GIMAP6 deficiency in kidney disease remain to be clarified.

Collectively, these findings establish autophagy as a critical safeguard of kidney EC function, limiting oxidative stress and inflammatory signalling, and highlight its importance in chronic and acute kidney disease.

3.3 Central nervous system

Under physiological conditions, the endothelial blood brain barrier (BBB) maintains central nervous system (CNS) homeostasis by protecting the CNS parenchyma from the constantly changing milieu of the bloodstream. Brain microvascular ECs are interconnected by continuous and highly specialised tight junctions that restrict paracellular diffusion of solutes and water-soluble molecules, a property that is compromised during neuroinflammatory states (Mapunda et al., 2022; Liebner et al., 2018).

A growing body of work implicates autophagy as a key regulator of BBB integrity and permeability, exerting both protective and deleterious effects, particularly in the context of stroke and ischaemic injury (Kim et al., 2018). Beyond barrier regulation, EC autophagy also controls phagocytic functions within the CNS (Zhou et al., 2019). In models of spinal cord injury (SCI) and experimental autoimmune encephalomyelitis (EAE),

CNS microvascular ECs engulf myelin debris and degrade it via autophagy. This process is coupled to the induction of cytokines, chemokines and adhesion molecules by ECs, thereby promoting macrophage accumulation and exacerbating disease progression (Zhou et al., 2019). In this context, EC autophagy plays a detrimental role by linking debris clearance to vascular inflammation and contributes to the pathology of demyelinating disorders.

Notably, non-canonical autophagy pathways may also operate in the CNS vasculature. Retinal pigment epithelial cells, for example, rely on LAP for photoreceptor degradation (Kim et al., 2013). Although the contribution of non-canonical autophagy to brain EC phagocytic activity remains unexplored, these findings raise the possibility that LAP or related processes may similarly regulate myelin clearance in ECs. In addition, non-canonical autophagy pathways have been implicated in extracellular vesicle (EV) secretion (Leidal & Debnath, 2020; Gardner et al., 2023). In line with this, cerebral EC–derived EVs (CEC-EVs) have recently been shown to modulate microglial behaviour during neuroinflammation (Wang et al., 2023). Specifically, CEC-EVs carrying miR-672–5p restored autophagy flux in microglia following LPS-induced blockade. As autophagy exerts a protective role in LPS-driven neuroinflammation, miR-672–5p–mediated restoration of autophagy skewed microglia towards an anti-inflammatory phenotype, characterised by increased IL-10 and IL-4 expression and reduced production of IL-1β, IL-6 and TNF-α (Wang et al., 2023). These results suggest that non-canonical autophagy in ECs may play a protective role by dampening neuroinflammation via EV secretion.

Together, these findings extend the immunomodulatory repertoire of brain ECs beyond barrier maintenance to include autophagy-dependent control of myelin phagocytosis and EV-mediated regulation of glial inflammatory responses.

3.4 Bone marrow

ECs also play a central role in supporting haematopoiesis within the bone marrow (BM). Silencing of the autophagy master regulator Beclin1 in BM ECs impaired their ability to support haematopoiesis in vitro, an effect reproduced by pharmacological inhibition of autophagy in BM ECs from patients with poor graft function following allogeneic hematopoietic stem cell (HSC) transplantation (Lyu et al., 2020). These findings indicate that BM EC autophagy exerts a beneficial effect by sustaining HSC homeostasis and differentiation. Recent work has further uncovered an autophagy/

CD200R/eNOS axis in HSC regulation (Furuhashi et al., 2025). Specifically, a subset of highly immune-privileged and regenerative HSCs (NO^{hi} HSCs) relies on autophagy to localise to BM capillaries enriched in the immune checkpoint molecule CD200. Within this niche, the ciliary protein IFT20, CD200 signalling, endothelial nitric oxide synthase, and autophagy cooperate to support NO^{hi} HSC regenerative function and immunoprotection (Furuhashi et al., 2025).

Collectively, these studies highlight BM EC autophagy as a key regulator of HSC maintenance and immune privilege within the vascular niche.

4. Other emerging immunomodulatory functions of EC autophagy

4.1 Xenophagy

Autophagy contributes directly to innate immunity by capturing and degrading intracellular microbes through xenophagy (Deretic, 2021; Sharma et al., 2018). ECs employ this process as a host-defence mechanism against invading pathogens, including *Mycobacterium tuberculosis* in LECs (Lerner et al., 2018) and *Salmonella typhimurium* in HUVECs (Spinnenhirn et al., 2014). In addition, the *Mycobacterium ulcerans* exotoxin mycolactone was shown to induce autophagy in human dermal microvascular ECs (Hall et al., 2022). However, not all pathogens are effectively cleared by this route: while Group A Streptococcus (GAS) undergoes xenophagic degradation in other cell types, in ECs it evades this pathway (Lu et al., 2017). Instead, GAS infection preferentially activates LC3-associated phagocytosis (LAP), while suppressing bacteriostatic xenophagy (Cheng et al., 2019). Mechanistically, GAS-mediated secretion of the pore-forming toxin streptolysin O activates β1-integrin signalling in ECs, leading to NOX2-dependent ROS production. ROS in turn promotes LAP and stimulates PI3K/AKT and MEK/ERK pathways, culminating in mTOR activation and xenophagy inhibition. Consequently, GAS is recruited to single-membrane $LC3^{+}$ compartments that fail to clear bacteria due to insufficient acidification following lysosomal fusion (Cheng et al., 2019). A follow-up study revealed that VEGF treatment can restore xenophagic activity in ECs, enabling efficient clearance of intracellular GAS in vitro and reducing mortality in murine models of GAS sepsis (Lu et al., 2022). Collectively, this evidence suggests that canonical autophagy plays a beneficial role in ECs by promoting the efficient clearance of specific pathogens.

4.2 Phagocytosis and efferocytosis

Although classically attributed to professional phagocytes (macrophages, neutrophils and DCs), ECs can also exhibit phagocytic activity (Uribe-Querol & Rosales, 2020). As mentioned above, in the CNS, microvascular ECs have been shown to phagocytose myelin debris. This phenomenon occurs in early SCI, a stage when microglia and astrocytic clearance are absent (Wu et al., 2024a). Phagocytosis depends on the autophagy–lysosome pathway and IgG-mediated opsonisation, and while it supports debris removal, it also triggers the release of pro-inflammatory cytokines, aggravating secondary tissue damage (Zhou et al., 2019). By contrast, the HDAC6 inhibitor Tubastatin-A has been shown to enhance microvascular EC-mediated phagocytosis of myelin debris while attenuating inflammatory cytokine release, in part by modulating IgG/IgM balance and restoring autophagic flux and degradation capacity (Wu et al., 2024a). These findings highlight the importance of autophagy in regulating EC phagocytic activity and point to a dual role, whereby autophagy is beneficial in some contexts by dampening inflammatory responses, yet detrimental in others by promoting inflammatory signalling.

Beyond clearing myelin debris, ECs also eliminate apoptotic cells through efferocytosis, a specialised form of phagocytosis that is non-inflammatory (Henson, 2017; Morioka et al., 2019; Liu et al., 2023). This pathway is closely linked to LAP, which promotes anti-inflammatory signalling (Peña-Martinez et al., 2022; Cunha et al., 2018). Indeed, blood-separated tissues rely on LAP machinery downstream of the efferocytic receptor MerTK for apoptotic corpse clearance (Yefimova et al., 2021). In ECs, silencing MerTK destabilises junctions, reduces VE-cadherin expression and enhances neutrophil TEM and vascular permeability in vitro (Wu et al., 2024b). *In vivo*, global MerTK deficiency results in heightened neutrophil infiltration and pulmonary leakage during *Pseudomonas aeruginosa* pneumonia (Li et al., 2019) and accelerates atherosclerosis progression (Wu et al., 2024b). Consistently, an emerging work suggests that EC-specific MerTK deficiency renders mice more susceptible to atherosclerosis, accompanied by increased inflammatory markers in the aortic arch (Liu et al., 2025). In line with this, enhancing efferocytosis in aortic ECs attenuates endothelial dysfunction and inflammatory signalling in aged mice (Liu et al., 2023). Therapeutically, induction of autophagy has been shown to favour efferocytic pathways. In hyperlipidaemic mice, the Gasdermin D inhibitor disulfiram reduced inflammation and conferred atheroprotection by promoting both autophagy and efferocytosis in macrophages.

Although EC efferocytosis was not directly assessed, these findings support the possibility that autophagy likewise enhances efferocytic capacity in ECs (Traughber et al., 2024). Collectively, these observations suggest that defective efferocytosis in ECs enhances leukocyte trafficking and vascular inflammation, whereas autophagy induction (potentially through LAP) has a beneficial effect by coupling apoptotic clearance with anti-inflammatory signalling and favouring homeostasis.

4.3 Antigen recognition

The recognition and presentation of antigens to T lymphocytes is a central step in adaptive immunity. While professional antigen-presenting cells (APCs) such as DCs and macrophages have long been considered the main players, it is now clear that ECs, particularly lymphatic ECs (LECs), also present antigens and shape immune responses in lymphoid organs and inflamed tissues. LECs present peripheral tissue antigens via MHC-II molecules to autoreactive T cells, thereby enforcing peripheral tolerance (Saxena et al., 2019; Nadafi et al., 2020). Autophagy provides an additional layer of regulation in antigen presentation. In professional APCs, autophagy facilitates MHC-II presentation by delivering cytosolic and endogenous antigens to MHC-II loading compartments, enhancing $CD4^+$ T cell stimulation (Lee et al., 2010; Ho et al., 2021; Germic et al., 2019). Conversely, loss of autophagy (for example, in ATG5-deficient APCs) prolongs antigen storage, favours MHC-I cross-presentation, and increases MHC-I surface levels by impairing internalisation and degradation pathways (Lee et al., 2010; Ho et al., 2021; Loi et al., 2016).

Emerging evidence suggests that similar mechanisms extend to LECs, where autophagy plays a beneficial role by dampening inflammation in the context of autoimmune disorders. In collagen-induced arthritis, autophagy is induced in LECs and regulates the frequency and phenotype of regulatory T (Treg) cells in draining LNs (Harlé et al., 2021). Deletion of ATG5-dependent autophagy in LECs impaired Treg proliferation, reduced expression of the integrin CD103 that anchors Tregs at inflamed sites, and exacerbated arthritis severity. Strikingly, these effects were recapitulated in mice lacking MHC-II specifically in LECs, directly linking LEC autophagy to MHC-II–restricted antigen presentation and Treg-mediated immune tolerance (Harlé et al., 2021).

Beyond tolerance, the contribution of EC autophagy to antigen presentation in other contexts remains largely unexplored. However, given the established role of autophagy in directing antigens to MHC-II compartments and regulating surface MHC-I expression in APCs (Lee et al., 2010; Ho et al., 2021; Germic et al., 2019; Loi et al., 2016), it is reasonable to hypothesise that vascular ECs and LECs may exploit similar mechanisms to influence lymphocyte activation and trafficking during inflammation and disease.

4.4 Vascular ageing

Ageing is the primary risk factor for cardiovascular and chronic inflammatory disease (Donato et al., 2018). Among the many molecular changes associated with old age, compromised autophagy pathways have recently emerged as a hallmark of ageing across diverse species (Kaushik et al., 2021; Aman et al., 2021; López-Otín et al., 2023). Within the vasculature, defective autophagy intersects with endothelial dysfunction, barrier breakdown, cellular senescence, and dysregulated chemokine expression—defining features of vascular ageing that critically shape inflammatory outcomes (Reglero-Real et al., 2025; Abdellatif et al., 2023). In line with this, impaired autophagy has been detected in the aged cardiovascular system of mice and humans (Donato et al., 2018; Sasaki et al., 2017; LaRocca et al., 2012). Importantly, pharmacological or genetic restoration of autophagy ameliorates endothelial dysfunction, normalises vascular tone, delays arterial stiffening and restricts neutrophil recruitment and vascular inflammation (LaRocca et al., 2013; Lesniewski et al., 2017). These findings strongly implicate impaired autophagy as a causal driver of vascular ageing.

EC senescence is another defining hallmark of vascular ageing and is mechanistically linked to autophagy decline. In *C. elegans*, loss of EC autophagy accelerates senescence and shortens lifespan (Zhang et al., 2023). In mice, selective deletion of *Atg5* in ECs induces a senescent and pro-inflammatory phenotype characterised by mitochondrial dysfunction, increased adhesion molecule expression, and secretion of inflammatory mediators, changes that accelerate atherosclerosis (Vion et al., 2017). Recent work from Nourshargh's laboratory further demonstrated that senescent microvascular ECs exert direct effects on neutrophil trafficking. In inflamed postcapillary venules, progerin-expressing senescent ECs promoted excessive neutrophil adhesion to the luminal surface and enhanced neutrophil-driven vascular permeability, resulting in pathogenic patterns of neutrophil trafficking (Rolas et al., 2024). Furthermore, aged microvascular ECs display enhanced junctional localisation of the chemokine CXCL1, which drives reverse transendothelial migration (rTEM) of neutrophils back into the circulation (Barkaway et al., 2021). Neutrophils undergoing rTEM accumulate in remote tissues, including the lung, where they align with sites of vascular leakage. Together, these studies identify endothelial ageing and senescence (closely linked to impaired autophagy) as central regulators of neutrophil trafficking dynamics, with the potential to amplify systemic inflammation.

Collectively, these findings underscore the close interplay between autophagy decline and endothelial senescence, positioning impaired EC autophagy as a central mechanism in vascular ageing. Therapeutic strategies that restore EC autophagy may therefore offer powerful means to limit endothelial ageing, preserve vascular homeostasis, and counteract inflammaging.

5. Concluding remarks and future avenues

Autophagy within ECs is increasingly recognised as a central regulator of immune cell behaviour, shaping both innate and adaptive responses. Genetic and imaging studies now position EC autophagy as a critical checkpoint in leukocyte trafficking: intact autophagy restrains excessive infiltration of neutrophils, monocytes, and lymphocytes across multiple models of acute and chronic inflammation. These findings highlight the therapeutic potential of boosting EC autophagy to limit tissue injury and restore immune balance.

Beyond regulating leukocyte entry, EC autophagy influences a broader repertoire of immune functions, including antigen presentation, mechanosensing, phagocytic capacity, and the stability of endothelial junctions. The marked heterogeneity of ECs across tissues, together with the diverse roles of canonical and non-canonical autophagy pathways, raises the possibility that distinct autophagy programmes contribute to specialised EC phenotypes. This is illustrated by tissue-specific contexts—such as the liver, where LAP may underpin the high endocytic and antigen-presenting activity of LSECs, or the CNS, where differential EC autophagy may influence immune privilege versus surveillance.

A related and emerging concept is that EC autophagy may participate in unconventional secretory pathways. In other cell types, secretory autophagy encompasses LC3-dependent routes of extracellular release, including EV biogenesis and autophagosome–lysosome or endosome–plasma membrane fusion, that enable secretion of bioactive proteins, lipids, metabolites, organelles, and EV-associated cargo (Piletic et al., 2023). Such pathways place autophagy at a strategic intersection between intracellular trafficking and intercellular communication, raising the exciting possibility that EC autophagy shapes immune cell behaviour through regulated secretory cues. Defining whether tissue-specific EC subsets use these mechanisms to deliver "angiocrine" immune instructions, particularly under stress or infection, represents a key priority for the field.

Emerging evidence also links impaired EC autophagy with cellular senescence, collectively reshaping vascular–immune interactions during ageing and chronic inflammation. Rather than reiterating its established contribution to vascular ageing, the central challenge now is to determine how autophagy status modulates senescence-associated signalling and secretory programmes in ECs, and how these, in turn, influence leukocyte recruitment, activation, and persistence within tissues.

Taken together, these findings identify EC autophagy as a mechanism that orchestrates immune cell trafficking, phenotype, and function. A deeper understanding of these pathways across tissues and disease states will clarify how autophagy integrates vascular and immune homeostasis and may reveal new therapeutic opportunities for inflammatory and ageing-associated diseases.

Acknowledgments

Figures were created in Biorender. Mayor, F. (2026) (https://BioRender.com/fxll95b, https://BioRender.com/7stdlct, https://BioRender.com/biqbyoi).

This work was principally funded by "Generación de Conocimiento" projects program of the Spanish Ministry of Innovation and Science (MCIN) and Spanish State Research Agency (AEI) (PID2022–137552OA-100). N. Reglero-Real is a Ramón y Cajal awardee (RYC2021–031221-I) funded by MCIN/AEI, cofunded by "NextGenerationEU". G. Marcos-Zapatero is additionally supported by PREP2022–000816, part of the project PID2022–137552OA-100, funded by MCIN/AEI/10.13039/501100011033/FEDER, EU. I. Carriles-Moar is additionally funded by the "Community of Madrid" under the project Research Assistant and Laboratory Technicians CM 2023 (ref: PEJ-2023-AI/SAL-GL-27400). This contract is co-financed at 40 % by the European Social Fund Plus (ESF+) 2021–2027.

References

Abdellatif, M., Rainer, P. P., Sedej, S., & Kroemer, G. (2023). Hallmarks of cardiovascular ageing. *Nature Reviews Cardiology, 20*(11), 754–777.

Aman, Y., Schmauck-Medina, T., Hansen, M., Morimoto, R. I., Simon, A. K., Bjedov, I., et al. (2021). Autophagy in healthy aging and disease. *Nature Aging, 1*(8), 634–650.

Amersfoort, J., Eelen, G., & Carmeliet, P. (2022). Immunomodulation by endothelial cells — partnering up with the immune system? *Nature Reviews. Immunology, 22*(9), 576–588.

Ballesteros, I., Rubio-Ponce, A., Genua, M., Lusito, E., Kwok, I., Fernández-Calvo, G., et al. (2020). Co-option of neutrophil fates by tissue environments. *Cell, 183*(5), 1282–1297.e18.

Barkaway, A., Rolas, L., Joulia, R., Bodkin, J., Lenn, T., Owen-Woods, C., et al. (2021). Age-related changes in the local milieu of inflamed tissues cause aberrant neutrophil trafficking and subsequent remote organ damage. *Immunity, 54*(7), 1494–1510.e7.

Bauer, A., Tatliadim, H., & Halin, C. (2022). Leukocyte trafficking in lymphatic vessels. *Cold Spring Harbor Perspectives in Medicine, 12*(10), a041186.

Blazer, A., Qian, Y., Schlegel, M. P., Algasas, H., Buyon, J. P., Cadwell, K., et al. (2022). APOL1 variant-expressing endothelial cells exhibit autophagic dysfunction and mitochondrial stress. *Frontiers in Genetics, 13*, 769936.

Cerezo-Wallis, D., Rubio-Ponce, A., Richter, M., Pitino, E., Kwok, I., Marteletto, G., et al. (2025). Architecture of the neutrophil compartment. *Nature.*

Chen, M. L., Yi, L., Jin, X., Liang, X. yu, Zhou, Y., Zhang, T., et al. (2013). Resveratrol attenuates vascular endothelial inflammation by inducing autophagy through the cAMP signaling pathway. *Autophagy, 9*(12), 2033–2045.

Cheng, Y. L., Kuo, C. F., Lu, S. L., Omori, H., Wu, Y. N., Hsieh, C. L., et al. (2019). Group a Streptococcus induces LAPosomes via SLO/β1 Integrin/NOX2/ROS pathway in endothelial cells that are ineffective in bacterial killing and suppress xenophagy. *mBio, 10*(5).

Claude-Taupin, A., Codogno, P., & Dupont, N. (2021). Links between autophagy and tissue mechanics. *Journal of Cell Science, 134*(17).

Colom, B., Bodkin, J. V., Beyrau, M., Woodfin, A., Ody, C., Rourke, C., et al. (2015). Leukotriene B4-neutrophil elastase axis drives neutrophil reverse transendothelial cell migration in vivo. *Immunity, 42*(6), 1075–1086.

Coon, B. G., Timalsina, S., Astone, M., Zhuang, Z. W., Fang, J., Han, J., et al. (2022). A mitochondrial contribution to anti-inflammatory shear stress signaling in vascular endothelial cells. *Journal of Cell Biology, 221*(7).

Cunha, L. D., Yang, M., Carter, R., Guy, C., Harris, L., Crawford, J. C., et al. (2018). LC3-associated phagocytosis in myeloid cells promotes tumor immune tolerance. *Cell, 175*(2), 429–441.e16.

Cyster, J. G., & Schwab, S. R. (2012). Sphingosine-1-phosphate and lymphocyte egress from lymphoid organs. *Annual Review of Immunology, 30*(1), 69–94.

De Filippo, K., Dudeck, A., Hasenberg, M., Nye, E., van Rooijen, N., Hartmann, K., et al. (2013). Mast cell and macrophage chemokines CXCL1/CXCL2 control the early stage of neutrophil recruitment during tissue inflammation. *Blood, 121*(24), 4930–4937.

Dejana, E., Tournier-Lasserve, E., & Weinstein, B. M. (2009). The control of vascular integrity by endothelial cell junctions: Molecular basis and pathological implications. *Developmental Cell, 16*(2), 209–221.

Deretic, V. (2021). Autophagy in inflammation, infection, and immunometabolism. *Immunity, 54*(3), 437–453.

Deretic, V., & Levine, B. (2018). Autophagy balances inflammation in innate immunity. *Autophagy, 14*(2), 243–251.

Donato, A. J., Machin, D. R., & Lesniewski, L. A. (2018). Mechanisms of dysfunction in the aging vasculature and role in age-related disease. *Circulation Research, 123*(7), 825–848.

Dong, W., He, B., Qian, H., Liu, Q., Wang, D., Li, J., et al. (2018). RAB26-dependent autophagy protects adherens junctional integrity in acute lung injury. *Autophagy, 14*(10), 1677–1692.

Dumas, S. J., Meta, E., Borri, M., Luo, Y., Li, X., Rabelink, T. J., et al. (2021). Phenotypic diversity and metabolic specialization of renal endothelial cells. *Nature Reviews Nephrology, 17*(7), 441–464.

Feng, Y., He, D., Yao, Z., & Klionsky, D. J. (2014). The machinery of macroautophagy. *Cell Research, 24*(1), 24–41.

Ficht, X., & Iannacone, M. (2020). Immune surveillance of the liver by T cells. *Science Immunology, 5*(51).

Fielhaber, J. A., Carroll, S. F., Dydensborg, A. B., Shourian, M., Triantafillopoulos, A., Harel, S., et al. (2012). Inhibition of mammalian target of rapamycin augments lipopolysaccharide-induced lung injury and apoptosis. *The Journal of Immunology, 188*(9), 4535–4542.

Fletcher, K., Ulferts, R., Jacquin, E., Veith, T., Gammoh, N., Arasteh, J. M., et al. (2018). The WD40 domain of ATG16L1 is required for its non-canonical role in lipidation of LC3 at single membranes. *The EMBO Journal, 37*(4).

Florey, O., Kim, S. E., Sandoval, C. P., Haynes, C. M., & Overholtzer, M. (2011). Autophagy machinery mediates macroendocytic processing and entotic cell death by targeting single membranes. *Nature Cell Biology, 13*(11), 1335–1343.

Fowell, D. J., & Kim, M. (2021). The spatio-temporal control of effector T cell migration. *Nature Reviews. Immunology, 21*(9), 582–596.

Furman, D., Campisi, J., Verdin, E., Carrera-Bastos, P., Targ, S., Franceschi, C., et al. (2019). Chronic inflammation in the etiology of disease across the life span. *Nature Medicine, 25*(12), 1822–1832.

Furuhashi, K., Kakiuchi, M., Ueda, R., Oda, H., Ummarino, S., Ebralidze, A. K., et al. (2025). Bone marrow niches orchestrate stem-cell hierarchy and immune tolerance. *Nature, 638*(8049), 206–215.

Galluzzi, L., Baehrecke, E. H., Ballabio, A., Boya, P., Bravo-San Pedro, J. M., Cecconi, F., et al. (2017). Molecular definitions of autophagy and related processes. *The EMBO Journal, 36*(13), 1811–1836.

Galluzzi, L., & Green, D. R. (2019). Autophagy-independent functions of the autophagy machinery. *Cell, 177*(7), 1682–1699.

Gardner, J. O., Leidal, A. M., Nguyen, T. A., & Debnath, J. (2023). LC3-dependent EV loading and secretion (LDELS) promotes TFRC (transferrin receptor) secretion via extracellular vesicles. *Autophagy, 19*(5), 1551–1561.

Geldhof, V., de Rooij, L. P. M. H., Sokol, L., Amersfoort, J., De Schepper, M., Rohlenova, K., et al. (2022). Single cell atlas identifies lipid-processing and immunomodulatory endothelial cells in healthy and malignant breast. *Nature Communications, 13*(1), 5511.

Germic, N., Frangez, Z., Yousefi, S., & Simon, H. U. (2019). Regulation of the innate immune system by autophagy: Monocytes, macrophages, dendritic cells and antigen presentation. *Cell Death and Differentiation, 26*(4), 715–727.

Girbl, T., Lenn, T., Perez, L., Rolas, L., Barkaway, A., Thiriot, A., et al. (2018). Distinct compartmentalization of the chemokines CXCL1 and CXCL2 and the atypical receptor ACKR1 determine discrete stages of neutrophil diapedesis. *Immunity, 49*(6), 1062–1076.e6.

Guixé-Muntet, S., de Mesquita, F. C., Vila, S., Hernández-Gea, V., Peralta, C., García-Pagán, J. C., et al. (2017). Cross-talk between autophagy and KLF2 determines endothelial cell phenotype and microvascular function in acute liver injury. *Journal of Hepatology, 66*(1), 86–94.

Hall, B. S., Dos Santos, S. J., Hsieh, L. T. H., Manifava, M., Ruf, M. T., Pluschke, G., et al. (2022). Inhibition of the SEC61 translocon by mycolactone induces a protective autophagic response controlled by EIF2S1-dependent translation that does not require ULK1 activity. *Autophagy, 18*(4), 841–859.

Hammoutene, A., Biquard, L., Lasselin, J., Kheloufi, M., Tanguy, M., Vion, A. C., et al. (2020). A defect in endothelial autophagy occurs in patients with non-alcoholic steatohepatitis and promotes inflammation and fibrosis. *Journal of Hepatology, 72*(3), 528–538.

Hampton, H. R., & Chtanova, T. (2019). Lymphatic migration of immune cells. *Frontiers in Immunology, 10*.

Harlé, G., Kowalski, C., Dubrot, J., Brighouse, D., Clavel, G., Pick, R., et al. (2021). Macroautophagy in lymphatic endothelial cells inhibits T cell–mediated autoimmunity. *Journal of Experimental Medicine, 218*(6).

Harris, R. C., & Neilson, E. G. (2006). Toward a unified theory of renal progression. *Annual Review of Medicine, 57*(1), 365–380.

Heckmann, B. L., & Green, D. R. (2019). LC3-associated phagocytosis at a glance. *Journal of Cell Science, 132*(5).

Heckmann, B. L., Teubner, B. J. W., Tummers, B., Boada-Romero, E., Harris, L., Yang, M., et al. (2019). LC3-Associated endocytosis facilitates β-amyloid clearance and mitigates neurodegeneration in murine Alzheimer's disease. *Cell, 178*(3), 536–551.e14.

Henson, P. M. (2017). Cell removal: Efferocytosis. *Annual Review of Cell and Developmental Biology, 33*(1), 127–144.

Herzig, S., & Shaw, R. J. (2018). AMPK: Guardian of metabolism and mitochondrial homeostasis. *Nature Reviews. Molecular Cell Biology, 19*(2), 121–135.

Ho, N. I., Camps, M. G. M., Verdoes, M., Münz, C., & Ossendorp, F. (2021). Autophagy regulates long-term cross-presentation by murine dendritic cells. *European Journal of Immunology, 51*(4), 835–847.

Houbaert, D., Nikolakopoulos, A. P., Jacobs, K. A., Meçe, O., Roels, J., Shankar, G., et al. (2024). An autophagy program that promotes T cell egress from the lymph node controls responses to immune checkpoint blockade. *Cell Rep, 43*(4), 114020.

Humphrey, J. D., & Schwartz, M. A. (2021). Vascular mechanobiology: Homeostasis, adaptation, and disease. *Annual Review of Biomedical Engineering, 23*(1), 1–27.

Hunter, M. C., Teijeira, A., & Halin, C. (2016). T Cell trafficking through lymphatic vessels. *Frontiers in Immunology, 7*.

Kadatane, S. P., Satariano, M., Massey, M., Mongan, K., & Raina, R. (2023). The role of inflammation in CKD. *Cells, 12*(12), 1581.

Kalucka, J., de Rooij, L., Goveia, J., Rohlenova, K., Dumas, S. J., Meta, E., et al. (2020). Single-cell transcriptome atlas of murine endothelial cells. *Cell, 180*(4), 764–779.e20.

Kaushik, S., Tasset, I., Arias, E., Pampliega, O., Wong, E., Martinez-Vicente, M., et al. (2021). Autophagy and the hallmarks of aging. *Ageing Research Reviews, 72*, 101468.

Kim, J. W., Nam, S. A., Koh, E. S., Kim, H. W., Kim, S., Woo, J. J., et al. (2024). The impairment of endothelial autophagy accelerates renal senescence by ferroptosis and NLRP3 inflammasome signaling pathways with the disruption of endothelial barrier. *Antioxidants, 13*(8), 886.

Kim, J. Y., Zhao, H., Martinez, J., Doggett, T. A., Kolesnikov, A. V., Tang, P. H., et al. (2013). Noncanonical autophagy promotes the visual cycle. *Cell, 154*(2), 365–376.

Kim, K. A., Shin, D., Kim, J. H., Shin, Y. J., Rajanikant, G. K., Majid, A., et al. (2018). Role of autophagy in endothelial damage and blood–brain barrier disruption in ischemic stroke. *Stroke; a Journal of Cerebral Circulation, 49*(6), 1571–1579.

Klionsky, D. J., Petroni, G., Amaravadi, R. K., Baehrecke, E. H., Ballabio, A., Boya, P., et al. (2021). Autophagy in major human diseases. *The EMBO Journal, 40*(19).

Lange, L., Morgan, M., & Schambach, A. (2021). The hemogenic endothelium: A critical source for the generation of PSC-derived hematopoietic stem and progenitor cells. *Cellular and Molecular Life Sciences, 78*(9), 4143–4160.

LaRocca, T. J., Gioscia-Ryan, R. A., Hearon, C. M., & Seals, D. R. (2013). The autophagy enhancer spermidine reverses arterial aging. *Mechanisms of Ageing and Development, 134*(7–8), 314–320.

LaRocca, T. J., Henson, G. D., Thorburn, A., Sindler, A. L., Pierce, G. L., & Seals, D. R. (2012). Translational evidence that impaired autophagy contributes to arterial ageing. *The Journal of Physiology, 590*(14), 3305–3316.

Le, N. T., Sandhu, U. G., Quintana-Quezada, R. A., Hoang, N. M., Fujiwara, K., & Abe, J. ichi (2017). Flow signaling and atherosclerosis. *Cellular and Molecular Life Sciences, 74*(10), 1835–1858.

Lee, H. K., Mattei, L. M., Steinberg, B. E., Alberts, P., Lee, Y. H., Chervonsky, A., et al. (2010). In vivo requirement for Atg5 in antigen presentation by dendritic cells. *Immunity, 32*(2), 227–239.

Leidal, A. M., & Debnath, J. (2020). LC3-dependent extracellular vesicle loading and secretion (LDELS). *Autophagy, 16*(6), 1162–1163.

Lerner, T. R., Queval, C. J., Fearns, A., Repnik, U., Griffiths, G., & Gutierrez, M. G. (2018). Phthiocerol dimycocerosates promote access to the cytosol and intracellular burden of mycobacterium tuberculosis in lymphatic endothelial cells. *BMC Biology, 16*(1), 1.

Lesniewski, L. A., Seals, D. R., Walker, A. E., Henson, G. D., Blimline, M. W., Trott, D. W., et al. (2017). Dietary rapamycin supplementation reverses age-related vascular dysfunction and oxidative stress, while modulating nutrient-sensing, cell cycle, and senescence pathways. *Aging Cell, 16*(1), 17–26.

Levine, B., & Kroemer, G. (2019). Biological functions of autophagy genes: A disease perspective. *Cell, 176*(1–2), 11–42.

Ley, K., Laudanna, C., Cybulsky, M. I., & Nourshargh, S. (2007). Getting to the site of inflammation: The leukocyte adhesion cascade updated. *Nature Reviews. Immunology,* 7(9), 678–689.

Li, Y., Suo, L., Fu, Z., Li, G., & Zhang, J. (2021). Pivotal role of endothelial cell autophagy in sepsis. *Life Sciences, 276*, 119413.

Li, Y., Wittchen, E. S., Monaghan-Benson, E., Hahn, C., Earp, H. S., Doerschuk, C. M., et al. (2019). The role of endothelial MERTK during the inflammatory response in lungs. *PLoS One, 14*(12), e0225051.

Liebner, S., Dijkhuizen, R. M., Reiss, Y., Plate, K. H., Agalliu, D., & Constantin, G. (2018). Functional morphology of the blood–brain barrier in health and disease. *Acta Neuropathologica, 135*(3), 311–336.

Lin, S. C., & Hardie, D. G. (2018). AMPK: Sensing glucose as well as cellular energy status. *Cell Metabolism, 27*(2), 299–313.

Liu, G. Y., & Sabatini, D. M. (2020). mTOR at the nexus of nutrition, growth, ageing and disease. *Nature Reviews. Molecular Cell Biology, 21*(4), 183–203.

Liu, S., Wu, J., Stolarz, A., Zhang, H., Boerma, M., Byrum, S. D., et al. (2023). PCSK9 attenuates efferocytosis in endothelial cells and promotes vascular aging. *Theranostics, 13*(9), 2914–2929.

Liu, S., Yao, J., Huang, H., Wu, J., Banerjee, O., Xue, B., et al. (2025). Impairment of endothelial MerTK accelerates atherosclerosis development. *medRxiv [Preprint].*

Loi, M., Müller, A., Steinbach, K., Niven, J., Barreira da Silva, R., Paul, P., et al. (2016). Macroautophagy proteins control MHC class I levels on dendritic cells and shape antiviral CD8 + T cell responses. *Cell Reports, 15*(5), 1076–1087.

López-Otín, C., Blasco, M. A., Partridge, L., Serrano, M., & Kroemer, G. (2023). Hallmarks of aging: An expanding universe. *Cell, 186*(2), 243–278.

Lu, S. L., Kawabata, T., Cheng, Y. L., Omori, H., Hamasaki, M., Kusaba, T., et al. (2017). Endothelial cells are intrinsically defective in xenophagy of Streptococcus pyogenes. *PLoS Pathogens, 13*(7) e1006444.

Lu, S. L., Omori, H., Zhou, Y., Lin, Y. S., Liu, C. C., Wu, J. J., et al. (2022). VEGF-mediated augmentation of autophagic and lysosomal activity in endothelial cells defends against intracellular Streptococcus pyogenes. *mBio, 13*(4).

Lyu, Z. S., Cao, X. N., Wen, Q., Mo, X. D., Zhao, H. Y., Chen, Y. H., et al. (2020). Autophagy in endothelial cells regulates their haematopoiesis-supporting ability. *EBioMedicine, 53*, 102677.

Macian, F. (2019). Autophagy in T cell function and aging. *Frontiers in Cell and Developmental Biology, 7*.

Mamdouh, Z., Chen, X., Pierini, L. M., Maxfield, F. R., & Muller, W. A. (2003). Targeted recycling of PECAM from endothelial surface-connected compartments during diapedesis. *Nature, 421*(6924), 748–753.

Mameli, E., Martello, A., & Caporali, A. (2022). Autophagy at the interface of endothelial cell homeostasis and vascular disease. *The FEBS Journal, 289*(11), 2976–2991.

Mannam, P., Zhang, X., Shan, P., Zhang, Y., Shinn, A. S., Zhang, Y., et al. (2013). Endothelial MKK3 is a critical mediator of lethal murine endotoxemia and acute lung injury. *The Journal of Immunology, 190*(3), 1264–1275.

Mapunda, J. A., Tibar, H., Regragui, W., & Engelhardt, B. (2022). How does the immune system enter the brain? *Frontiers in Immunology, 13*.

Martinez-Martin, N., Maldonado, P., Gasparrini, F., Frederico, B., Aggarwal, S., Gaya, M., et al. (2017). A switch from canonical to noncanonical autophagy shapes B cell responses. *Science (1979), 355*(6325), 641–647.

Medzhitov, R. (2008). Origin and physiological roles of inflammation. *Nature, 454*(7203), 428–435.

Moreno-Cañadas, R., Luque-Martín, L., & Arroyo, A. G. (2021). Intravascular crawling of patrolling monocytes: A lèvy-like motility for unique search functions? *Frontiers in Immunology, 12*.

Morioka, S., Maueröder, C., & Ravichandran, K. S. (2019). Living on the edge: Efferocytosis at the interface of homeostasis and pathology. *Immunity, 50*(5), 1149–1162.

Morishita, H., & Mizushima, N. (2019). Diverse cellular roles of autophagy. *Annual Review of Cell and Developmental Biology, 35*(1), 453–475.

Nadafi, R., Gago de Graça, C., Keuning, E. D., Koning, J. J., de Kivit, S., Konijn, T., et al. (2020). Lymph node stromal cells generate antigen-specific regulatory T cells and control autoreactive T and B cell responses. *Cell Reports, 30*(12), 4110–4123.e4.

Nourshargh, S., & Alon, R. (2014). Leukocyte migration into inflamed tissues. *Immunity, 41*(5), 694–707.

Nourshargh, S., Hordijk, P. L., & Sixt, M. (2010). Breaching multiple barriers: Leukocyte motility through venular walls and the interstitium. *Nature Reviews. Molecular Cell Biology, 11*(5), 366–378.

Nussenzweig, S. C., Verma, S., & Finkel, T. (2015). The role of autophagy in vascular biology. *Circulation Research, 116*(3), 480–488.

Paik, D. T., Tian, L., Williams, I. M., Rhee, S., Zhang, H., Liu, C., et al. (2020). Single-cell RNA sequencing unveils unique transcriptomic signatures of organ-specific endothelial cells. *Circulation, 142*(19), 1848–1862.

Pang, R., Dong, L., Liu, J., Ji, X., Zhuang, H., & Duan, M. (2023). The study on role of endothelial cell autophagy in rats with sepsis-induced acute kidney injury. *Heliyon, 9*(3), e13796.

Pankratz, F., Hohnloser, C., Bemtgen, X., Jaenich, C., Kreuzaler, S., Hoefer, I., et al. (2018). MicroRNA-100 suppresses chronic vascular inflammation by stimulation of endothelial autophagy. *Circulation Research, 122*(3), 417–432.

Patella, F., Neilson, L. J., Athineos, D., Erami, Z., Anderson, K. I., Blyth, K., et al. (2016). In-depth proteomics identifies a role for autophagy in controlling reactive oxygen species mediated endothelial permeability. *Journal of Proteome Research, 15*(7), 2187–2197.

Peña-Martinez, C., Rickman, A. D., & Heckmann, B. L. (2022). Beyond autophagy: LC3-associated phagocytosis and endocytosis. *Science Advances, 8*(43).

Piletic, K., Alsaleh, G., & Simon, A. K. (2023). Autophagy orchestrates the crosstalk between cells and organs. *EMBO Reports, 24*(9), e57289.

Pober, J. S., & Sessa, W. C. (2007). Evolving functions of endothelial cells in inflammation. *Nature Reviews. Immunology, 7*(10), 803–815.

Pober, J. S., & Tellides, G. (2012). Participation of blood vessel cells in human adaptive immune responses. *Trends in Immunology, 33*(1), 49–57.

Poisson, J., Lemoinne, S., Boulanger, C., Durand, F., Moreau, R., Valla, D., et al. (2017). Liver sinusoidal endothelial cells: Physiology and role in liver diseases. *Journal of Hepatology, 66*(1), 212–227.

Prame Kumar, K., Nicholls, A. J., & Wong, C. H. Y. (2018). Partners in crime: Neutrophils and monocytes/macrophages in inflammation and disease. *Cell and Tissue Research, 371*(3), 551–565.

Proebstl, D., Voisin, M. B., Woodfin, A., Whiteford, J., D'Acquisto, F., Jones, G. E., et al. (2012). Pericytes support neutrophil subendothelial cell crawling and breaching of venular walls in vivo. *Journal of Experimental Medicine, 209*(6), 1219–1234.

Reglero-Real, N., Colom, B., Bodkin, J. V., & Nourshargh, S. (2016). Endothelial cell junctional adhesion molecules. *Arteriosclerosis, Thrombosis, and Vascular Biology, 36*(10), 2048–2057.

Reglero-Real, N., Marcos-Ramiro, B., & Millán, J. (2012). Endothelial membrane reorganization during leukocyte extravasation. *Cellular and Molecular Life Sciences, 69*(18), 3079–3099.

Reglero-Real, N., Pérez-Gutiérrez, L., & Nourshargh, S. (2021b). Endothelial cell autophagy keeps neutrophil trafficking under control. *Autophagy, 17*(12), 4509–4511.

Reglero-Real, N., Pérez-Gutiérrez, L., Yoshimura, A., Rolas, L., Garrido-Mesa, J., Barkaway, A., et al. (2021a). Autophagy modulates endothelial junctions to restrain neutrophil diapedesis during inflammation. *Immunity, 54*(9), 1989–2004 e9.

Reglero-Real, N., Rolas, L., & Nourshargh, S. (2025). Aging microvasculature: Effects on immune cell trafficking and inflammatory diseases. *Journal of Experimental Medicine, 222*(7).

Rolas, L., Stein, M., Barkaway, A., Reglero-Real, N., Sciacca, E., Yaseen, M., et al. (2024). Senescent endothelial cells promote pathogenic neutrophil trafficking in inflamed tissues. *EMBO Reports, 25.*

Ruart, M., Chavarria, L., Campreciós, G., Suárez-Herrera, N., Montironi, C., Guixé-Muntet, S., et al. (2019). Impaired endothelial autophagy promotes liver fibrosis by aggravating the oxidative stress response during acute liver injury. *Journal of Hepatology, 70*(3), 458–469.

Sachdev, U., & Lotze, M. T. (2017). Perpetual change: Autophagy, the endothelium, and response to vascular injury. *Journal of Leukocyte Biology, 102*(2), 221–235.

Sanjuan, M. A., Dillon, C. P., Tait, S. W. G., Moshiach, S., Dorsey, F., Connell, S., et al. (2007). Toll-like receptor signalling in macrophages links the autophagy pathway to phagocytosis. *Nature, 450*(7173), 1253–1257.

Santambrogio, L., Berendam, S. J., & Engelhard, V. H. (2019). The antigen processing and presentation machinery in lymphatic endothelial cells. *Frontiers in Immunology, 10.*

Santovito, D., Egea, V., Bidzhekov, K., Natarelli, L., Mourão, A., Blanchet, X., et al. (2020). Noncanonical inhibition of caspase-3 by a nuclear microRNA confers endothelial protection by autophagy in atherosclerosis. *Science Translational Medicine, 12*(546).

Sasaki, Y., Ikeda, Y., Iwabayashi, M., Akasaki, Y., & Ohishi, M. (2017). The impact of autophagy on cardiovascular senescence and diseases. *International Heart Journal, 58*(5), 666–673.

Saxena, V., Li, L., Paluskievicz, C., Kasinath, V., Bean, A., Abdi, R., et al. (2019). Role of lymph node stroma and microenvironment in T cell tolerance. *Immunological Reviews, 292*(1), 9–23.

Schaaf, M. B., Houbaert, D., Meçe, O., & Agostinis, P. (2019). Autophagy in endothelial cells and tumor angiogenesis. *Cell Death and Differentiation, 26*(4), 665–679.

Schupp, J. C., Adams, T. S., Cosme, C., Raredon, M. S. B., Yuan, Y., Omote, N., et al. (2021). Integrated single-cell atlas of endothelial cells of the human lung. *Circulation, 144*(4), 286–302.

Sharma, V., Verma, S., Seranova, E., Sarkar, S., & Kumar, D. (2018). Selective autophagy and xenophagy in infection and disease. *Frontiers in Cell and Developmental Biology, 6*.
Shim, H. B., Deniset, J. F., & Kubes, P. (2022). Neutrophils in homeostasis and tissue repair. *International Immunology, 34*(8), 399–407.
Slavin, S. A., Leonard, A., Grose, V., Fazal, F., & Rahman, A. (2018). Autophagy inhibitor 3-methyladenine protects against endothelial cell barrier dysfunction in acute lung injury. *American Journal of Physiology-Lung Cellular and Molecular Physiology, 314*(3), L388–L396.
Spinnenhirn, V., Farhan, H., Basler, M., Aichem, A., Canaan, A., & Groettrup, M. (2014). The ubiquitin-like modifier FAT10 decorates autophagy targeted Salmonella and contributes to resistance of mice. *Journal of Cell Science*.
Sreeramkumar, V., Adrover, J. M., Ballesteros, I., Cuartero, M. I., Rossaint, J., Bilbao, I., et al. (2014). Neutrophils scan for activated platelets to initiate inflammation. *Science (1979), 346*(6214), 1234–1238.
Torisu, K., Singh, K. K., Torisu, T., Lovren, F., Liu, J., Pan, Y., et al. (2016). Intact endothelial autophagy is required to maintain vascular lipid homeostasis. *Aging Cell, 15*(1), 187–191.
Torisu, T., Torisu, K., Lee, I. H., Liu, J., Malide, D., Combs, C. A., et al. (2013). Autophagy regulates endothelial cell processing, maturation and secretion of von Willebrand factor. *Nature Medicine, 19*(10), 1281–1287.
Traughber, C. A., Timinski, K., Prince, A., Bhandari, N., Neupane, K., Khan, M. R., et al. (2024). Disulfiram reduces atherosclerosis and enhances efferocytosis, autophagy, and atheroprotective gut microbiota in hyperlipidemic mice. *Journal of the American Heart Association, 13*(8).
Uribe-Querol, E., & Rosales, C. (2020). Phagocytosis: Our current understanding of a universal biological process. *Frontiers in Immunology, 11*.
Vargas, J. N. S., Hamasaki, M., Kawabata, T., Youle, R. J., & Yoshimori, T. (2023). The mechanisms and roles of selective autophagy in mammals. *Nature Reviews. Molecular Cell Biology, 24*(3), 167–185.
Verhoeven, J., Baelen, J., Agrawal, M., & Agostinis, P. (2021). Endothelial cell autophagy in homeostasis and cancer. *FEBS Letters, 595*(11), 1497–1511.
Verhoeven, J., Jacobs, K. A., Rizzollo, F., Lodi, F., Hua, Y., Poźniak, J., et al. (2023). Tumor endothelial cell autophagy is a key vascular-immune checkpoint in melanoma. *EMBO Molecular Medicine, 15*(12).
Vestweber, D. (2015). How leukocytes cross the vascular endothelium. *Nature Reviews. Immunology, 15*(11), 692–704.
Vion, A. C., Kheloufi, M., Hammoutene, A., Poisson, J., Lasselin, J., Devue, C., et al. (2017). Autophagy is required for endothelial cell alignment and atheroprotection under physiological blood flow. *Proceedings of the National Academy of Sciences, 114*(41).
Voisin, M. B., & Nourshargh, S. (2013). Neutrophil transmigration: Emergence of an adhesive cascade within venular walls. *Journal of Innate Immunity, 5*(4), 336–347.
Voisin, M. B., Pröbstl, D., & Nourshargh, S. (2010). Venular basement membranes ubiquitously express matrix protein low-expression regions. *The American Journal of Pathology, 176*(1), 482–495.
Wang, C., Feng, L., Zhu, L., Wu, L., Chen, B., Cui, C., et al. (2023). Cerebral endothelial cell-derived extracellular vesicles regulate microglial polarization and promote autophagy via delivery of miR-672-5p. *Cell Death & Disease, 14*(9), 643.
Wang, X., Su, W., Ma, M., Zhu, L., Gao, R., Qin, C., et al. (2022). The KLF4–p62 axis prevents vascular endothelial cell injury via the mTOR/S6K pathway and autophagy in diabetic kidney disease. *Endokrynologia Polska, 73*(5), 837–845.
Woodfin, A., Voisin, M. B., Beyrau, M., Colom, B., Caille, D., Diapouli, F. M., et al. (2011). The junctional adhesion molecule JAM-C regulates polarized transendothelial migration of neutrophils in vivo. *Nature Immunology, 12*(8), 761–769.

Wu, C., Pan, Y., Wang, L., Liu, M., Tu, P., Chen, S., et al. (2024a). Inhibition of HDAC6 promotes microvascular endothelial cells to phagocytize myelin debris and reduces inflammatory response to accelerate the repair of spinal cord injury. *CNS Neuroscience & Therapeutics, 30*(3).

Wu, J., Liu, S., Banerjee, O., Shi, H., Xue, B., & Ding, Z. (2024b). Disturbed flow impairs MerTK-mediated efferocytosis in aortic endothelial cells during atherosclerosis. *Theranostics, 14*(6), 2427–2441.

Wu, J., Ma, Z., Raman, A., Beckerman, P., Dhillon, P., Mukhi, D., et al. (2021). APOL1 risk variants in individuals of African genetic ancestry drive endothelial cell defects that exacerbate sepsis. *Immunity, 54*(11), 2632–2649.e6.

Wu, W., Xu, H., Wang, Z., Mao, Y., Yuan, L., Luo, W., et al. (2015). PINK1-parkin-mediated mitophagy protects mitochondrial integrity and prevents metabolic stress-induced endothelial injury. *PLoS One, 10*(7), e0132499.

Yao, Y., Du Jiang, P., Chao, B. N., Cagdas, D., Kubo, S., Balasubramaniyam, A., et al. (2022). GIMAP6 regulates autophagy, immune competence, and inflammation in mice and humans. *Journal of Experimental Medicine, 219*(6).

Yau, J. W., Singh, K. K., Hou, Y., Lei, X., Ramadan, A., Quan, A., et al. (2017). Endothelial-specific deletion of autophagy-related 7 (ATG7) attenuates arterial thrombosis in mice. *The Journal of Thoracic and Cardiovascular Surgery, 154*(3), 978–988.e1.

Yefimova, M. G., Ravel, C., Rolland, A. D., Bourmeyster, N., & Jégou, B. (2021). MERTK-mediated LC3-associated phagocytosis (LAP) of apoptotic substrates in blood-separated tissues: Retina, testis, ovarian follicles. *Cells, 10*(6), 1443.

Yu, L., Chen, Y., & Tooze, S. A. (2018). Autophagy pathway: Cellular and molecular mechanisms. *Autophagy, 14*(2), 207–215.

Zhang, L., Yang, P., Chen, J., Chen, Z., Liu, Z., Feng, G., et al. (2023). CD44 connects autophagy decline and ageing in the vascular endothelium. *Nature Communications, 14*(1), 5524.

Zhang, X., Ramírez, C. M., Aryal, B., Madrigal-Matute, J., Liu, X., Diaz, A., et al. (2020). Cav-1 (Caveolin-1) deficiency increases autophagy in the endothelium and attenuates vascular inflammation and atherosclerosis. *Arteriosclerosis, Thrombosis, and Vascular Biology, 40*(6), 1510–1522.

Zhao, H., Chen, H., Xiaoyin, M., Yang, G., Hu, Y., Xie, K., et al. (2019). Autophagy activation improves lung injury and inflammation in sepsis. *Inflammation, 42*(2), 426–439.

Zhao, H., Wu, L., Yan, G., Chen, Y., Zhou, M., Wu, Y., et al. (2021). Inflammation and tumor progression: Signaling pathways and targeted intervention. *Signal Transduction and Targeted Therapy, 6*(1), 263.

Zhou, T., Zheng, Y., Sun, L., Badea, S. R., Jin, Y., Liu, Y., et al. (2019). Microvascular endothelial cells engulf myelin debris and promote macrophage recruitment and fibrosis after neural injury. *Nature Neuroscience, 22*(3), 421–435.

CHAPTER FIVE

Viral trafficking: The role of autophagy in the release of virus-loaded extracellular vesicles

Laura Vipraio Tiberi[a,b,1], Angelica Di Matteo[a,c,1], Fatima Domenica Elisa De Palma[a,b], Manuela Lizarralde-Guerrero[a,d], Mojgan Djavaheri-Mergny[a], Maria Vitale[e], Filippo Scialó[b,e], Lucio Pastore[b,e], Maria Chiara Maiuri[a,b,*], and Jonathan Pol[a,*]

[a]Team "Metabolism, Cancer & Immunity", Équipe labellisée par la Ligue contre le cancer, Centre de Recherche des Cordeliers, UMRS1138, Inserm, Université Paris Cité, Sorbonne Université, Paris, France
[b]Department of Molecular Medicine and Medical Biotechnologies, University of Napoli Federico II, Naples, Italy
[c]Department of Precision Medicine, University of Campania "L. Vanvitelli", Naples, Italy
[d]Université Paris-Saclay, Faculté de Médecine, Kremlin-Bicêtre, France
[e]CEINGE-Biotecnologie Avanzate Franco Salvatore, Naples, Italy
*Corresponding authors. e-mail address: maria-chiara.maiuri@inserm.fr; pol_jonathan@yahoo.fr

Contents

[1] Equal contribution

International Review of Cell and Molecular Biology, Volume 406
ISSN 1937-6448, https://doi.org/10.1016/bs.ircmb.2025.12.002

Abstract

Over the past decade, the interplay between autophagy process and viral replication, alongside the mechanisms by which viruses evade immune surveillance, has been extensively investigated. Increasing evidence indicates that viruses hijack the autophagic machinery to facilitate the release of extracellular vesicles (EVs) carrying viral components. This involvement highlights autophagy as a central hub in viral trafficking, bridging intracellular remodelling with extracellular dissemination. In this review, we explore how viruses exploit autophagy and EV pathways to promote replication and enable non-lytic release. We highlight the molecular interactions underlying EV biogenesis and secretory autophagy and discuss the emerging potential of EVs as biomarkers and therapeutic platforms, despite current technical challenges. Finally, we examine strategies to target the autophagy-EV interface as an antiviral strategy.

1. Introduction

Viruses depend entirely on host cellular machinery to complete their life cycle. During evolution, they acquired mechanisms to remodel intracellular processes, notably autophagy and vesicular trafficking, in order to evade immunity, replicate, and disseminate (Paul & Münz, 2016). In this context, the autophagy machinery is not only involved in the degradative process that maintains cellular homeostasis (Ginini et al., 2022; Leidal & Debnath, 2021) but is also implicated into the unconventional secretory pathway that releases cytoplasmic material, including viral components. Extracellular vesicles (EVs) have emerged as versatile mediators of cell-to-cell communication and pathogen dissemination. Multiple viruses manipulate autophagy and EV pathways to achieve non-lytic egress and immune evasion, using EVs as vectors for infectious material and mediators of intercellular communication (Miller & Krijnse-Locker, 2008).

Macroautophagy, and more specifically xenophagy, is widely recognized as a key defense mechanism against intracellular pathogens. In turn, viruses have evolved diverse strategies to bypass or redirect the autophagy process to support their replication (Münz et al., 2025). Viral evasion often involves blocking autophagosome formation or maturation (Jassey & Jackson, 2024).

RNA viruses like picornaviruses, flaviviruses and coronaviruses, inhibit the fusion of autophagosomes with lysosomes to prevent degradation of their own components, and reroute autophagosomes to secretory pathways to promote non-lytic viral dissemination and immune evasion. This manipulation is a widely shared evolutionary-conserved strategy (Chu et al., 2024; Li et al., 2024; Wang et al., 2022).

This review dissects the mechanisms by which viruses reshape the autophagy-EV axis to enhance infection. It outlines the main molecular components involved in the interaction between autophagy and exocytotic trafficking, in clinically relevant RNA and DNA viruses. Finally, we discuss the therapeutic prospects associated with the selective inhibition of these pathways, with the aim of preventing viral dissemination and enhancing the host's immune response. Understanding these mechanisms will guide the development of innovative antiviral interventions for complex clinical contexts.

2. Host and viral EVs

EVs are lipid-bilayer particles released from cells. By packaging proteins, lipids and nucleic acids that reflect their cell of origin, they coordinate intercellular communication across diverse physiological and pathological contexts, including cancer, neurodegeneration, and infectious diseases (Théry et al., 2018). In the latter context, viruses hijack EVs to disseminate, modulate immunity, and bypass autophagy-based defense (Bello-Morales & López-Guerrero, 2018; Caobi et al., 2020; Pleet, et al., 2018).

2.1 General classification and biogenesis of EVs

EVs are typically categorized based on their size, biogenesis, content, type of secreting cells and ultimate function. They comprise exosomes (30–150 nm), microvesicles (100 nm-1 μm), and apoptotic bodies (1–5 μm). The exosomes and microvesicles are released from healthy cells whereas the apoptotic bodies are produced by cells undergoing apoptosis (Colombo et al., 2014; Rey-Cadilhac et al., 2023; van Niel et al., 2018).

Exosomes originate from a highly regulated endosomal pathway. The inward budding of endosomal membranes gives rise to intraluminal vesicles (ILVs) regrouped within multivesicular bodies (MVBs). This budding process is tightly orchestrated by the endosomal sorting complex required for transport (ESCRT) machinery composed by ESCRT-0, -I, -II, and -III. More precisely,

ESCRT-0 recognizes ubiquitinated cargo proteins, ESCRT-I and -II promote membrane deformation, and ESCRT-III completes vesicle scission. Accessory proteins, such as ALG-2 interacting protein X (ALIX) and tumor susceptibility gene 101 (TSG101), regulate the process by stabilizing the ESCRT complexes and directing specific cargo into the budding vesicles (Colombo et al., 2014; Lau & Yam, 2023; van Niel et al., 2018; Williams & Urbé, 2007). Exosome formation can also proceed via ESCRT-independent pathways, involving lipid-modifying enzymes such as neutral sphingomyelinase 2 (nSMase2), and tetraspanins (CD9, CD63, CD81) that cluster cargo and create curvature for ILVs (Hadizadeh et al., 2022; Tkach & Théry, 2016). After maturation, MVBs either fuse with lysosomes for degradation, or fuse with the plasma membrane to release ILVs as exosomes. Rab GTPases, particularly Rab27a/b, guide MVB trafficking to the cell surface, while specific SNARE (soluble N-ethylmaleimide-sensitive-factor attachment protein receptor) proteins mediate membrane fusion (Colombo et al., 2014). Through this regulated process, exosomes deliver selected proteins, RNAs and lipids, that modulate neighbouring or distant cells (van Niel et al., 2018).

In contrast with exosomes biogenesis, microvesicles arise from direct outward budding and fission of the plasma membrane. Small GTPases, such as ARF6 and some Rho family members (RhoA, Rac1, Cdc42), promote localized actin remodelling and membrane curvature, while myosin light chain kinase (MLCK) induces actomyosin contraction that facilitates membrane blebbing. Cargo selection and budding sites are influenced by the externalization of phosphatidylserine (PS), cholesterol enrichment, and specific adhesion molecules (e.g., integrins, selectins) in local membrane domains. Therefore, microvesicles reflect the immediate physiological state of the plasma membrane, providing a rapid mechanism for cells to signal or dispose of material under stress (Tkach & Théry, 2016; van Niel et al., 2018).

Apoptotic bodies are the largest EVs. They derive from an active process driven by caspase-mediated cleavage of cytoskeletal and nuclear proteins, resulting in membrane blebbing and fragmentation of the dying cell into vesicles. These vesicles can enclose organelles and chromatin fragments. Surface-exposed PS serves as "eat-me" signals for phagocytes, ensuring the safe clearance of cellular debris and limiting inflammation, while the enclosed cargo can still influence neighbouring cells (Akers et al., 2013).

2.2 Viral EVs

Viral extracellular vesicles (vEVs) are host-derived vesicles whose biogenesis, cargo, and functions are reprogrammed by viruses to enhance their replication, persistence, and immune evasion. Unlike physiological EVs that mainly contain host macromolecules (Théry et al., 2002), vEVs are often enriched in viral proteins, lipids, or nucleic acids that support viral propagation.

Enveloped viruses, such as the human immunodeficiency virus 1 (HIV-1), exploit the ESCRT machinery during exosome formation: the Gag polyprotein contains PTAP and YPXnL motifs that interact with TSG101 and ALIX, recruiting ESCRT-III and the ATPase VPS4 (vacuolar protein sorting 4) to mediate non-lytic release (Feng et al., 2013; Kalamvoki et al., 2014). Herpesviruses reroute MVB traffic to package viral miRNAs into exosomes, thereby silencing antiviral genes in recipient cells (Kalamvoki et al., 2014).

Non-enveloped viruses exploit similar pathways. For instance, hepatitis A virus (HAV) is secreted as quasi-enveloped particles containing tetraspanins such as CD9 and ESCRT-III components but lacking the autophagy marker LC3 (microtubule-associated protein 1 light chain 3), indicating an ESCRT-dependent but autophagy-independent MVB release (Feng et al., 2013; McKnight & Lemon, 2018). Hepatitis E virus (HEV) exits cells in exosome-like vesicles containing ORF3, dependent on TSG101 and VPS4 (Nagashima et al., 2017). Adenoviruses (AdVs) exploit exosome-like vesicles to achieve receptor-independent entry, thus broadening cell tropism (Chatterjee et al., 2024; Ipinmoroti et al., 2021; Crenshaw et al., 2019) (Fig. 1).

Three main types of vEV cargo can be distinguished:

1. *Virions*. Naked viruses (e.g., AdVs, picornaviruses) can be packaged in PS-rich EVs via virion-ESCRT interactions or diversion of secretory autophagosomes (Chatterjee et al., 2024; Feng et al., 2013; Rey-Cadilhac et al., 2023; Robinson et al., 2014). Some enveloped viruses can acquire an additional host membrane before release, generating EV-like virions enriched in PS. For instance, hepatitis B virus (HBV) buds into MVBs in an ESCRT- and VPS4-dependent manner. Dengue virus (DENV) recruits TSG101/ALIX to exit in exosome-like particles (Agudelo et al., 2025; Chatterjee et al., 2024; Li et al., 2025).
2. *Viral proteins*. Structural and functional viral proteins can enter EVs through interactions with ESCRT or tetraspanin scaffolds (Rey-Cadilhac et al., 2023). They can consist of envelope glycoproteins (e.g., influenza virus, SARS-CoV-2), or immune-modulatory proteins (e.g., flavivirus NS1) (Barberis et al., 2021; Reyes-Ruiz et al., 2020; Stertz et al., 2007).

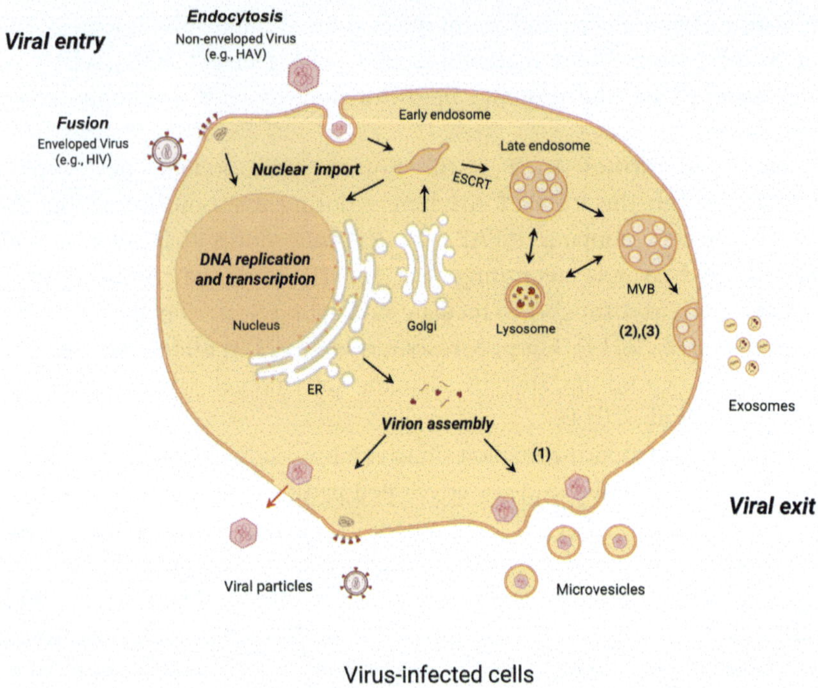

Fig. 1 Viral EV biogenesis in infected cells. Viruses exploit host-derived EV pathways to package and release viral cargo, thus supporting replication, persistence, and immune evasion. Viral EVs can incorporate: (1) virions, both naked (e.g., AdVs, picornaviruses) and enveloped (e.g., HBV, DENV), released within PS-enriched vesicles, or as quasi-enveloped HAV and HEV particles dependent on ESCRT machinery; (2) viral proteins, including structural envelope glycoproteins that facilitate cell entry (e.g., influenza virus, SARS-CoV-2), and non-structural proteins that modulate host immunity (e.g., flavivirus NS1). Incorporation occurs via direct interactions with ESCRT components or tetraspanin-enriched membrane domains; and (3) viral nucleic acids, ranging from complete genomes (e.g., enteroviruses) to regulatory RNAs (e.g., EBV miRNAs) selectively sorted by RNA-binding proteins (e.g., hnRNPA2B1, SYNCRIP, Ago2) that are linked to EV biogenesis pathways. Examples of virus-specific mechanisms include HIV-1 Gag motifs (PTAP, YPXnL) that recruit TSG101 and ALIX to hijack ESCRT for budding, and herpesvirus-mediated rerouting of MVB trafficking to load viral miRNAs into exosomes. Collectively, EVs disseminate viral particles, genomes and proteins, modulate immune signaling, and facilitate viral spread beyond canonical infection routes. AdV, adenovirus; EBV, Epstein-Barr virus; ESCRT, endosomal sorting complex required for transport; EV, extracellular vesicle; HAV, hepatitis A virus; HBV, hepatitis B virus; HEV, hepatitis E virus; HIV-1, human immunodeficiency virus 1; DENV, dengue virus; MVB, multivesicular body; PS, phosphatidylserine; SARS-CoV-2, severe acute respiratory syndrome coronavirus 2. *The figure was created with BioRender.com.*

3. *Viral nucleic acids.* These nucleic acids range from complete viral genomes (e.g., enteroviruses) to subgenomic mRNAs and regulatory miRNAs (e.g., herpesviruses). Sorting can occur through two pathways: (i) direct interaction with ESCRT components endowed with RNA-binding properties or (ii) indirect association via RNA-binding proteins (RBPs) associated with EV biogenesis, such as hnRNPA2B1, SYNCRIP, and Ago2 (Santangelo et al., 2016; Villarroya-Beltri et al., 2013; Xia et al., 2021). For example, hepatitis C virus (HCV) genomic RNA associates with miR-122 and Ago2 in a complex stabilized by HSP90 to promote EV loading (Bukong et al., 2014).

Altogether, these selective cargo-loading strategies transform EVs into Trojan carriers that expand viral tropism, hide infectious particles from immune recognition and neutralization, modulate host defenses, and amplify viral spread.

2.3 Impact of vEVs on the viral life cycle

Once released, vEVs deliver cargo to recipient cells, modulating several biological processes. Their functional impact depends on the virus species, the transported viral elements, the type of EVs and the recipient cell. vEVs demonstrate dual roles, facilitating viral replication and immune evasion while also mediating antiviral defense by transferring host-derived restriction factors, immune stimulators, and viral antigens (Nolte-'t Hoen et al., 2016; Rey-Cadilhac et al., 2023; Théry et al., 2002; Urbanelli et al., 2019; Xia et al., 2021) (Fig. 2).

2.3.1 Pro-viral effects of vEVs

vEVs can support infection via five main interconnected mechanisms:

1. *Viral particle-dependent transmission.* The encapsulation of fully formed virions within EVs creates "Trojan horses" that shield viral particles from neutralizing antibodies and innate immune sensors (Meckes & Raab-Traub, 2011). HAV, HCV, SARS-CoV-2, and human polyomavirus (JCPyV) utilize this strategy (Feng et al., 2013; Robinson et al., 2014). For non-enveloped viruses such as coxsackievirus B3 (CVB3), EV-mediated egress enables non-lytic release with high multiplicity of infection (Robinson et al., 2014).
2. *Viral particle-independent transmission.* vEVs deliver viral nucleic acids and cofactors to initiate infection upon cytoplasmic delivery, often accompanied by replication enhancers (Kouwaki et al., 2017). This strategy

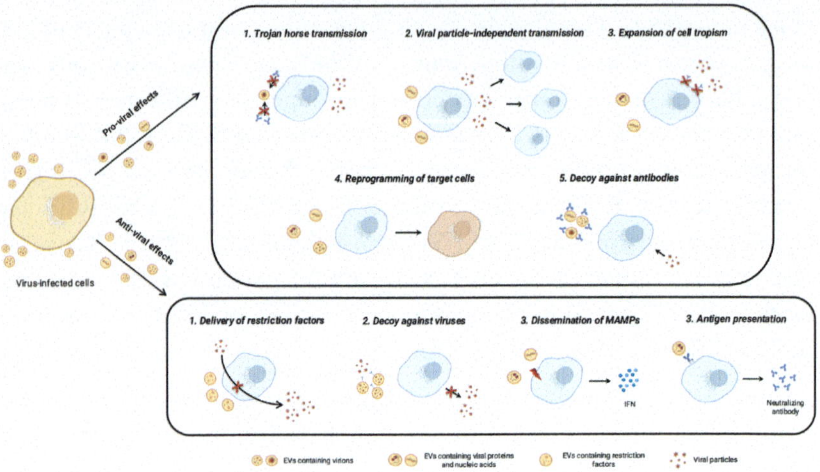

Fig. 2 Pro- and anti-viral effects of vEVs. vEVs deliver viral or host-derived cargo to recipient cells, exerting context-dependent pro- or anti-viral effects based on cargo composition and target cell type. Pro-viral mechanisms include: (1) Trojan horse transmission, where virions encapsulated within EVs escape antibody neutralization (e.g., HAV, HCV, CVB3, AdVs); (2) viral particle-independent transmission, in which EVs deliver viral nucleic acids and cofactors to initiate infection in new cells (e.g., DENV RNA); (3) expansion of cell tropism, enabling infection of non-permissive cells lacking canonical entry receptors (e.g., JCPyV in glial cells); (4) reprogramming of target cells, through delivery of viral proteins (e.g., HIV Nef, EBV LMP-1, HBV HBx) or RNAs (HIV TAR, DENV sfRNA) that suppress antiviral responses; (5) antibody decoy activity via surface-displayed viral antigens (e.g., SARS-CoV-2 Spike, ZIKV E protein, Ebola GP) that sequester neutralizing antibodies. Conversely, antiviral mechanisms comprise: (1) delivery of restriction factors, such as APOBEC3G, reducing viral replication; (2) Decoy against virus infection that expose receptors for viral entry (e.g. ACE2) thus interfering with virus infection; (3) dissemination of MAMPs, including viral RNAs, that activate PRRs (e.g., TLRs) and induce NF-κB/IFN responses; (4) Transfer of viral antigens to APCs, thereby stimulating $CD4^+$ and $CD8^+$ T cell responses. These findings illustrate the dual role of vEVs as dynamic vehicles that can either promote viral spread and immune evasion or support antiviral immunity and infection control. AdV, adenovirus; APC, antigen-presenting cell; CVB3, coxsackievirus B3; DENV, Dengue virus; EBV, Epstein-Barr virus; EV, extracellular vesicles; HAV, hepatitis A virus; HBV, hepatitis B virus; HCV, hepatitis C virus; HIV, human immunodeficiency virus 1; IFN, interferon; JCPyV, JC polyomavirus (JCPyV); MAMP, *microbe-associated molecular pattern;* PRR, pattern recognition receptor; SARS-CoV-2, severe acute respiratory syndrome coronavirus 2; TLR, toll-like receptor; vEVs, viral EVs; ZIKV, Zika virus. *The figure was created with BioRender.com.*

allows viruses, such as DENV, to bypass neutralizing antibodies and persist under immune pressure (Latanova et al., 2024; Urbanelli et al., 2019).

3. *Expansion of viral cell tropism.* vEVs enable infection of non-permissive tissues lacking canonical entry receptors (Urbanelli et al., 2019). For instance, AdVs exploit exosome-like vesicles to infect cells lacking the coxsackie and adenovirus receptor (CAR) (Hambo & Harb, 2023; Ipinmoroti et al., 2021; Crenshaw et al., 2019). Similarly, JCPyV can infect glial cells through EV-mediated transport (O'Hara et al., 2020). This ability facilitates barrier crossing (e.g., blood-brain barrier) and systemic dissemination.
4. *Host cell reprogramming.* vEVs deliver viral factors that make recipient cells more susceptible to infection. Viral proteins such as HIV-1 Nef, HBV HBx, and Epstein-Barr virus (EBV) LMP-1, alter signaling, favoring viral pathogenesis (Campbell et al., 2008; Canitano et al., 2013; Kapoor et al., 2017; Lenassi et al., 2010). Similarly, viral RNAs, including HIV trans-activation response (TAR) RNA, herpes simplex virus 1 (HSV-1) miRNAs, and DENV subgenomic flavivirus RNA (sfRNA), suppress immunity or promote latency (Bidet et al., 2014; Manokaran et al., 2015; Narayanan et al., 2013).
5. *Decoy against humoral immunity.* vEVs can act as molecular decoys against humoral immunity when viral antigens are displayed on their surface. This antigen exposure allows vEVs to sequester neutralizing antibodies, thus preserving circulating infectious virions. For instance, epitopes derived from SARS-CoV-2 Spike protein, Ebola virus glycoprotein (GP), and Zika virus envelope (E) protein can be exposed on vEVs surface thus promoting this decoy function (Nehls et al., 2019; Troyer et al., 2021; Zhao et al., 2023).

2.3.2 Antiviral effects of vEVs

Conversely, vEVs can enhance antiviral defense through four main mechanisms (Nolte-'t Hoen et al., 2016; Théry et al., 2002; Urbanelli et al., 2019):

1. *Transfer of restriction factors.* One notable example is the EV-mediated delivery of APOBEC3G, a cytidine deaminase that restricts retroviral replication by inducing G→A hypermutations during reverse transcription (Khatua et al., 2009).
2. *Decoy against virus infection.* EVs can expose on their surface receptors for viral entry (e.g. ACE2, PS), thus interfering with virus infection.

Circulating ACE2-displaying extracellular vesicles (evACE2), such as ACE2+ exosomes produced via ATG16L1-dependent mechanisms, can neutralize SARS-CoV-2 infection by directly blocking viral entry (Ching et al., 2022; El-Shennawy et al., 2022). Similarly, phosphatidylserine (PS)-exposing extracellular vesicles (PS-rich EVs) can interfere with viral attachment and entry into host cells by blocking PS receptors (PSR) targeted by viruses like Zika virus (ZIKV), dengue virus (DENV) and other viruses that rely on apoptotic mimicry strategies (Groß et al., 2024).

3. *Dissemination of microbe-associated molecular patterns (MAMPs).* The dissemination of MAMPs, such as viral RNAs, within vEVs can augment antiviral state of neighbouring non-infected cells. For instance, HCV RNA and HIV miRNAs can engage pattern recognition receptors (PRRs), including Toll-like receptors (TLRs), and trigger innate immune signaling cascades culminating in type I interferon (IFN-I) production and NF-κB activation (Bernard et al., 2014; Dreux et al., 2012).
4. *Priming of adaptive antiviral immunity.* vEVs can deliver viral antigens to antigen-presenting cells (APCs), such as dendritic cells. Once internalized, these antigens can be processed and loaded onto major histocompatibility complex (MHC) molecules, and initiate virus-specific $CD4^+$ and $CD8^+$ T cell responses (Nolte-'t Hoen et al., 2016; Pegtel & Gould, 2019; Théry et al., 2002).

3. Autophagy in viral life cycle: From degradation to secretion

Autophagy maintains cellular homeostasis by selectively eliminating cytoplasmic material, including damaged organelles, aggregated proteins and intracellular pathogens, through the lysosomal pathway (Zaffagnini & Martens, 2016). However, recent studies revealed a functional plasticity of autophagy. Under certain pathological conditions, such as viral infections, oxidative stress, or inhibition of autophagic flux, autophagosomes can function as secretory platforms facilitating the release of cytosolic components, including RNA, proteins and viral particles. This dual function, degradative or secretory, makes autophagy a favoured viral target, as viruses can dynamically exploit both modes for replication and immune evasion (Deretic et al., 2013; Galluzzi et al., 2017; Jassey & Jackson, 2024).

3.1 Canonical degradative autophagy in viral infection

Canonical autophagy is a tightly regulated and evolutionarily conserved multistep process that degrades cytoplasmic components. Upon viral infection, a selective form of autophagy called xenophagy contributes to cellular defense by sequestering viral elements into double-membrane autophagosomes and delivering them to lysosomes for degradation (Deretic et al., 2013; Levine & Kroemer, 2019; Orvedahl & Levine, 2008; Randow & Youle, 2014). This process unfolds in five steps described below with mechanistic differences between DNA and RNA viruses (Table 1).

3.1.1 Viral recognition and initiation

MAMPs of viral origin engage PRRs. In case of DNA viruses, cyclic GMP-AMP synthase (cGAS)-stimulator of interferon genes (STING) and TLR9 sense viral DNA in the cytosol and endosomes, respectively (Chen et al., 2016). For RNA viruses, recognition relies on cytosolic RNA sensors like retinoic acid-inducible gene-I (RIG-I) and melanoma differentiation-associated protein 5 (MDA5), which both belong to the RIG-I-like receptor family (RLR), or on endosomal RNA sensors such as TLR3 and TLR7. Once activated, these receptors initiate STING or mitochondrial antiviral signalling protein (MAVS) pathways that trigger both IFN-I production and autophagy (Gui et al., 2019; Sun et al., 2016).

Upon activation via cGAS-mediated recognition of viral DNA, STING translocates from the endoplasmic reticulum (ER) to the Golgi apparatus, where it activates TANK-binding kinase 1 (TBK1) and IRF3, ultimately inducing IFN-I responses. Moreover, STING recruits autophagy-related proteins such as LC3 to facilitate pathogen sequestration and autophagosome formation, effectively linking innate viral sensing to the selective degradation of viral material (Gui et al., 2019; Huang et al., 2025).

MAVS, which operates downstream of RIG-I, forms aggregates on the mitochondrial outer membrane, triggering TBK1-IRF3 signalling and subsequent IFN-I production. The mitochondrial localization of MAVS also intersects with mitophagy, a selective autophagy targeting damaged mitochondria. Altogether, these processes contribute to clear viral components and counter viral strategies aimed at disrupting mitochondrial antiviral signalling (Liu et al., 2015; Sun et al., 2016).

TBK1 serves as a pivotal kinase linking nucleic acid sensing to both interferon production and autophagy by phosphorylating IRF3 to drive IFN-β transcription and, at the same time, sequestosome 1 (p62/SQSTM1), nuclear dot protein 52 (NDP52), and optineurin (OPTN) to enhance

Table 1 Canonical autophagy-mediated degradation during DNA and RNA virus infections. The table summarises the main differences across the five steps of the xenophagy process occurring in host cells during viral infection, distinguishing between DNA and RNA viruses. AdV, adenovirus; HIV-1, human immunodeficiency virus 1; HSV-1, herpes simplex virus 1; KSHV, Kaposi's sarcoma-associated herpesvirus; RLR, RIG-I-like receptor.

Steps of degradative xenophagy	DNA virus	RNA virus	DNA vs RNA virus differences
1. Viral recognition and initiation	Viral DNA is sensed in the cytosol via the cGAS-STING pathway, which activates TRIM23 → TBK1 and triggers antiviral macroautophagy (e.g., HSV-1 and KSHV) (Acharya et al., 2025; Pino-Belmar et al., 2024). Upon entry, endosomal membrane damage exposes glycans recognised by galectin-8, which recruits NDP52/p62 to initiate degradative xenophagy (e.g., AdVs, KSHV) (Montespan et al., 2017).	Viral RNA and replicative intermediates are sensed via RIG-I/MDA5 → TBK1, which phosphorylates selective autophagy receptors (OPTN, p62 and TAX1BP1), thereby enhancing xenophagy (White et al., 2025). TRIM receptors, such as TRIM5α, directly recognise capsids (e.g., HIV-1) and target them to LC3 for degradative autophagy, while p62 recognises ubiquitinated capsids (e.g., Sindbis virus) (Mandell et al., 2014; Orvedahl et al., 2010).	Initiation is often cGAS-STING-dependent for DNA viruses and more RLR/TBK1-dependent for RNA viruses. Non-enveloped DNA viruses (e.g., AdVs) also use galectin-8 to sense membrane damage (Montespan et al., 2017; Pied et al., 2022).

2. Phagophore nucleation	TBK1 supports phagophore initiation by phosphorylating STX17 and coordinating the ULK1-FIP200-ATG13 complex. The cGAS-STING-TRIM23 axis is triggered in response to DNA viruses, such as HSV-1 and KSHV (Acharya et al., 2025; Kumar et al., 2019).	TBK1, which is activated by RLRs or TRIM5α, promotes nucleation. Selective receptors (OPTN, p62 and TAX1BP1) recruit LC3 to nascent phagophores around capsids or cargoes (e.g. HIV-1 via TRIM5 and Sindbis virus via p62) (Mandell et al., 2014; Orvedahl et al., 2010).	The ULK1-Beclin1/PI3KC3 machinery is shared, but the upstream trigger differs (cGAS-STING vs RLR/TRIM). STX17-TBK1 is a common initiation control point (Kumar et al., 2019).
3. Phagophore expansion and cargo recognition	The addition of ubiquitin tags and galectin-8 is linked to the occurrence of endosomal damage caused by AdVs or KSHV. This damage exposes glycans, which then lead to Gal-8 recruiting NDP52/p62. This, in turn, results in the binding of LC3 and the subsequent capture of virions for degradation (Montespan et al., 2017; Pied et al., 2022).	Selective autophagy receptors facilitate targeted viral degradation through several distinct mechanisms. p62 binds to Sindbis virus capsid proteins and delivers them to LC3. TRIM5α acts as a selective receptor for HIV-1, linking directly to LC3/GABARAP. TBK1 phosphorylates OPTN/p62, enhancing Ub-chain binding and cargo recruitment (Mandell et al., 2014; Orvedahl et al., 2010; White et al., 2025).	DNA viruses often use galectin-8→NDP52/p62, which sense damage to the entry membrane. RNA viruses typically involve direct capsid/protein recognition via p62/OPTN/TRIM (Montespan et al., 2017).

(continued)

Table 1 Canonical autophagy-mediated degradation during DNA and RNA virus infections. The table summarises the main differences across the five steps of the xenophagy process occurring in host cells during viral infection, distinguishing between DNA and RNA viruses. AdV, adenovirus; HIV-1, human immunodeficiency virus 1; HSV-1, herpes simplex virus 1; KSHV, Kaposi's sarcoma-associated herpesvirus; RLR, RIG-I-like receptor. (*cont'd*)

Steps of degradative xenophagy	DNA virus	RNA virus	DNA vs RNA virus differences
4. Autophagosome maturation	TBK1 is essential for the maturation of autophagosomes into degradative compartments and facilitates progression to lysosome-competent vesicles. This has been demonstrated for pathogens like HSV-1 (Pilli et al., 2012; Yakoub & Shukla, 2015).	A similar mechanism applies to RNA viruses, whose clearance also depends on TBK1 activity and proper LC3/ ATG8 lipidation. Inhibiting key adapters such as p62 or OPTN, or disrupting LC3/ATG8 function, delays viral degradation. For example, blocking p62 or ATG5 can lead to the accumulation of Sindbis virus capsids[11].	TBK1 serves as a hub for maturation in both cases; the receptors/cargo differ (Pilli et al., 2012).
5. Autophagosome-lysosome fusion and degradation	Autophagosome-lysosome fusion relies on STX17, whose activity is regulated by TBK1, enabling the delivery of cargo for degradation. This process contributes to	Lysosomal degradation effectively reduces viral components. For instance, autophagy mediated by p62 lead to reduced capsid accumulation and	The molecular mechanisms of lysosome fusion and degradation are shared across DNA and RNA viruses, but they differ as follows: DNA viruses primarily eliminate mislocalised entry particles,

	functional viral restriction, as observed with HSV-1 when autophagy is active, and also with vaccinia virus when xenophagy receptors are overexpressed (Kumar et al., 2019; Yakoub & Shukla, 2015; Krause et al., 2024).	cytotoxicity in Sindbis virus whereas TRIM5α directs HIV-1 cores LC3/ATG8-dependent degradation (Mandell et al., 2014; Orvedahl et al., 2010).	whereas RNA viruses mainly target cytosolic capsids and viral proteins (Montespan et al., 2017).

pathogens recognition and LC3 recruitment (Pilli et al., 2012). These pathways converge on the activation of Unc-51 like autophagy activating kinase 1 (ULK1) to initiate the formation of phagophores that sequester viral components (Deretic et al., 2013).

3.1.2 Phagophore nucleation

Once activated, the ULK1 kinase complex initiates phagophore nucleation by phosphorylating components of the class III phosphatidylinositol 3-kinase (PI3K-III) complex, which consists of Beclin-1 (BECN1), vacuolar protein sorting-associated protein 34 (VPS34), vacuolar protein sorting-associated protein 15 (VPS15) and autophagy-related protein 14-like protein 1 (ATG14L).

Specifically, BECN1 acts as a central scaffold that recruits additional regulatory proteins to initiate autophagosome formation (Tran et al., 2021). Its phosphorylation by ULK1 enhances the activity of the associated lipid kinase VPS34 which is the catalytic subunit of the complex. VPS34, stabilized by the regulatory adaptor VPS15, generates phosphatidylinositol 3-phosphate (PI3P) at ER level (Axe et al., 2008). PI3P serves as a crucial signaling lipid involved in the recruitment of downstream effectors such as WD repeat domain phosphoinositide-interacting protein 2 (WIPI2) and double FYVE domain-containing protein 1 (DFCP1) that remodel membranes and stabilize the nascent phagophore. ATG14L, an autophagy-specific regulatory subunit, specifically directs the PI3K-III complex to the ER, where the phagophore membrane originates. The spatially restricted generation of PI3P defines the site of autophagosome formation and is a critical step in segregating cytoplasmic components or pathogens for degradation (Mizushima, 2018; Russell et al., 2014).

Altogether, these proteins establish a PI3P-enriched microdomain, which marks the site for phagophore initiation and provides a platform for the recruitment of proteins required for membrane expansion.

3.1.3 Phagophore expansion and cargo recognition

The elongation of the phagophore membrane is orchestrated by the ATG5-ATG12-ATG16L1 complex, an ubiquitin-like conjugation system. In particular, ATG12 is covalently conjugated to ATG5 through the sequential action of ATG7 (E1-like enzyme) and ATG10 (E2-like enzyme). The ATG12-ATG5 conjugate associates non-covalently with ATG16L1, forming a multimeric complex that acts as an E3-like enzyme to facilitate the lipidation of cytosolic LC3 (i.e., LC3-I), with phosphatidylethanolamine

(PE) to generate membrane-anchored LC3 (i.e., LC3-II) (Mizushima et al., 2011). LC3-II stably associates with the growing phagophore membrane, driving its curvature and elongation, and is widely used as a marker for autophagosomes (Axe et al., 2008; Klionsky et al., 2021; Nakatogawa, 2020; Zaffagnini & Martens, 2016). Importantly, selective autophagy receptors including p62/SQSTM1, NDP52, and OPTN act as adaptors that recognize ubiquitinated substrates or viral proteins. They contain LC3-interacting regions (LIR motifs) that bind directly to LC3-II, thereby linking specific cargo to the autophagic machinery. This cargo recognition process ensures the selective degradation of viral proteins or entire pathogens.

3.1.4 Autophagosome formation and maturation

During the elongation phase, the phagophore progressively extends to engulf viral material or cytoplasmic components marked for degradation. Once the opposing edges of the phagophore converge, they undergo a tightly regulated closure process that results in the formation of a mature double-membrane autophagosome. This process is coordinated by membrane remodelling factors and components of the ESCRT machinery (Longatti & Tooze, 2009; Takahashi et al., 2019), which ensure proper membrane sealing and vesicle integrity. Once formed, autophagosomes are transported along microtubules toward the lysosome-rich perinuclear region of the cell. This trafficking is facilitated by dynein motor proteins, which mediate retrograde transport toward the microtubule-organizing center (Kimura et al., 2008). During this step, the autophagosome undergoes progressive maturation, acquiring fusion-competent machinery and eliminating inhibitory factors that would otherwise prevent lysosomal docking (Birgisdottir & Johansen, 2020). A key aspect of this maturation process is the stable incorporation of LC3-II into the autophagosomal membrane, which maintains vesicle integrity and provides docking sites for autophagy receptors and adaptor proteins. Additional ATG proteins are recruited to enhance the vesicles stability and facilitate its interaction with fusion complexes (Longatti & Tooze, 2009).

3.1.5 Autophagosome-lysosome fusion and degradation

The final step involves autophagosome-lysosome fusion to form autolysosome, where the enclosed viral cargo is degraded by lysosomal hydrolases. This process is coordinated by several molecular machineries. Rab7, a small GTPase localized into late endosomes and autophagosomes, regulates their trafficking, maturation and fusion competence (Ke, 2024;

Walia et al., 2024). In addition, Rab7 recruits the homotypic fusion and protein sorting (HOPS) complex, which bridges autophagosomes and lysosomes, facilitating membrane proximity and ensuring the proper orientation for fusion (Jiang et al., 2014; Ke, 2024). Subsequently, the SNARE proteins, including Syntaxin 17 (STX17) on autophagosomes, the adaptor SNAP29, and VAMP8 on lysosomes, form a trans-SNARE complex that drives membrane fusion, generating a degradative autolysosome (Song et al., 2024; Jian et al., 2025). Once fusion occurs, the autophagosome inner membrane and its contents are exposed to the acidic environment and hydrolytic enzymes of the lysosome, including proteases, lipases, and nucleases. These enzymes efficiently degrade viral proteins and nucleic acids, completing the xenophagic process and contributing to host defense (Itakura et al., 2012; Takáts et al., 2014; Liu et al., 2025).

3.2 Non-canonical secretory autophagy in viral infection

Under specific conditions, autophagy can divert from degradation to secretion, releasing cytosolic contents including viral particles. This adaptation is particularly prominent during viral infections where many viruses block the fusion of autophagosomes and lysosomes to exploit secretory pathways (Table 2).

Secretory autophagy initiates through the same PRR recognition and early steps (phagophore formation, nucleation, elongation) as canonical autophagy. However, instead of fusing with lysosomes, autophagosomes containing sequestered viral components are rerouted toward plasma membrane or endosomal compartments. This redirection creates amphisomes, which are hybrid vesicles displaying both autophagic (LC3) and endosomal markers (Ganesan & Cai, 2021). Secretory trafficking employs a different group of Rab GTPases (Rab8a, Rab11, and Rab27a) and additional SNAREs (Syntaxin-3, Syntaxin-4, and Sec22b) that mediate fusion with plasma membrane or MVBs, enabling extracellular cargo release (Lin et al., 2024; Moreau et al., 2013; Zhao et al., 2025).

In addition to classical secretory autophagy, recent studies have identified an alternative mechanism of LC3 conjugation to pre-existing single membranes, referred as Conjugation of ATG8s to Single Membranes (CASM) (Durgan et al., 2021; Fletcher et al., 2018). Unlike canonical macroautophagy, CASM is characterised by the lipidation of ATG8/LC3 proteins onto single membranes of endosomes, lysosomes, or phagosomes, independently of the ULK1 or VPS34 complexes (Durgan et al., 2021; Hooper et al., 2022). This process relies on the ATG5–ATG12–ATG16L1

Table 2 Non-canonical autophagy-mediated secretion during DNA and RNA virus infections. The table summarises the main differences across the five steps of secretory autophagy in host cells during viral infection, distinguishing between DNA viruses (HSV-1, AdVs, EBV) and RNA viruses (poliovirus, CVB3, EV71). AdV, adenovirus; AWOL, autophagosome-mediated exit without lysis; CVB3, coxsackievirus B3; EBV, Epstein-Barr virus; EV71, enterovirus A71; HSV-1, herpes simplex virus 1.

Steps of secretory autophagy	DNA virus	RNA virus	DNA vs RNA differences
1. Viral recognition and initiation	DNA viruses rarely trigger secretory autophagy. However, some herpesviruses (e.g., EBV and HSV-1) can modulate ATG proteins and redirect autophagic membranes towards exocytosis rather than degradation (e.g. in neurons for HSV-1 spread) (Zhai et al., 2023)	Many RNA viruses hijack the pathway. For example, poliovirus, coxsackieviruses and EV71 induce autophagy via dsRNA intermediates, but viral proteins (2BC and 3A) block fusion and divert autophagosomes to the secretory pathway (Jackson, 2015; Lai et al., 2016).	RNA viruses more frequently initiate autophagy to exploit secretion, while DNA viruses mostly re-route existing autophagosomes (Jackson, 2015; Zhai et al., 2023).
2. Phagophore nucleation	In the life cycle of herpesviruses such as HSV-1 and EBV, the nucleation of autophagosomes via the Beclin1/PI3KC3 complex occurs but their maturation is stalled. Instead, the resulting membranes are repurposed for secretion. For example, in HSV-1 infection, they are found along axons (English et al., 2009).	Enteroviruses, including poliovirus and CVB3, actively trigger nucleation via the PI3KC3/ULK1 pathway to generate double-membrane vesicles. These vesicles provide platforms for viral replication and also serve as precursors for secretory autophagosomes (Jackson, 2015; Lai et al., 2016).	Nucleation tends to be passive in DNA viruses, but active in RNA viruses (English et al., 2009; Jackson, 2015).

(continued)

Table 2 Non-canonical autophagy-mediated secretion during DNA and RNA virus infections. The table summarises the main differences across the five steps of secretory autophagy in host cells during viral infection, distinguishing between DNA viruses (HSV-1, AdVs, EBV) and RNA viruses (poliovirus, CVB3, EV71). AdV, adenovirus; AWOL, autophagosome-mediated exit without lysis; CVB3, coxsackievirus B3; EBV, Epstein-Barr virus; EV71, enterovirus A71; HSV-1, herpes simplex virus 1. (*cont'd*)

Steps of secretory autophagy	DNA virus	RNA virus	DNA vs RNA differences
3. Phagophore expansion and cargo recruitment	Some DNA viruses exploit LC3 lipidation to package viral proteins into vesicles. For example, EBV LMP1 is incorporated into LC3-positive compartments for exosomal export (Huang & Yue, 2020).	RNA viruses rely on LC3 and ATG5/7 to enwrap viral particles or RNA replication complexes (Huang & Yue, 2020; Lai et al., 2016). For example, poliovirus particles are loaded into LC3-positive vesicles that are subsequently released via non-lytic pathways (Jackson, 2015; Lai et al., 2016).	RNA viruses tend to use autophagy-derived membranes directly for virion enwrapping, while DNA viruses primarily utilise these compartments for the secretion of viral proteins or oncoproteins (Huang & Yue, 2020; Jackson, 2015).
4. Autophagosome maturation (non-degradative route)	In DNA viruses such as HSV-1 and EBV, maturation is stalled: autophagosomes do not fuse with lysosomes but are redirected toward plasma membrane trafficking pathways involving Rab11, Rab27 (English et al., 2009; Huang & Yue, 2020).	Enteroviruses block lysosomal fusion through viral proteins such as 2BC and 3 A, thereby diverting autophagosomes towards the secretory autophagy pathway. These vesicles are then transported to the plasma membrane via Rab8a and GRASP55 (Jackson, 2015; Lai et al., 2016; Zhang & Wang, 2018).	Both DNA and RNA viruses actively block degradation, but RNA viruses show a stronger dependence on viral proteins that arrest fusion (English et al., 2009; Jackson, 2015; Lai et al., 2016).

5. Fusion with plasma membrane and secretion	HSV-1 uses secretory autophagy in neurons for anterograde axonal spread. This involves the fusion of vesicles at the plasma membrane via syntaxin-4 and Rab27a. Similarly, EBV packages its oncogenic protein LMP1 into LC3-positive secretory vesicles that are released in exosomes (English et al., 2009; Huang & Yue, 2020; Zhai et al., 2023).	Virions of enteroviruses, including poliovirus, CVB3, and EV71, are released in bulk within LC3-positive vesicles via the AWOL pathway (Jackson, 2015; Lai et al., 2016).	DNA viruses primarily utilise secretory autophagy for protein secretion or neuronal trafficking, while RNA viruses exploit it to release fully formed virions, enabling non-lytic egress (Huang & Yue, 2020; Jackson, 2015; Zhang & Wang, 2018).

complex, but the WD40C-terminal domain of ATG16L1 plays a distinctive role by directing the lipidation machinery to endolysosomal membranes (Fletcher et al., 2018). In certain contexts, CASM can also be mediated by Tectonin Beta-Propeller Repeat Containing 1 (TECPR1), which interacts with ATG5 and phosphatidylinositol 3-phosphate (PI3P) on damaged endolysosomal membranes, acting as a platform for LC3 conjugation (Fletcher et al., 2018; Kaur et al., 2024).

Mechanistically, CASM activation often follows perturbations in the vacuolar-type H^+-ATPase (V-ATPase) complex, which senses lysosomal pH alterations and recruits ATG16L1 to single membranes (Hooper et al., 2022). V-ATPase–ATG16L1 interaction represents a key molecular switch that distinguishes CASM from canonical autophagy. Viral ion channel proteins such as Influenza A virus M2 disrupt lysosomal acidification, thereby promoting V-ATPase disassembly and triggering LC3 lipidation onto endosomal membranes (Gannagé et al., 2009; Hooper et al., 2022). Similarly, SARS-CoV-2 ORF3a localizes to late endosomes and lysosomes, where it impairs autolysosome maturation and enhances LC3 decoration of single membranes, a phenotype consistent with CASM activation (Chen et al., 2021; Miao et al., 2021).

Importantly, CASM is also linked to innate immune signalling. The cyclic GMP-AMP synthase (cGAS)–stimulator of interferon genes (STING) pathway, which senses cytosolic viral DNA, has been shown to promote ATG16L1-dependent LC3 lipidation on endolysosomal membranes independently of double-membrane autophagosome formation (Fischer et al., 2020; Gui et al., 2019). Upon activation, STING translocates from the endoplasmic reticulum (ER) to the Golgi apparatus, where it interacts with the TANK-binding kinase 1 (TBK1) and the ATG5–ATG12–ATG16L1 complex, leading to non-canonical LC3 lipidation (Fischer et al., 2020; Gui et al., 2019). This link between innate immune sensing and CASM establishes a crosstalk between antiviral defense and membrane trafficking.

CASM activation goes beyond degradation. By decorating endolysosomal membranes with LC3, CASM can promote the loading and secretion of LC3-dependent extracellular vesicles (LDELS), a process in which LC3 marks single-membrane compartments for cargo selection (e.g. RNA, protein) and release (Gardner et al., 2023; Leidal & Debnath, 2020). This mechanism is amplified under viral stress thus enhancing cell-to-cell communication and immune evasion (Bare et al., 2025; Ma et al., 2024; Towers et al., 2020). During viral infection, CASM may therefore serve as a mechanistic bridge between immune activation, membrane remodeling, and the release of LC3-positive EVs carrying viral proteins, RNA, or immunomodulatory molecules. Although the precise contribution of CASM to EV biogenesis remains incomplete, its involvement in virus-induced modulation of endolysosomal trafficking strongly supports its inclusion in models of autophagy-dependent EV secretion (Figueras-Novoa et al., 2024; Fischer et al., 2020; Hooper et al., 2022).

3.3 Autophagy machinery in virion packaging and release

Several viruses directly exploit components of the autophagy machinery to facilitate virion assembly and release, highlighting a functional overlap between secretory autophagy and viral morphogenesis. Components of the autophagy pathway, including LC3 and gamma-aminobutyric acid receptor-associated proteins (GABARAPs), have been shown to contribute structurally to virion assembly rather than to degradation.

In HIV-1, GABARAP family proteins regulate the packaging of genomic RNA into virions and assist in proper virion assembly (Palaric et al., 2025). Disruption of GABARAP expression significantly impairs RNA incorporation and reduces the production of infectious particles, underscoring the direct role of autophagy-related proteins in HIV-1 biogenesis.

In EBV, a DNA virus, autophagy proteins including LC3 and other macroautophagy components can be incorporated into newly formed virions. These proteins assist in viral particle formation and intracellular transport, suggesting that autophagy machinery is co-opted for structural support during viral morphogenesis (Nowag et al., 2014).

Finally, certain RNA viruses, such as IAV, exploit LC3 conjugation to stabilize virions. Specifically, the M2 protein of IAV contains an LC3-interacting motif that redirects LC3-positive membranes toward sites of virion assembly, maintaining particle stability and enhancing viral exit (Beale et al., 2014).

Collectively, these findings highlight that the autophagy machinery acts as an integral mediator of viral biogenesis, providing vesicular platforms that facilitate virion assembly, genomic RNA packaging, and extracellular release.

3.4 Vesicular fate: Degradation or secretion?

The fate of autophagosomes, degradative or secretory, depends on molecular regulators and environmental signals that viruses frequently manipulate (Table 1). During persistent infections or when autophagic flux is blocked, cells activate compensatory pathways such as SALI (secretory autophagy during lysosome inhibition) or MAD (membrane-Atg8-dependent secretion), which redirect autophagosomes towards exocytosis. These routes allow the removal of undegraded toxic material and can be exploited by viruses to achieve non-lytic release and silent dissemination (Ahat et al., 2022; Debnath & Leidal, 2022; Feng et al., 2022).

TRIM16 and GRASP55 are regulatory proteins that act as molecular "switches", directing vesicles toward secretion. TRIM16 recognizes secretory cargo like IL-1β and facilitates LC3-positive membrane delivery via Sec22b, while GRASP55 undergoes phosphorylation changes that relocate it to autophagosomes and MVBs, enabling cargo secretion (Di Rienzo et al., 2020; Dupont et al., 2011; Noh et al., 2022; Nüchel et al., 2021). ESCRT complexes and specific Rab GTPases (Rab27a for exosomes, Rab8a for autophagy-associated proteins like p62/SQSTM1) orchestrate the release of EVs containing viral components and other immunomodulatory factors (Blanc & Vidal, 2018; Colombo et al., 2014; Zhao et al., 2025).

4. Viral strategies for evading xenophagy

Viruses have evolved several strategies to counteract xenophagy: (i) direct interference with pathogen recognition, autophagosome formation, or lysosomal fusion and (ii) indirect exploitation of EVs to transfer viral factors and evade immune detection (Table 3) (Alenquer & Amorim, 2015).

4.1 Direct interference with xenophagy

RNA viruses can disrupt innate signaling pathways (STING, MAVS, and TBK1 pathways) that trigger xenophagy. For instance, HCV NS3/4 A protease cleaves MAVS, dampening RIG-I signaling (Li et al., 2005). DNA viruses can inhibit autophagosome formation. For instance, HSV-1 ICP34.5 binds BECN1, blocking autophagosome formation (Orvedahl et al., 2007), while AdV capsid protein VI contains a PPxY motif that interacts with the E3 ubiquitin ligase NEDD4 to restrict autophagosome formation (Montespan et al., 2017).

Some viruses, notably herpesviruses, target xenophagy receptors. The human cytomegalovirus (CMV) produces TRS1 protein which impairs the recognition of viral components by p62/SQSTM1, NDP52, or OPTN

Table 3 Viral strategies to evade xenophagy. The table summarizes the main approaches used by DNA and RNA viruses to inhibit pathogen recognition, autophagosome formation, lysosomal fusion, and cargo degradation (direct strategies), as well as to exploit host-derived EVs to transfer viral components without recognition by xenophagy (indirect strategy). AdV, adenovirus; CMV, human cytomegalovirus; DENV, Dengue virus; EBV, Epstein-Barr virus; EV, extracellular vesicle; HAV, hepatitis A virus; HCV, hepatitis C virus; HIV-1, human immunodeficiency virus 1; HSV-1, herpes simplex virus 1; IAV, influenza A virus.

	Steps of autophagy	DNA viruses	RNA viruses
Direct strategies	**Inhibition of autophagosome formation**	HSV-1 ICP34.5 protein binds BECN1 blocking autophagosome formation (Orvedahl et al., 2007) CMV TRS1 protein interferes with autophagy initiation (Chaumorcel et al., 2012) AdV capsid protein VI restricts autophagosome formation (Montespan et al., 2017)	Poliovirus and DENV hijack autophagic membranes for replication (Abernathy et al., 2019; Jackson, 2015). IAV M2 protein manipulates autophagosomes for viral replication (Gannagé et al., 2009).
	Interference with Xenophagy Receptors	CMV TRS1 protein, interferes with selective autophagy receptors like p62/SQSTM1, NDP52, OPTN to avoid xenophagy (Chaumorcel et al., 2012) EBV modulates ubiquitination of viral proteins to reduce selective autophagy receptor recognition (Dantuma & Masucci, 2003; Granato et al., 2014)	HCV protease-mediated MAVS cleavage indirectly reduces activation of autophagy receptors (Li et al., 2005).
	Prevention of Lysosomal Fusion	HSV-1 and CMV can also block fusion indirectly via selective	IAV, poliovirus, and DENV prevent fusion, using

(continued)

Table 3 Viral strategies to evade xenophagy. The table summarizes the main approaches used by DNA and RNA viruses to inhibit pathogen recognition, autophagosome formation, lysosomal fusion, and cargo degradation (direct strategies), as well as to exploit host-derived EVs to transfer viral components without recognition by xenophagy (indirect strategy). AdV, adenovirus; CMV, human cytomegalovirus; DENV, Dengue virus; EBV, Epstein-Barr virus; EV, extracellular vesicle; HAV, hepatitis A virus; HCV, hepatitis C virus; HIV-1, human immunodeficiency virus 1; HSV-1, herpes simplex virus 1; IAV, influenza A virus. (*cont'd*)

	Steps of autophagy	DNA viruses	RNA viruses
		autophagy receptor interference (Orvedahl et al., 2007; Chaumorcel et al., 2012).	double-membrane autophagosomal structures as replication scaffolds (Abernathy et al., 2019; Gannagé et al., 2009; Jackson, 2015).
Indirect strategy	**Exploitation of EVs**	EBV BART miRNAs carried by EVs downregulate ATG5/ATG7 in recipient cells (Pegtel et al., 2010)	HIV-1 EVs carry Nef protein, inhibiting autophagy in neighboring cells (Castro-Gonzalez et al., 2021); HAV is secreted quasi-enveloped form in EVs, evading LC3 recruitment and ubiquitination (Feng et al., 2013).
Outcome		Evade recognition by xenophagic mechanisms, block autophagosome formation and prevent fusion with lysosomes degradation, thus reducing viral clearance.	Hijack autophagic membranes for replication, prevent lysosomal degradation, modulate xenophagy in neighboring cells to enhance spread and immune evasion.

(Chaumorcel et al., 2012), while EBV regulates protein ubiquitination to decrease recognition by xenophagy receptors (Dantuma & Masucci, 2003).

Other viruses block the fusion of autophagosomes with lysosomes. The matrix protein 2 (M2) of influenza A virus (IAV) manipulates autophagosomal membranes for replication while avoiding lysosomal destruction (Gannagé et al., 2009). Similarly, poliovirus and DENV hijack autophagosome-like membranes as scaffolds for viral RNA replication that resist fusion with lysosomes (Abernathy et al., 2019; Jackson, 2015).

4.2 Indirect EV-mediated evasion

Viruses exploit exosomes and microvesicles to manipulate xenophagy in neighbouring cells. For instance, EBV BART miRNAs downregulate ATG5 and ATG7 in recipient cells, reducing xenophagy efficiency (Pegtel et al., 2010). Cells infected by HIV-1 secrete EVs containing Nef protein, which inhibits autophagy in recipient cells (Castro-Gonzalez et al., 2021).

In addition, EVs can shield viral particles. HAV quasi-enveloped forms evade LC3-mediated recruitment and ubiquitination, hiding from the autophagic machinery (Feng et al., 2013).

Overall, these host-pathogen interactions highlight the importance of autophagy in antiviral immunity and may inform novel therapeutic strategies targeting xenophagy to control viral infections.

5. Intersection of secretory autophagy and EV pathways in viral infection

Viruses exploit the convergence between EV biogenesis and autophagy, which respectively contribute to intercellular communication and intracellular degradation, to control vesicle release, remodel host communication, and calibrate immune responses, thereby enhancing persistence and dissemination.

RNA viruses such as poliovirus, CVB3, and enterovirus 71 (EV71) can be loaded within vesicles that display both classical EV markers, such as CD63 and flotillin-1, and the autophagy marker LC3-II. During secretory autophagy, autophagosomes, instead of fusing with lysosomes, merge with the plasma membrane to secrete their cargo while shielding virions from immune recognition (Pleet et al., 2018). A recent study confirms that DENV similarly hijacks autophagy vesicles for secretory pathways, with 26 % of DC-derived EVs co-expressing LC3, and depletion of LC3-positive EVs reducing viral transmission (Cloherty et al., 2024).

Other viruses bypass autophagy and exploit the MVB-exosome pathway for release. HAV and HEV acquire host-derived membranes via the ESCRT machinery and Rab GTPases, without detectable LC3, generating quasi-enveloped particles through endosomal rather than autophagic trafficking (Feng et al., 2013; Nagashima et al., 2014).

The amphisome represents direct pathway convergence, when autophagosomes fuse with MVBs before lysosomal degradation (Fader & Colombo, 2009). Under suppressed autophagic flux, amphisomes divert toward secretion, releasing vesicular cargo. This crosstalk means that modulation of autophagy profoundly affects both the quantity and composition of EVs, allowing viruses to amplify spread by tipping the balance toward secretion.

Some enveloped viruses integrate both ESCRT-dependent EV biogenesis and autophagy into their life cycle. For instance, HIV-1 Gag associates with LC3 for trafficking while Nef inhibits autophagosome maturation, preventing degradation and enhancing release (Ojha et al., 2017). Similarly, HCV and various herpesviruses employ these pathways to disseminate viral components with immunomodulatory effects (Meckes & Raab-Traub, 2011).

The intersection presents multiple therapeutic intervention points. Strategies to restore autophagic degradation, inhibit vesicle biogenesis, or prevent amphisome diversion, could disrupt viral dissemination and re-establish immune control. The challenge lies in targeting these pathways without compromising essential physiological functions, but their shared vulnerability makes this attractive for antiviral research (van Niel et al., 2018).

6. Therapeutic opportunities and future directions in EV-targeted interventions

The processes that govern the biogenesis, loading and release of EVs are connected with central nodes of viral biology, such as vesicular trafficking, membrane remodeling and secretory autophagy, creating potential host-targeted therapeutic opportunities (van Niel et al., 2022; O'Donogue & Krachler, 2021). These interconnected processes offer advantages over virus-specific treatments, notably reduced viral resistance, broader antiviral spectrum, and potential EV-based therapeutic delivery. However, clinical translation faces challenges including off-target toxicity due to shared human pathways, compensatory pathway activation, and insufficient drug specificity. Success requires balancing between the treatment's safety and antiviral effectiveness, potentially through tissue-specific or time-limited interventions (Urbanelli et al., 2023).

6.1 Molecular targets

Among the most studied molecular targets, Rab27a holds a prominent position. It is a small GTPase essential for EV trafficking and fusion with the plasma membrane, whose role in viral propagation has been extensively documented for pathogens like IAV, HIV-1, HCV, HEV, enterovirus A71 and CMV (Ostrowski et al., 2010; Kowal et al., 2014). Rab27a's activity depends on interaction with effectors like SYTL4, whose depletion has proven even more protective than Rab27a inhibition itself in IAV models (Chen et al., 2025). The protein KIBRA is also a target of interest as it stabilizes Rab27a and prevents its degradation. This interaction modulates the size and number of MVBs, consequently modulating EV secretion (Song et al., 2019). However, the essential role of Rab27a in intercellular communication and regulation of the tumor microenvironment requires careful consideration.

The ESCRT system represents another point of intervention considering its functions in both exosome biogenesis and viral budding (e.g., HIV-1, the filoviruses Ebola and Marburg) (Votteler & Sundquist, 2013; Carlton & Martin-Serrano, 2009). Interfering with the ESCRT system could prevent vEV release but risks disrupting cellular processes like cell division and breaking down faulty proteins.

Secretory autophagy components offer less-explored targets: modulating Sec22b or Rab8a could enhance viral release frequency while preserving canonical autophagy (Jeppesen et al., 2019). However, these proteins participate in fundamental immune signaling, necessitating treatments that avoid interfering with antiviral immunity (ten Broeke et al., 2013). Emerging solutions include targeted nanocarriers that specifically deliver substances to block cell actions, or systems that can alter genes in specific organs or tissues to minimize systemic toxicity (Gao et al., 2018; Lu et al., 2023).

6.2 EVs as biomarkers: Potential applications and limitations

EVs not only help viruses to spread, but also provide useful information for disease diagnosis through their cargo composition (Kalluri & LeBleu, 2020; Théry et al., 2018). Their accessibility in body fluids such as plasma, urine, and saliva enables non-invasive monitoring with reduced procedural risks and discomfort compared to conventional diagnostic methods. In oncology, EV-based biomarkers have demonstrated utility for the early detection of malignancies, prediction of recurrence, and identification of occult infections (Chen et al., 2018).

However, EV-based diagnosis faces some technical challenges. Sample collection methods can influence both EV quality and cargo composition, introducing analytical variability (Kumar et al., 2024). EV heterogeneity also complicates their characterization due to overlapping markers between EV subtypes and cellular processes of origin (e.g., CD63, ALIX, LC3). Moreover, EV instability ex vivo requires standardized collection and storage procedures to maintain cargo integrity (Willms et al., 2018). To address these challenges, it is essential to adhere to standardized protocols, conduct multi-site studies, and apply advanced analytical methodologies, including machine learning integration of multiple indicators to enhance diagnostic accuracy (Albanese et al., 2021; Shao et al., 2012).

6.3 Engineered EVs: Advanced platforms for therapeutic applications

Engineered EVs offer unique advantages as precision therapeutic delivery systems due to their endogenous membrane composition, which confers biocompatibility and reduced immunogenicity compared to synthetic carriers (Armstrong & Stevens, 2018). In addition to carrying conventional cargos, such as mRNA and siRNA, these vesicles can be functionalized with targeting ligands, fusogenic proteins, or cell-penetrating domains for enhanced cellular uptake and specificity (Vader et al., 2016).

Recent advances demonstrated that EVs can incorporate complete CRISPR-Cas systems, enabling precise genome editing applications (O'Brien et al., 2020). Notably, engineered exosomes have shown success as antiviral delivery platforms, successfully delivering siRNA against HCV and CRISPR systems for HBV targeting (Owliaee et al., 2023; Schwarz et al., 2025; Zeng et al., 2024). EVs can also display viral antigens and mimic the virus itself, potentially eliciting stronger immune responses than traditional vaccine platforms (Liang et al., 2021). Moreover, EVs can expose on their surface receptors such as ACE2/TMPRSS2 (SARS-CoV-2) serving as a decoy to trap viruses and reduce infection (Ching et al., 2022; El-Shennawy, et al., 2022). Nevertheless, caution is warranted, as under certain conditions this therapeutic decoy strategy may enhance infection rather than inhibit it (Tey et al., 2022).

Clinical translation requires addressing scalable production with strict quality control to reduce batch effects, improving lyophilisation techniques for long-term stability, and standardized potency assays (Lamichhane & Jay, 2018). Regulatory approval will necessitate comprehensive studies of EV safety, efficacy, and pharmacokinetics. Despite these challenges, engineered EVs have the potential to revolutionize the development of vaccines and other therapies in genetic and oncological diseases (Huang et al., 2023; Mendt et al., 2019) (Fig. 3).

6.4 Emerging technologies for the study of EVs and autophagosomes

The characterization of EVs and autophagosomes represents a complex technological challenge due to size heterogeneity and overlap between the types of vesicles. While EVs generally range in size from 30 to 200 nanometres, autophagosomes can reach sizes between 300 and 1500 nm,

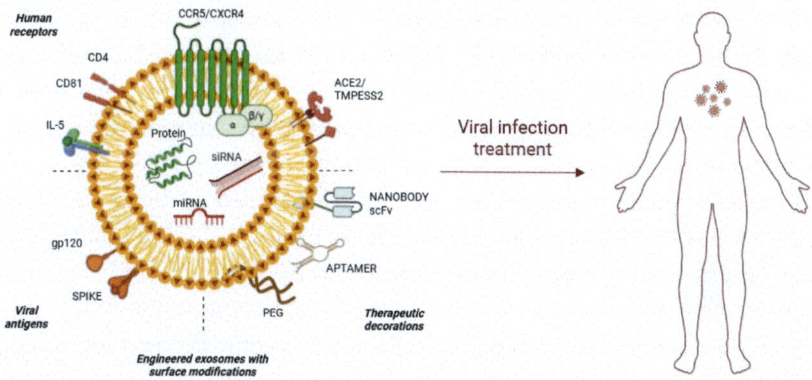

Fig. 3 Exosomal surface engineering for precision targeting in viral infection. Schematic representation of EV-based antiviral strategies. EVs can be engineered to display viral entry receptors on their surface, acting as decoys that capture and neutralize free virions by preventing them from interacting with target cells. Examples include EVs that expose ACE2/TMPRSS2 (SARS-CoV-2), CD4/CCR5/CXCR4 (HIV), CD81, SR-B1, CLDN1, and OCLN (HCV), or sialic acid residues (influenza virus). Alternatively, EVs can be designed to deliver antiviral agents to infected or immune cells, thereby enhancing therapeutic efficacy. This can be achieved by decorating the surface with ligands, aptamers, or antibody fragments to direct the EVs towards infected cells [e.g., anti-Spike EVs for SARS-CoV-2 or anti-gp120 EVs for HIV] or immune cells (e.g., EVs carrying CD40/CD86 ligands or IL-15 to stimulate dendritic, NK, or T cells). EV, extracellular vesicles; HCV, hepatitis C virus; HIV, human immunodeficiency virus; SARS-CoV-2, severe acute respiratory syndrome coronavirus 2. *The figure was created with BioRender.com.*

with significant overlap from fragments and amphisomes (Zhang et al., 2018). This complexity means that traditional isolation methods, such as ultracentrifugation, are often inadequate as they do not always guarantee complete purity. This exposes samples to the risk of contamination by membrane fragments, lipoproteins, or organelles, compromising subsequent analyses (Jeppesen et al., 2019).

To overcome these difficulties, the scientific community is adopting a combined approach that integrates physical separation techniques with advanced imaging technologies. In particular, cryo-electron microscopy (cryo-EM) and cryo-electron tomography (cryo-ET) enable direct visualization of membrane architecture and protein interactions at nanometric level without chemical fixation artefacts (Wan & Briggs, 2016). These methods, together with correlative light-electron microscopy (CLEM) techniques, offer dynamic insights into fusion, internalization, and trafficking processes, using bioorthogonal labeling (Scorrano et al., 2019).

Advances in proteomics such as data-independent acquisition (DIA) method and its variant known as sequential window acquisition of all theoretical mass spectra (SWATH-MS) have improved reproducibility and detection sensitivity. The development of single-particle proteomics, which uses microchip-based systems or high-sensitivity immunofluorescence methods, allows the identification of EV subsets, including vEVs, that would be masked in bulk analyses (Jiang et al., 2017; Leidal & Debnath, 2021).

Despite these improvements, challenges remain including standardized isolation procedures, ion suppression in mass spectrometry, and variability in analyzing single vesicles. Nevertheless, clinical applications appear promising with single-vesicle molecular profiling potentially revolutionizing early viral infection diagnosis and enabling targeted therapeutic interventions (Melo et al., 2015; Zhang et al., 2019).

7. Conclusions

Autophagy has emerged as a critical process exploited by viruses to coordinate replication, facilitate non-lytic release, and evade immunosurveillance via EV-mediated dissemination (Paul & Bartenschlager, 2013). Rather than just modifying individual components like Rab27a or ESCRT, viruses create sophisticated intracellular environments that simultaneously support viral replication while minimizing cellular damage and immune detection (Wan & Briggs, 2016).

The discovery that EVs can carry immunomodulatory proteins and viral RNAs opens novel therapeutic and diagnostic opportunities, including EV-based vaccines, delivery of antiviral RNAs, and EV biomarker development (Albanese et al., 2021; Urbanelli et al., 2019).

However, there are still major technical obstacles to overcome, notably for the purification and enrichment of EVs, uneasy due to EV heterogeneity (overlap between EV and autophagosomes markers), contamination (separate pure structures from cellular fragments), and resolution limits of conventional analytical techniques. Modern technologies are contributing to advance our understanding of these complex mechanisms, notably real-time analysis, single-vesicle proteomics, and super-resolution imaging (Li et al., 2015).

Future work must determine whether the release of virus-loaded EVs can be selectively blocked by targeting secretory autophagy without compromising canonical autophagy. Promising strategies include the use of allosteric inhibitors, tissue-targeted nanocarriers, or CRISPR-Cas9 and PROTAC-based interventions, designed to act reversibly and locally (Lu et al., 2023). Concurrently, EVs themselves could be used as diagnostic biomarkers to predict relapses or as therapeutic platforms, carrying antigens or antiviral molecules (Gao et al., 2025).

In conclusion, understanding the interface between autophagy and EVs not only enriches fundamental biology but also provides a strong foundation for developing next-generation antiviral treatments and diagnostic tools. Bridging cell biology, molecular engineering, and omics technologies will be essential to turn these mechanisms into precision medicine tools against viral diseases.

Funding

JP is supported by the ANRS | Maladies infectieuses émergentes (ANRS MIE); SIRIC Cancer Research and Personalized Medicine (CARPEM); Domaine de recherche et d'Innovation Majeur – Immunothérapies, Auto-immunité et Cancer (DIM-ITAC, Région Île-de-France); Institut National du Cancer (INCa); and Multi-Organism Institute (ITMO) Aviesan Cancer (National Alliance for Life Sciences and Health).

Conflicts of interest

MCM and FDEDP are the inventors of a patent covering therapeutic targeting of cancer, not relevant with this study. JP is the inventor of patents covering the diagnosis, prognosis, and treatment of cancers, including patents on oncolytic viruses licensed to Turnstone Biologics. LP is a scientific co-founder of Kimera srl. The funders had no role in the design of the study; in the writing of the manuscript, or in the decision to publish the results.

Authors' contributions

MCM conceived and designed the study. JP and MCM supervised the project. LVT and ADM wrote the first version of the manuscript and prepared the figures. MCM, JP, MDM, LP, MLG, MV, FS and FDEDP provided intellectual input and edited the paper. All authors read and approved the final manuscript.

References

Abernathy, E., Mateo, R., Majzoub, K., van Buuren, N., Bird, S. W., Carette, J. E., & Kirkegaard, K. (2019). Differential and convergent utilization of autophagy components by positive-strand RNA viruses. *PLoS Biology, 17*(1), e2006926.

Acharya, D., Sayyad, Z., Hoenigsperger, H., Hirschenberger, M., Zurenski, M., Balakrishnan, K., ... Gack, M. U. (2025). Publisher Correction: TRIM23 mediates cGAS-induced autophagy in anti-HSV defense. *Nature Communications, 16*(1), 6114.

Agudelo, J. S. H., Pereira, G., & Fernandes, C. J. D. C. (2025). Role of extracellular vesicles in severe dengue: Virus-host interactions and biomarker potential. *Viruses, 17*(6), 807.

Ahat, E., Bui, S., Zhang, J., da Veiga Leprevost, F., Sharkey, L., Reid, W., ... Wang, Y. (2022). GRASP55 regulates the unconventional secretion and aggregation of mutant huntingtin. *The Journal of Biological Chemistry, 298*(8), 102219.

Akers, J. C., Gonda, D., Kim, R., Carter, B. S., & Chen, C. C. (2013). Biogenesis of extracellular vesicles (EV): Exosomes, microvesicles, retrovirus-like vesicles, and apoptotic bodies. *Journal of Neuro-oncology, 113*(1), 1–11.

Albanese, M., Chen, Y. A., Hüls, C., Gärtner, K., Tagawa, T., Mejias-Perez, E., ... Hammerschmidt, W. (2021). MicroRNAs are minor constituents of extracellular vesicles that are rarely delivered to target cells. *PLoS Genetics, 17*(12), e1009951.

Alenquer, M., & Amorim, M. J. (2015). Exosome biogenesis, regulation, and function in viral infection. *Viruses, 7*(9), 5066–5083.

Armstrong, J. P. K., & Stevens, M. M. (2018). Strategic design of extracellular vesicle drug delivery systems. *Advanced Drug Delivery Reviews, 130*, 12–16.

Axe, E. L., Walker, S. A., Manifava, M., Chandra, P., Roderick, H. L., Habermann, A., ... Ktistakis, N. T. (2008). Autophagosome formation from membrane compartments enriched in phosphatidylinositol 3-phosphate and dynamically connected to the endoplasmic reticulum. *The Journal of Cell Biology, 182*(4), 685–701.

Barberis, E., Vanella, V. V., Falasca, M., Caneapero, V., Cappellano, G., Raineri, D., ... Manfredi, M. (2021). Circulating exosomes are strongly involved in SARS-CoV-2 infection. *Frontiers in Molecular Biosciences, 8*, 632290.

Bare, Y., Defourny, K., Bretou, M., Van Niel, G., Nolte-'t Hoen, E., & Gaudin, R. (2025). The endoplasmic reticulum as a cradle for virus and extracellular vesicle secretion. *Trends in Cell Biology, 35*(4), 282–293.

Beale, R., Wise, H., Stuart, A., Ravenhill, B. J., Digard, P., & Randow, F. (2014). A LC3-interacting motif in the influenza A virus M2 protein is required to subvert autophagy and maintain virion stability. *Cell Host & Microbe, 15*(2), 239–247. https://doi.org/10.1016/j.chom.2014.01.006 PMID: 24528869; PMCID: PMC3991421.

Bello-Morales, R., & López-Guerrero, J. A. (2018). Extracellular vesicles in herpes viral spread and immune evasion. *Frontiers in Microbiology, 9*, 2572.

Bernard, M. A., Zhao, H., Yue, S. C., Anandaiah, A., Koziel, H., & Tachado, S. D. (2014). Novel HIV-1 miRNAs stimulate TNFα release in human macrophages via TLR8 signaling pathway. *PLoS One, 9*(9), e106006.

Bidet, K., Dadlani, D., & Garcia-Blanco, M. A. (2014). G3BP1, G3BP2 and CAPRIN1 are required for translation of interferon stimulated mRNAs and are targeted by a dengue virus non-coding RNA. *PLoS Pathogens, 10*(7), e1004242.

Birgisdottir, Å. B., & Johansen, T. (2020). Autophagy and endocytosis - interconnections and interdependencies. *Journal of Cell Science, 133*(10), jcs228114.

Blanc, L., & Vidal, M. (2018). New insights into the function of Rab GTPases in the context of exosomal secretion. *Small GTPases, 9*(1-2), 95–106.

Bukong, T. N., Momen-Heravi, F., Kodys, K., Bala, S., & Szabo, G. (2014). Exosomes from hepatitis C infected patients transmit HCV infection and contain replication competent viral RNA in complex with Ago2-miR122-HSP90. *PLoS Pathogens, 10*(10), e1004424.

Campbell, T. D., Khan, M., Huang, M. B., Bond, V. C., & Powell, M. D. (2008). HIV-1 Nef protein is secreted into vesicles that can fuse with target cells and virions. *Ethnicity & Disease, 18*(2), S2 14-9.

Canitano, A., Venturi, G., Borghi, M., Ammendolia, M. G., & Fais, S. (2013). Exosomes released in vitro from Epstein–Barr virus (EBV)-infected cells contain EBV-encoded latent phase mRNAs. *Cancer Letters, 337*(2), 193–199.

Caobi, A., Nair, M., & Raymond, A. D. (2020). Extracellular vesicles in the pathogenesis of viral infections in humans. *Viruses, 12*(10), 1200.

Carlton, J. G., & Martin-Serrano, J. (2009). The ESCRT machinery: New functions in viral and cellular biology. *Biochemical Society Transactions, 37*, 195–199 Pt 1.

Castro-Gonzalez, S., Shi, Y., Colomer-Lluch, M., Song, Y., Mowery, K., Almodovar, S., ... Serra-Moreno, R. (2021). HIV-1 Nef counteracts autophagy restriction by enhancing the association between BECN1 and its inhibitor BCL2 in a PRKN-dependent manner. *Autophagy, 17*(2), 553–577.

Chatterjee, S., Kordbacheh, R., & Sin, J. (2024). Extracellular vesicles: A novel mode of viral propagation exploited by enveloped and non-enveloped viruses. *Microorganisms, 12*(2), 274.

Chaumorcel, M., Lussignol, M., Mouna, L., Cavignac, Y., Fahie, K., Cotte-Laffitte, J., ... Esclatine, A. (2012). The human cytomegalovirus protein TRS1 inhibits autophagy via its interaction with Beclin 1. *Journal of Virology, 86*(5), 2571–2584.

Chen, G., Huang, A. C., Zhang, W., Zhang, G., Wu, M., Xu, W., ... Guo, W. (2018). Exosomal PD-L1 contributes to immunosuppression and is associated with anti-PD-1 response. *Nature, 560*(7718), 382–386.

Chen, T., Ouyang, A., Zou, J., Feng, Y., Wu, C., Jiang, M., ... Zhou, H. (2025). Rab27a regulates the transport of influenza virus membrane proteins to the plasma membrane. *Nature Communications, 16*(1), 6271.

Chen, Q., Sun, L., & Chen, Z. J. (2016). Regulation and function of the cGAS-STING pathway of cytosolic DNA sensing. *Nature Immunology, 17*(10), 1142–1149.

Chen, D., Zheng, Q., Sun, L., Ji, M., Li, Y., Deng, H., & Zhang, H. (2021). ORF3a of SARS-CoV-2 promotes lysosomal exocytosis-mediated viral egress. *Developmental Cell, 56*(23), 3250–3263. https://doi.org/10.1016/j.devcel.2021.10.006 Epub 2021 Oct 11. PMID: 34706264; PMCID: PMC8502680.

Ching, K. L., de Vries, M., Gago, J., Dancel-Manning, K., Sall, J., Rice, W. J., ... Cadwell, K. (2022). ACE2-containing defensosomes serve as decoys to inhibit SARS-CoV-2 infection. *PLoS Biology, 20*(9), e3001754. https://doi.org/10.1371/journal.pbio.3001754 PMID: 36099266; PMCID: PMC9469972.

Chu, Y. D., Chen, M. C., Yeh, C. T., & Lai, M. W. (2024). Hijacking host extracellular vesicle machinery by hepatotropic viruses: Current understandings and future prospects. *Journal of Biomedical Science, 31*(1), 97.

Cloherty, A. P. M., Rader, A. G., Patel, K. S., Eisden, T. T. H. D., van Piggelen, S., Schreurs, R. R. C. E., & Ribeiro, C. M. S. (2024). Dengue virus exploits autophagy vesicles and secretory pathways to promote transmission by human dendritic cells. *Frontiers in Immunology, 15*, 1260439.

Colombo, M., Raposo, G., & Théry, C. (2014). Biogenesis, secretion, and intercellular interactions of exosomes and other extracellular vesicles. *Annual Review of Cell and Developmental Biology, 30*, 255–289.

Crenshaw, B. J., Jones, L. B., Bell, C. R., Kumar, S., & Matthews, Q. L. (2019). Perspective on adenoviruses: Epidemiology, pathogenicity, and gene therapy. *Biomedicines*, 7(3), 61.

Dantuma, N. P., & Masucci, M. G. (2003). The ubiquitin/proteasome system in Epstein-Barr virus latency and associated malignancies. *Seminars in Cancer Biology, 13*(1), 69–76.

Debnath, J., & Leidal, A. M. (2022). Secretory autophagy during lysosome inhibition (SALI). *Autophagy, 18*(10), 2498–2499.

Deretic, V., Saitoh, T., & Akira, S. (2013). Autophagy in infection, inflammation and immunity. *Nature Reviews. Immunology, 13*(10), 722–737.

Di Rienzo, M., Romagnoli, A., Antonioli, M., Piacentini, M., & Fimia, G. M. (2020). TRIM proteins in autophagy: Selective sensors in cell damage and innate immune responses. *Cell Death and Differentiation, 27*(3), 887–902.

Dreux, M., Garaigorta, U., Boyd, B., Décembre, E., Chung, J., Whitten-Bauer, C., ... Chisari, F. V. (2012). Short-range exosomal transfer of viral RNA from infected cells to plasmacytoid dendritic cells triggers innate immunity. *Cell Host & Microbe, 12*(4), 558–570.

Dupont, N., Jiang, S., Pilli, M., Ornatowski, W., Bhattacharya, D., & Deretic, V. (2011). Autophagy-based unconventional secretory pathway for extracellular delivery of IL-1β. *The EMBO Journal, 30*(23), 4701–4711.

Durgan, J., Lystad, A. H., Sloan, K., Carlsson, S. R., Wilson, M. I., Marcassa, E., ... Florey, O. (2021). Non-canonical autophagy drives alternative ATG8 conjugation to phosphatidylserine. *Molecular Cell, 81*(9), 2031–2040.e8. https://doi.org/10.1016/j.molcel.2021.03.020 Epub 2021 Apr 27. PMID: 33909989; PMCID: PMC8122138.

El-Shennawy, L., Hoffmann, A. D., Dashzeveg, N. K., McAndrews, K. M., Mehl, P. J., Cornish, D., ... Liu, H. (2022). Circulating ACE2-expressing extracellular vesicles block broad strains of SARS-CoV-2. *Nature Communications, 13*(1), 405. https://doi.org/10.1038/s41467-021-27893-2 PMID: 35058437; PMCID: PMC8776790.

English, L., Chemali, M., Duron, J., Rondeau, C., Laplante, A., Gingras, D., ... Desjardins, M. (2009). Autophagy enhances the presentation of endogenous viral antigens on MHC class I molecules during HSV-1 infection. *Nature Immunology, 10*(5), 480–487.

Fader, C. M., & Colombo, M. I. (2009). Autophagy and multivesicular bodies: Two closely related partners. *Cell Death and Differentiation, 16*(1), 70–78.

Feng, Z., Hensley, L., McKnight, K. L., Hu, F., Madden, V., Ping, L., ... Lemon, S. M. (2013). A pathogenic picornavirus acquires an envelope by hijacking cellular membranes. *Nature, 496*(7445), 367–371.

Feng, H., Wang, N., Zhang, N., & Liao, H. H. (2022). Alternative autophagy: Mechanisms and roles in different diseases. *Cell Communication and Signaling: CCS, 20*(1), 43.

Figueras-Novoa, C., Timimi, L., Marcassa, E., Ulferts, R., & Beale, R. (2024). Conjugation of ATG8s to single membranes at a glance. *Journal of Cell Science, 137*(15), jcs261031. https://doi.org/10.1242/jcs.261031 Epub 2024 Aug 15. PMID: 39145464; PMCID: PMC11361636.

Fischer, T. D., Wang, C., Padman, B. S., Lazarou, M., & Youle, R. J. (2020). STING induces LC3B lipidation onto single-membrane vesicles via the V-ATPase and ATG16L1-WD40 domain. *The Journal of Cell Biology, 219*(12), e202009128. https://doi.org/10.1083/jcb.202009128 PMID: 33201170; PMCID: PMC7716379.

Fletcher, K., Ulferts, R., Jacquin, E., Veith, T., Gammoh, N., Arasteh, J. M., ... Florey, O. (2018). The WD40 domain of ATG16L1 is required for its non-canonical role in lipidation of LC3 at single membranes. *The EMBO Journal, 37*(4), e97840. https://doi.org/10.15252/embj.201797840 Epub 2018 Jan 9. PMID: 29317426; PMCID: PMC5813257.

Galluzzi, L., Baehrecke, E. H., Ballabio, A., Boya, P., Bravo-San Pedro, J. M., Cecconi, F., ... Kroemer, G. (2017). Molecular definitions of autophagy and related processes. *The EMBO Journal, 36*(13), 1811–1836.

Ganesan, D., & Cai, Q. (2021). Understanding amphisomes. *The Biochemical Journal, 478*(10), 1959–1976.

Gannagé, M., Dormann, D., Albrecht, R., Dengjel, J., Torossi, T., Rämer, P. C., ... Münz, C. (2009). Matrix protein 2 of influenza A virus blocks autophagosome fusion with lysosomes. *Cell Host & Microbe, 6*(4), 367–380.

Gao, W., Chen, Y., Zhang, Y., Zhang, Q., & Zhang, L. (2018). Nanoparticle-based local antimicrobial drug delivery. *Advanced Drug Delivery Reviews, 127*, 46–57.

Gao, Q., Zhan, Y., Zhang, J., Sun, D., Xiang, H., Bao, C., Gao, Q., Zhang, M., Wang, J., & Mao, L. (2025). The diagnostic potential of extracellular vesicles in virus-related diseases. *Frontiers in Cellular and Infection Microbiology, 15*, 1641405.

Gardner, J. O., Leidal, A. M., Nguyen, T. A., & Debnath, J. (2023). LC3-dependent EV loading and secretion (LDELS) promotes TFRC (transferrin receptor) secretion via extracellular vesicles. *Autophagy, 19*(5), 1551–1561.

Ginini, L., Billan, S., Fridman, E., & Gil, Z. (2022). Insight into extracellular vesicle-cell communication: From cell recognition to intracellular fate. *Cells, 11*(9), 1375.

Granato, M., Santarelli, R., Farina, A., Gonnella, R., Lotti, L. V., Faggioni, A., & Cirone, M. (2014). Epstein-barr virus blocks the autophagic flux and appropriates the autophagic machinery to enhance viral replication. *Journal of Virology, 88*(21), 12715–12726.

Groß, R., Reßin, H., von Maltitz, P., Albers, D., Schneider, L., Bley, H., ... Müller, J. A. (2024). Phosphatidylserine-exposing extracellular vesicles in body fluids are an innate defence against apoptotic mimicry viral pathogens. *Nature Microbiology, 9*(4), 905–921. https://doi.org/10.1038/s41564-024-01637-6 Epub 2024 Mar 25. PMID: 38528146; PMCID: PMC10994849.

Gui, X., Yang, H., Li, T., Tan, X., Shi, P., Li, M., Du, F., & Chen, Z. J. (2019). Autophagy induction via STING trafficking is a primordial function of the cGAS pathway. *Nature, 567*(7747), 262–266.

Hadizadeh, N., Bagheri, D., Shamsara, M., Hamblin, M. R., Farmany, A., Xu, M., ... Hashemi, E. (2022). Extracellular vesicles biogenesis, isolation, manipulation and genetic engineering for potential in vitro and in vivo therapeutics: An overview. *Frontiers in Bioengineering and Biotechnology, 10*, 1019821.

Hambo, S., & Harb, H. (2023). Extracellular vesicles and their role in lung infections. *International Journal of Molecular Sciences, 24*(22), 16139.

Hooper, K. M., Jacquin, E., Li, T., Goodwin, J. M., Brumell, J. H., Durgan, J., & Florey, O. (2022). V-ATPase is a universal regulator of LC3-associated phagocytosis and non-canonical autophagy. *The Journal of Cell Biology, 221*(6), e202105112. https://doi.org/10.1083/jcb.202105112 Epub 2022 May 5. PMID: 35511089; PMCID: PMC9082624.

Huang, T., Sun, C., Du, F., & Chen, Z. J. (2025). STING-induced noncanonical autophagy regulates endolysosomal homeostasis. *Proceedings of the National Academy of Sciences of the United States of America, 122*(8), e2415422122.

Huang, L., Wu, E., Liao, J., Wei, Z., Wang, J., & Chen, Z. (2023). Research advances of engineered exosomes as drug delivery carrier. *ACS Omega, 8*(46), 43374–43387.

Huang, L., & Yue, J. (2020). The interplay of autophagy and enterovirus. *Seminars in Cell & Developmental Biology, 101*, 12–19.

Ipinmoroti, A. O., Crenshaw, B. J., Pandit, R., Kumar, S., Sims, B., & Matthews, Q. L. (2021). Human adenovirus serotype 3 infection modulates the biogenesis and composition of lung cell-derived extracellular vesicles. *Journal of Immunological Research*, 2958394.

Itakura, E., Kishi-Itakura, C., & Mizushima, N. (2012). The hairpin-type tail-anchored SNARE syntaxin 17 targets to autophagosomes for fusion with endosomes/lysosomes. *Cell, 151*(6), 1256–1269.

Jackson, W. T. (2015). Viruses and the autophagy pathway. *Virology, 479-480*, 450–456.
Jassey, A., & Jackson, W. T. (2024). Viruses and autophagy: Bend, but don't break. *Nature Reviews. Microbiology, 22*(5), 309–321.
Jeppesen, D. K., Fenix, A. M., Franklin, J. L., Higginbotham, J. N., Zhang, Q., Zimmerman, L. J., ... Coffey, R. J. (2019). Reassessment of exosome composition. *Cell, 177*(2), 428–445.e18.
Jiang, P., Nishimura, T., Sakamaki, Y., Itakura, E., Hatta, T., Natsume, T., & Mizushima, N. (2014). The HOPS complex mediates autophagosome-lysosome fusion through interaction with syntaxin 17. *Molecular Biology of the Cell, 25*(8), 1327–1337.
Jiang, L., Vader, P., & Schiffelers, R. M. (2017). Extracellular vesicles for nucleic acid delivery: Progress and prospects for safe RNA-based gene therapy. *Gene Therapy, 24*(3), 157–166.
Kalamvoki, M., Du, T., & Roizman, B. (2014). Cells infected with herpes simplex virus 1 export to uninfected cells exosomes containing STING, viral mRNAs, and microRNAs. *Proceedings of the National Academy of Sciences of the United States of America, 111*(46), E4991–E4996.
Kalluri, R., & LeBleu, V. S. (2020). The biology, function, and biomedical applications of exosomes. *Science (New York, N. Y.), 367*(6478), eaau6977.
Kapoor, N. R., Chadha, R., Kumar, S., Choedon, T., Reddy, V. S., & Kumar, V. (2017). The HBx gene of hepatitis B virus can influence hepatic microenvironment via exosomes by transferring its mRNA and protein. *Virus Research, 240*, 166–174.
Kaur, N., Carlsson, S. R., & Lystad, A. H. (2024). The separate axes of TECPR1 and ATG16L1 in CASM. *Autophagy, 20*(1), 214–215. https://doi.org/10.1080/15548627.2023.2255462 Epub 2023 Sep 7. PMID: 37676042; PMCID: PMC10761082.
Ke, P. Y. (2024). Molecular mechanism of autophagosome-lysosome fusion in mammalian cells. *Cells, 13*(6), 500 13.
Khatua, A. K., Taylor, H. E., Hildreth, J. E., & Popik, W. (2009). Exosomes packaging APOBEC3G confer human immunodeficiency virus resistance to recipient cells. *Journal of Virology, 83*(2), 512–521.
Kimura, S., Noda, T., & Yoshimori, T. (2008). Dynein-dependent movement of autophagosomes mediates efficient encounters with lysosomes. *Cell Structure and Function, 33*(1), 109–122.
Klionsky, D. J., et al. (2021). Guidelines for the use and interpretation of assays for monitoring autophagy (4th edition)[1]. *Autophagy, 17*(1), 1–382.
Kouwaki, T., Okamoto, M., Tsukamoto, H., Fukushima, Y., & Oshiumi, H. (2017). Extracellular vesicles deliver host and virus rna and regulate innate immune response. *International Journal of Molecular Sciences, 18*(3), 666.
Kowal, J., Tkach, M., & Théry, C. (2014). Biogenesis and secretion of exosomes. *Current Opinion in Cell Biology, 29*, 116–125.
Krause, M., Samolej, J., Yakimovich, A., Kriston-Vizi, J., Huttunen, M., Lara-Reyna, S., ... Mercer, J. (2024). Vaccinia virus subverts xenophagy through phosphorylation and nuclear targeting of p62. *The Journal of Cell Biology, 223*(6), e202104129.
Kumar, M. A., Baba, S. K., Sadida, H. Q., Marzooqi, S. A., Jerobin, J., Altemani, F. H., ... Bhat, A. A. (2024). Extracellular vesicles as tools and targets in therapy for diseases. *Signal Transduction and Targeted Therapy, 9*(1), 27.
Kumar, S., Gu, Y., Abudu, Y. P., Bruun, J. A., Jain, A., Farzam, F., ... Deretic, V. (2019). Phosphorylation of syntaxin 17 by TBK1 controls autophagy initiation. *Developmental Cell, 49*(1), 130–144 e6.
Lai, J. K., Sam, I. C., & Chan, Y. F. (2016). The autophagic machinery in enterovirus infection. *Viruses, 8*(2), 32.
Lamichhane, T. N., & Jay, S. M. (2018). Production of extracellular vesicles loaded with therapeutic cargo. *Methods in Molecular Biology, 1831*, 37–47.

Latanova, A., Karpov, V., & Starodubova, E. (2024). Extracellular vesicles in flaviviridae pathogenesis: Their roles in viral transmission, immune evasion, and inflammation. *International Journal of Molecular Sciences, 25*(4), 2144.

Lau, N. C. H., & Yam, J. W. P. (2023). From exosome biogenesis to absorption: Key takeaways for cancer research. *Cancers (Basel), 15*(7), 1992.

Leidal, A. M., & Debnath, J. (2020). LC3-dependent extracellular vesicle loading and secretion (LDELS). *Autophagy, 16*(6), 1162–1163. https://doi.org/10.1080/15548627.2020.1756557 Epub 2020 Apr 24. PMID: 32330402; PMCID: PMC7469500.

Leidal, A. M., & Debnath, J. (2021). Emerging roles for the autophagy machinery in extracellular vesicle biogenesis and secretion. *FASEB Bioadv, 3*(5), 377–386.

Lenassi, M., Cagney, G., Liao, M., Vaupotic, T., Bartholomeeusen, K., Cheng, Y., ... Peterlin, B. M. (2010). HIV Nef is secreted in exosomes and triggers apoptosis in bystander CD4+ T cells. *Traffic (Copenhagen, Denmark), 11*(1), 110–122.

Levine, B., & Kroemer, G. (2019). Biological functions of autophagy genes: A disease perspective. *Cell, 176*(1-2), 11–42.

Li, Q., Peng, G., Liu, H., Wang, L., Lu, R., & Li, L. (2024). Molecular mechanisms of secretory autophagy and its potential role in diseases. *Life Sciences, 347*, 122653.

Li, J., Prange, R., & Lu, M. (2025). ESCRT machinery in HBV life cycle: Dual roles in autophagy and membrane dynamics for viral pathogenesis. *Cells, 14*(8), 603.

Li, X. D., Sun, L., Seth, R. B., Pineda, G., & Chen, Z. J. (2005). Hepatitis C virus protease NS3/4A cleaves mitochondrial antiviral signaling protein off the mitochondria to evade innate immunity. *Proceedings of the National Academy of Sciences of the United States of America, 102*(49), 17717–17722.

Li, L., Tan, J., Miao, Y., Lei, P., & Zhang, Q. (2015). ROS and autophagy: Interactions and molecular regulatory mechanisms. *Cellular and Molecular Neurobiology, 35*(5), 615–621.

Liang, Y., Duan, L., Lu, J., & Xia, J. (2021). Engineered EVs and virus. *Theranostics, 11*(7), 3183–3195.

Lin, P. W., Chu, M. L., Liu, Y. W., Chen, Y. C., Shih, Y. H., Lan, S. H., ... Lee, Y. R. (2024). Revealing potential Rab proteins participate in regulation of secretory autophagy machinery. *The Kaohsiung Journal of Medical Sciences, 40*(7), 642–649.

Liu, S., Cai, X., Wu, J., Cong, Q., Chen, X., Li, T., ... Chen, Z. J. (2015). Phosphorylation of innate immune adaptor proteins MAVS, STING, and TRIF induces IRF3 activation. *Science (New York, N. Y.), 347*(6227), aaa2630.

Longatti, A., & Tooze, S. A. (2009). Vesicular trafficking and autophagosome formation. *Cell Death and Differentiation, 16*(7), 956–965.

Lu, Y., Godbout, K., Lamothe, G., & Tremblay, J. P. (2023). CRISPR-Cas9 delivery strategies with engineered extracellular vesicles. *Molecular Therapy Nucleic Acids, 34*, 102040.

Ma, W., Lu, Y., Jin, X., Lin, N., Zhang, L., & Song, Y. (2024). Targeting selective autophagy and beyond: From underlying mechanisms to potential therapies. *Journal of Advanced Research, 65*, 297–327.

Mandell, M. A., Kimura, T., Jain, A., Johansen, T., & Deretic, V. (2014). TRIM proteins regulate autophagy: TRIM5 is a selective autophagy receptor mediating HIV-1 restriction. *Autophagy, 10*(12), 2387–2388.

Manokaran, G., Finol, E., Wang, C., Gunaratne, J., Bahl, J., Ong, E. Z., ... Ooi, E. E. (2015). Dengue subgenomic RNA binds TRIM25 to inhibit interferon expression for epidemiological fitness. *Science (New York, N. Y.), 350*(6257), 217–221.

McKnight, K. L., & Lemon, S. M. (2018). Hepatitis A virus genome organization and replication strategy. *Cold Spring Harbor Perspectives in Medicine, 8*(12), a033480.

Meckes, D. G., Jr, & Raab-Traub, N. (2011). Microvesicles and viral infection. *Journal of Virology, 85*(24), 12844–12854.

Melo, S. A., Luecke, L. B., Kahlert, C., Fernandez, A. F., Gammon, S. T., Kaye, J., ... Kalluri, R. (2015). Glypican-1 identifies cancer exosomes and detects early pancreatic cancer. *Nature, 523*(7559), 177–182.

Mendt, M., Rezvani, K., & Shpall, E. (2019). Mesenchymal stem cell-derived exosomes for clinical use. *Bone Marrow Transplantation, 54*(2), 789–792.

Miao, G., Zhao, H., Li, Y., Ji, M., Chen, Y., Shi, Y., ... Zhang, H. (2021). ORF3a of the COVID-19 virus SARS-CoV-2 blocks HOPS complex-mediated assembly of the SNARE complex required for autolysosome formation. *Developmental Cell, 56*(4), 427–442. https://doi.org/10.1016/j.devcel.2020.12.010 Epub 2020 Dec 16. PMID: 33422265; PMCID: PMC7832235.

Mizushima, N. (2018). A brief history of autophagy from cell biology to physiology and disease. *Nature Cell Biology, 20*(5), 521–527.

Mizushima, N., Yoshimori, T., & Ohsumi, Y. (2011). The role of Atg proteins in autophagosome formation. *Annual Review of Cell and Developmental Biology, 27*, 107–132.

Montespan, C., Marvin, S. A., Austin, S., Burrage, A. M., Roger, B., Rayne, F., ... Wodrich, H. (2017). Multi-layered control of Galectin-8 mediated autophagy during adenovirus cell entry through a conserved PPxY motif in the viral capsid. *PLoS Pathogens, 13*(2), e1006217.

Montespan, C., Wiethoff, C. M., & Wodrich, H. (2017). A small viral PPxY peptide motif to control antiviral autophagy. *Journal of Virology, 91*(18), e00581-17.

Moreau, K., Renna, M., & Rubinsztein, D. C. (2013). Connections between SNAREs and autophagy. *Trends in Biochemical Sciences, 38*(2), 57–63.

Münz, C., Campbell, G. R., Esclatine, A., Faure, M., Labonte, P., Lussignol, M., ... Wodrich, H. (2025). Autophagy machinery as exploited by viruses. *Autophagy Reports, 4*(1), 2464986.

Nagashima, S., Jirintai, S., Takahashi, M., Kobayashi, T., Tanggis, Nishizaw, T., Kouki, T., Yashiro, T., & Okamoto, H. (2014). Hepatitis E virus egress depends on the exosomal pathway, with secretory exosomes derived from multivesicular bodies. *Journal of General Virology, 95*, 2166–2175 (Pt 10).

Nagashima, S., Takahashi, M., Kobayashi, T., Tanggis, Nishizawa, T., Nishiyama, T., ... Okamoto, H. (2017). Characterization of the quasi-enveloped hepatitis E virus particles released by the cellular exosomal pathway. *Journal of Virology, 91*(22) e00822-17.

Nakatogawa, H. (2020). Mechanisms governing autophagosome biogenesis. *Nature Reviews. Molecular Cell Biology, 21*(8), 439–458.

Narayanan, A., Iordanskiy, S., Das, R., Van Duyne, R., Santos, S., Jaworski, E., ... Kashanchi, F. (2013). Exosomes derived from HIV-1-infected cells contain trans-activation response element RNA. *The Journal of Biological Chemistry, 288*(27), 20014–20033.

Nehls, J., Businger, R., Hoffmann, M., Brinkmann, C., Fehrenbacher, B., Schaller, M., ... Schindler, M. (2019). Release of immunomodulatory ebola virus glycoprotein-containing microvesicles is suppressed by tetherin in a species-specific manner. *Cell Reports, 26*(7), 1841–1853 e6.

Noh, S. H., Kim, Y. J., & Lee, M. G. (2022). Autophagy-related pathways in vesicular unconventional protein secretion. *Frontiers in Cell and Developmental Biology, 10*, 892450.

Nolte-'t Hoen, E., Cremer, T., Gallo, R. C., & Margolis, L. B. (2016). Extracellular vesicles and viruses: Are they close relatives? *Proceedings of the National Academy of Sciences of the United States of America, 113*(33), 9155–9161.

Nowag, H., Guhl, B., Thriene, K., Romao, S., Ziegler, U., Dengjel, J., & Münz, C. (2014). Macroautophagy proteins assist epstein barr virus production and get incorporated into the virus particles. *EBioMedicine, 1*(2-3), 116–125. https://doi.org/10.1016/j.ebiom.2014.11.007 PMID: 26137519; PMCID: PMC4457436.

Nüchel, J., Tauber, M., Nolte, J. L., Mörgelin, M., Türk, C., Eckes, B., ... Plomann, M. (2021). An mTORC1-GRASP55 signaling axis controls unconventional secretion to reshape the extracellular proteome upon stress. *Molecular Cell, 81*(16), 3275–3293.e12.
O'Donoghue, E. J., & Krachler, A. M. (2021). Mechanisms of EV-mediated intercellular communication in viral infections. *Frontiers in Cell and Developmental Biology, 9*, 735386.
O'Brien, K., Breyne, K., Ughetto, S., Laurent, L. C., & Breakefield, X. O. (2020). RNA delivery by extracellular vesicles in mammalian cells and its applications. *Nature Reviews. Molecular Cell Biology, 21*(10), 585–606.
O'Hara, B. A., Morris-Love, J., Gee, G. V., Haley, S. A., & Atwood, W. J. (2020). JC Virus infected choroid plexus epithelial cells produce extracellular vesicles that infect glial cells independently of the virus attachment receptor. *PLoS Pathogens, 16*(3), e1008371.
Ojha, C. R., Lapierre, J., Rodriguez, M., Dever, S. M., Zadeh, M. A., DeMarino, C., ... El-Hage, N. (2017). Interplay between autophagy, exosomes and HIV-1 Associated neurological disorders: New insights for diagnosis and therapeutic applications. *Viruses, 9*(7), 176.
Orvedahl, A., Alexander, D., Tallóczy, Z., Sun, Q., Wei, Y., Zhang, W., ... Levine, B. (2007). HSV-1 ICP34.5 confers neurovirulence by targeting the Beclin 1 autophagy protein. *Cell Host & Microbe, 1*(1), 23–35.
Orvedahl, A., & Levine, B. (2008). Viral evasion of autophagy. *Autophagy, 4*(3), 280–285.
Orvedahl, A., MacPherson, S., Sumpter, R., Jr, Tallóczy, Z., Zou, Z., & Levine, B. (2010). Autophagy protects against Sindbis virus infection of the central nervous system. *Cell Host & Microbe, 7*(2), 115–127.
Ostrowski, M., Carmo, N. B., Krumeich, S., Fanget, I., Rapos, G., Savina, A., ... Thery, C. (2010). Rab27a and Rab27b control different steps of the exosome secretion pathway. *Nature Cell Biology, 12*(1), 19–30 sup pp 1-13.
Owliaee, I., Khaledian, M., Boroujeni, A. K., & Shojaeian, A. (2023). Engineered small extracellular vesicles as a novel platform to suppress human oncovirus-associated cancers. *Infectious Agents and Cancer, 18*(1), 69.
Palaric, M., Versapuech, M., Judith, D., Aubé, C., Leduc, M., Dutrieux, J., ... Berlioz-Torrent, C. (2025). GABARAP proteins regulate the packaging of HIV-1 genomic RNA into virions. *EMBO Reports*. https://doi.org/10.1038/s44319-025-00607-1 Epub ahead of print. PMID: 41174262.
Paul, D., & Bartenschlager, R. (2013). Architecture and biogenesis of plus-strand RNA virus replication factories. *World Journal of Virology, 2*(2), 32–48.
Pegtel, D. M., Cosmopoulos, K., Thorley-Lawson, D. A., van Eijndhoven, M. A., Hopmans, E. S., Lindenberg, J. L., ... Middeldorp, J. M. (2010). Functional delivery of viral miRNAs via exosomes. *Proceedings of the National Academy of Sciences of the United States of America, 107*(14), 6328–6333.
Pegtel, D. M., & Gould, S. J. (2019). Exosomes. *Annual Review of Biochemistry, 88*, 487–514.
Pied, N., Daussy, C. F., Denis, Z., Ragues, J., Faure, M., Iggo, R., ... Wodrich, H. (2022). TBK1 is part of a galectin 8 dependent membrane damage recognition complex and drives autophagy upon Adenovirus endosomal escape. *PLoS Pathogens, 18*(7), e1010736.
Pilli, M., Arko-Mensah, J., Ponpuak, M., Roberts, E., Master, S., Mandell, M. A., ... Deretic, V. (2012). TBK-1 promotes autophagy-mediated antimicrobial defense by controlling autophagosome maturation. *Immunity, 37*(2), 223–234.
Pino-Belmar, C., Aguilar, R., Valenzuela-Nieto, G. E., Cavieres, V. A., Cerda-Troncoso, C., Navarrete, V. C., ... Bustamante, H. A. (2024). An intrinsic host defense against HSV-1 relies on the activation of xenophagy with the active clearance of autophagic receptors. *Cells, 13*(15), 1256.
Pleet, M. L., Branscome, H., DeMarino, C., Pinto, D. O., Zadeh, M. A., Rodriguez, M., ... Kashanchi, F. (2018). Autophagy, EVs, and infections: A perfect question for a perfect time. *Frontiers in Cellular and Infection Microbiology, 8*, 362.

Randow, F., & Youle, R. J. (2014). Self and nonself: How autophagy targets mitochondria and bacteria. *Cell Host & Microbe, 15*(4), 403–411.

Rey-Cadilhac, F., Rachenne, F., Missé, D., & Pompon, J. (2023). Viral components trafficking with(in) extracellular vesicles. *Viruses, 15*(12), 2333.

Reyes-Ruiz, J. M., Osuna-Ramos, J. F., De Jesús-González, L. A., Palacios-Rápalo, S. N., Cordero-Rivera, C. D., Farfan-Morales, C. N., ... Del Ángel, R. M. (2020). The regulation of flavivirus infection by hijacking exosome-mediated cell-cell communication: New insights on virus-host interactions. *Viruses, 12*(7), 765.

Robinson, S. M., Tsueng, G., Sin, J., Mangale, V., Rahawi, S., McIntyre, L. L., ... Feuer, R. (2014). Coxsackievirus B exits the host cell in shed microvesicles displaying autophagosomal markers. *PLoS Pathogens, 10*(4), e1004045.

Russell, R. C., Yuan, H. X., & Guan, K. L. (2014). Autophagy regulation by nutrient signaling. *Cell Research, 24*(1), 42–57.

Santangelo, L., Giurato, G., Cicchini, C., Montaldo, C., Mancone, C., Tarallo, R., ... Tripodi, M. (2016). The RNA-binding protein SYNCRIP Is a component of the hepatocyte exosomal machinery controlling MicroRNA sorting. *Cell Reports, 17*(3), 799–808.

Schwarz, G., Ren, X., Xie, W., Guo, H., Jiang, Y., & Zhang, J. (2025). Engineered exosomes: A promising drug delivery platform with therapeutic potential. *Frontiers in Molecular Biosciences, 12*, 1583992.

Scorrano, L., De Matteis, M. A., Emr, S., Giordano, F., Hajnóczky, G., Kornmann, B., ... Schuldiner, M. (2019). Coming together to define membrane contact sites. *Nature Communications, 10*(1), 1287.

Shao, H., Chung, J., Balaj, L., Charest, A., Bigner, D. D., Carter, B. S., ... Lee, H. (2012). Protein typing of circulating microvesicles allows real-time monitoring of glioblastoma therapy. *Nature Medicine, 18*(12), 1835–1840.

Song, M. H., Sun, Y., & Qiu, X. B. (2024). Hijacking autophagy for infection by flaviviruses. *Virus Research, 347*, 199422.

Song, L., Tang, S., Han, X., Jiang, Z., Dong, L., Liu, C., ... Du, Y. (2019). KIBRA controls exosome secretion via inhibiting the proteasomal degradation of Rab27a. *Nature Communications, 10*(1), 1639.

Stertz, S., Reichelt, M., Spiegel, M., Kuri, T., Martínez-Sobrido, L., García-Sastre, A., ... Kochs, G. (2007). The intracellular sites of early replication and budding of SARS-coronavirus. *Virology, 361*(2), 304–315.

Sun, X., Sun, L., Zhao, Y., Li, Y., Lin, W., Chen, D., & Sun, Q. (2016). MAVS maintains mitochondrial homeostasis via autophagy. *Cell Discovery, 2*, 16024.

Takahashi, Y., Liang, X., Hattori, T., Tang, Z., He, H., Chen, H., ... Wang, H. G. (2019). VPS37A directs ESCRT recruitment for phagophore closure. *The Journal of Cell Biology, 218*(10), 3336–3354.

Takáts, S., Pircs, K., Nagy, P., Varga, Á., Kárpáti, M., Hegedűs, K., ... Juhász, G. (2014). Interaction of the HOPS complex with syntaxin 17 mediates autophagosome clearance in Drosophila. *Molecular Biology of the Cell, 25*(8), 1338–1354.

ten Broeke, T., Wubbolts, R., & Stoorvogel, W. (2013). MHC class II antigen presentation by dendritic cells regulated through endosomal sorting. *Cold Spring Harbor Perspectives in Biology, 5*(12), a016873.

Tey, S. K., Lam, H., Wong, S. W. K., Zhao, H., To, K. K., & Yam, J. W. P. (2022). ACE2-enriched extracellular vesicles enhance infectivity of live SARS-CoV-2 virus. *Journal of Extracellular Vesicles, 11*(5), e12231. https://doi.org/10.1002/jev2.12231 PMID: 35582880; PMCID: PMC9115585.

Théry, C., Witwer, K. W., Aikawa, E., et al. (2018). Minimal information for studies of extracellular vesicles 2018 (MISEV2018): A position statement of the international society for extracellular vesicles and update of the MISEV2014 guidelines. *Journal of Extracellular Vesicles*, 7(1), 1535750.

Théry, C., Zitvogel, L., & Amigorena, S. (2002). Exosomes: Composition, biogenesis and function. *Nature Reviews. Immunology, 2*(8), 569–579. https://doi.org/10.1038/nri855 PMID: 12154376.

Tkach, M., & Théry, C. (2016). Communication by extracellular vesicles: Where we are and where we need to go. *Cell, 164*(6), 1226–1232.

Towers, C. G., Wodetzki, D., & Thorburn, A. (2020). Autophagy and cancer: Modulation of cell death pathways and cancer cell adaptations. *The Journal of Cell Biology, 219*(1), e201909033.

Tran, S., Fairlie, W. D., & Lee, E. F. (2021). BECLIN1: Protein structure, function and regulation. *Cells, 10*(6), 1522.

Troyer, Z., Alhusaini, N., Tabler, C. O., Sweet, T., de Carvalho, K. I. L., Schlatzer, D. M., ... Tilton, J. C. (2021). Extracellular vesicles carry SARS-CoV-2 spike protein and serve as decoys for neutralizing antibodies. *Journal of Extracellular Vesicles, 10*(8), e12112.

Urbanelli, L., Buratta, S., Tancini, B., Sagini, K., Delo, F., Porcellati, S., & Emiliani, C. (2019). The role of extracellular vesicles in viral infection and transmission. *Vaccines (Basel), 7*(3), 102.

Urbanelli, L., Buratta, S., Tancini, B., Sagini, K., Delo, F., Porcellati, S., et al. (2023). Extracellular vesicles in viral infection: Pathophysiological roles and therapeutic perspectives. *Cells, 12*(1), 122.

Vader, P., Mol, E. A., Pasterkamp, G., & Schiffelers, R. M. (2016). Extracellular vesicles for drug delivery. *Advanced Drug Delivery Reviews, 106*(Pt A), 148–156.

van Niel, G., D'Angelo, G., & Raposo, G. (2022). Shedding light on the cell biology of extracellular vesicles. *Nature Reviews. Molecular Cell Biology, 23*(8), 467–484.

van Niel, G., D'Angelo, G., & Raposo, G. (2018). Shedding light on the cell biology of extracellular vesicles. *Nature Reviews. Molecular Cell Biology, 19*(4), 213–228.

Villarroya-Beltri, C., Gutiérrez-Vázquez, C., Sánchez-Cabo, F., Pérez-Hernández, D., Vázquez, J., Martin-Cofreces, N., ... Sánchez-Madrid, F. (2013). Sumoylated hnRNPA2B1 controls the sorting of miRNAs into exosomes through binding to specific motifs. *Nature Communications, 4*, 2980.

Votteler, J., & Sundquist, W. I. (2013). Virus budding and the ESCRT pathway. *Cell Host & Microbe, 14*(3), 232–241.

Walia, K., Sharma, A., Paul, S., Chouhan, P., Kumar, G., Ringe, R., ... Tuli, A. (2024). SARS-CoV-2 virulence factor ORF3a blocks lysosome function by modulating TBC1D5-dependent Rab7 GTPase cycle. *Nature Communications, 15*(1), 2053.

Wan, W., & Briggs, J. A. (2016). Cryo-electron tomography and subtomogram averaging. *Methods in Enzymology, 579*, 329–367.

Wang, T., Zhang, L., Liang, W., Liu, S., Deng, W., Liu, Y., ... Zhang, Y. (2022). Extracellular vesicles originating from autophagy mediate an antibody-resistant spread of classical swine fever virus in cell culture. *Autophagy, 18*(6), 1433–1449.

White, J., Choi, Y. B., Zhang, J., Vo, M. T., He, C., Shaikh, K., & Harhaj, E. W. (2025). Phosphorylation of the selective autophagy receptor TAX1BP1 by TBK1 and IKBKE/IKKi promotes ATG8-family protein-dependent clearance of MAVS aggregates. *Autophagy, 21*(1), 160–177.

Williams, R. L., & Urbé, S. (2007). The emerging shape of the ESCRT machinery. *Nature Reviews. Molecular Cell Biology, 8*(5), 355–368.

Willms, E., Cabañas, C., Mäger, I., Wood, M. J. A., & Vader, P. (2018). Extracellular vesicle heterogeneity: Subpopulations, isolation techniques, and diverse functions in cancer progression. *Frontiers in Immunology, 9*, 738.

Xia, X., Wang, Y., & Zheng, J. C. (2021). Emerging roles of extracellular vesicles in mediating RNA virus infection. *Fundamental Research, 1*(2), 179–185.

Yakoub, A. M., & Shukla, D. (2015). Autophagy stimulation abrogates herpes simplex virus-1 infection. *Scientific Reports, 5*, 9730.

Zaffagnini, G., & Martens, S. (2016). Mechanisms of selective autophagy. *Journal of Molecular Biology, 428*(9 Pt A), 1714–1724.
Zeng, W., Zheng, L., Li, Y., Yang, J., Mao, T., Zhang, J., ... Lu, F. (2024). Engineered extracellular vesicles for delivering functional Cas9/gRNA to eliminate hepatitis B virus cccDNA and integration. *Emerging Microbes & Infections, 13*(1), 2284286.
Zhai, H., Wang, T., Liu, D., Pan, L., Sun, Y., & Qiu, H. J. (2023). Autophagy as a dual-faced host response to viral infections. *Frontiers in Cellular and Infection Microbiology, 13*, 1289170.
Zhang, J., Lan, Y., Li, M. Y., Lamers, M. M., Fusade-Boyer, M., Klemm, E., ... Sanyal, S. (2018). Flaviviruses exploit the lipid droplet protein AUP1 to trigger lipophagy and drive virus production. *Cell Host & Microbe, 23*(6), 819–831 e5.
Zhang, Y., Liu, Y., Liu, H., & Tang, W. H. (2019). Exosomes: biogenesis, biologic function and clinical potential. *Cell & Bioscience, 9*, 19.
Zhang, X., & Wang, Y. (2018). GRASP55 facilitates autophagosome maturation under glucose deprivation. *Molecular and Cellular Oncology, 5*(4), e1494948.
Zhao, P., Tian, R., Song, D., Zhu, Q., Ding, X., Zhang, J., ... Sun, Q. (2025). Rab GTPases are evolutionarily conserved signals mediating selective autophagy. *The Journal of Cell Biology, 224*(5), e202410150.
Zhao, F., Xu, Y., Liu, N., Lv, D., Chen, Y., Liu, Z., ... Long, G. (2023). Extracellular vesicles from Zika virus-infected cells display viral E protein that binds ZIKV-neutralizing antibodies to prevent infection enhancement. *The EMBO Journal, 42*(6), e112096.

Further reading

Arakelyan, A., Fitzgerald, W., Zicari, S., Vanpouille, C., & Margolis, L. (2017). Extracellular vesicles carry HIV Env and facilitate Hiv infection of human lymphoid tissue. *Scientific Reports, 7*(1), 1695.
Ashraf Malik, M., Ishtiyaq Ali Mirza, J., Umar, M., & Manzoor, S. (2019). CD81 + Exosomes play a pivotal role in the establishment of hepatitis C persistent infection and contribute toward the progression of hepatocellular carcinoma. *Viral Immunology, 32*(10), 453–462.
Balakireva, L., Schoehn, G., Thouvenin, E., & Chroboczek, J. (2003). Binding of adenovirus capsid to dipalmitoyl phosphatidylcholine provides a novel pathway for virus entry. *Journal of Virology, 77*(8), 4858–4866.
Belov, G. A., Altan-Bonnet, N., Kovtunovych, G., Jackson, C. L., Lippincott-Schwartz, J., & Ehrenfeld, E. (2007). Hijacking components of the cellular secretory pathway for replication of poliovirus RNA. *Journal of Virology, 81*(2), 558–567.
Belov, G. A., & van Kuppeveld, F. J. (2012). (+)RNA viruses rewire cellular pathways to build replication organelles. *Current Opinion in Virology, 2*(6), 740–747.
Bestebroer, J., V'kovski, P., Mauthe, M., & Reggiori, F. (2013). Hidden behind autophagy: The unconventional roles of ATG proteins. *Traffic (Copenhagen, Denmark), 14*(10), 1029–1041.
Cao, W., Li, J., Yang, K., & Cao, D. (2021). An overview of autophagy: Mechanism, regulation and research progress. *Bulletin du Cancer, 108*(3), 304–322.
Chau, B. A., Chen, V., Cochrane, A. W., Parent, L. J., & Mouland, A. J. (2023). Liquid-liquid phase separation of nucleocapsid proteins during SARS-CoV-2 and HIV-1 replication. *Cell Reports, 42*(1), 111968.
Chen, Q., Sun, L., & Chen, Z. J. (2016). Regulation and function of the cGAS-STING pathway of cytosolic DNA sensing. *Nature Immunology, 17*(10), 1142–1149.

Cocozza, F., N é vo, N., Piovesana, E., Lahaye, X., Buchrieser, J., Schwartz, O., ... Martin-Jaular, L. (2020). Extracellular vesicles containing ACE2 efficiently prevent infection by SARS-CoV-2 Spike protein-containing virus. *Journal of Extracellular Vesicles, 10*(2), e12050.

Codogno, P., Mehrpour, M., & Proikas-Cezanne, T. (2011). Canonical and non-canonical autophagy: Variations on a common theme of self-eating? *Nature Reviews. Molecular Cell Biology, 13*(1), 7–12.

Crenshaw, B. J., Gu, L., Sims, B., & Matthews, Q. L. (2018). Exosome biogenesis and biological function in response to viral infections. *Open Virology Journal, 12*, 134–148.

den Boon, J. A., & Ahlquist, P. (2010). Organelle-like membrane compartmentalization of positive-strand RNA virus replication factories. *Annual Review of Microbiology, 64*, 241–256.

den Boon, J. A., Diaz, A., & Ahlquist, P. (2010). Cytoplasmic viral replication complexes. *Cell Host & Microbe, 8*(1), 77–85.

Deng, H., Cao, H., Wang, Y., Li, J., Dai, J., Li, L. F., ... Li, S. (2024). Viral replication organelles: The highly complex and programmed replication machinery. *Frontiers in Microbiology, 15*, 1450060.

Di Bonito, P., Ridolfi, B., Columba-Cabezas, S., Giovannelli, A., Chiozzini, C., Manfredi, F., ... Federico, M. (2015). HPV-E7 delivered by engineered exosomes elicits a protective $CD8^{+}$ T cell-mediated immune response. *Viruses, 7*(3), 1079–1099.

Garofalo, M., Villa, A., Rizzi, N., Kuryk, L., Rinner, B., Cerullo, V., ... Ciana, P. (2019). Extracellular vesicles enhance the targeted delivery of immunogenic oncolytic adenovirus and paclitaxel in immunocompetent mice. *Journal of Controlled Release: Official Journal of the Controlled Release Society, 294*, 165–175.

Habib, A., Liang, Y., & Zhu, N. (2023). Exosomes multifunctional roles in HIV-1: Insight into the immune regulation, vaccine development and current progress in delivery system. *Frontiers in Immunology, 14*, 1249133.

Herker, E. (2024). Lipid droplets in virus replication. *FEBS Letters, 598*(10), 1299–1300.

Hsu, N. Y., Ilnytska, O., Belov, G., Santiana, M., Chen, Y. H., Takvorian, P. M., ... Altan-Bonnet, N. (2010). Viral reorganization of the secretory pathway generates distinct organelles for RNA replication. *Cell, 141*(5), 799–811.

Jian, F., Wang, S., Tian, W., Chen, Y., Wang, S., Li, Y., ... Rong, Y. (2025). Deacetylated SNAP47 recruits HOPS to facilitate autophagosome-lysosome fusion independent of STX17. *Nature Communications, 16*(1), 543.

Liang, B., Peng, P., Chen, S., et al. (2020). Characterization and clinical value of exosomal microRNAs in ovarian cancer. *Oncology Reports, 44*(5), 2101–2112.

Liu, S., Yan, H., Diao, J., Zhang, S., & Zhong, Q. (2025). Spatio-temporal processes in autophagosome-lysosome fusion. *Medical Reviews (2021), 5*(4), 297–317.

Ma-Lauer, Y., Li, P., Niemeyer, D., Richter, A., Pusl, K., von Brunn, B., ... von Brunn, A. (2024). Oxysterole-binding protein targeted by SARS-CoV-2 viral proteins regulates coronavirus replication. *Frontiers in Cellular and Infection Microbiology, 14*, 1383917.

Melia, C. E., van der Schaar, H. M., de Jong, A. W. M., Lyoo, H. R., Snijder, E. J., Koster, A. J., ... Bárcena, M. (2018). The origin, dynamic morphology, and PI4P-independent formation of encephalomyocarditis virus replication organelles. *mBio, 9*(2), e00420-18.

Menzel, N., Waschke, J., Wolk, B., Papkalla, A., Pape, C., von, Hahn, T., et al. (2021). Whole lotta lipids—from HCV RNA replication to the mature viral particle. *International Journal of Molecular Sciences, 21*(8), 2888.

Miller, S., & Krijnse-Locker, J. (2008). Modification of intracellular membrane structures for virus replication. *Nature Reviews. Microbiology, 6*(5), 363–374.

Mizushima, N., & Komatsu, M. (2011). Autophagy: Renovation of cells and tissues. *Cell, 147*(4), 728–741.

Paul, P., & Münz, C. (2016). Autophagy and mammalian viruses: Roles in immune response, viral replication, and beyond. *Advances in Virus Research, 95*, 149–195.
Radke, J. R., & Cook, J. L. (2018). Human adenovirus infections: Update and consideration of mechanisms of viral persistence. *Current Opinion in Infectious Diseases, 31*(3), 251–256.
Reggiori, F., Monastyrska, I., Verheije, M. H., Calì, T., Ulasli, M., Bianchi, S., ... Molinari, M. (2010). Coronaviruses Hijack the LC3-I-positive EDEMosomes, ER-derived vesicles exporting short-lived ERAD regulators, for replication. *Cell Host & Microbe, 7*(6), 500–508.
Romero-Brey, I., & Bartenschlager, R. (2014). Membranous replication factories induced by plus-strand RNA viruses. *Viruses, 6*(7), 2826–2857.
Russell, R. C., Tian, Y., Yuan, H., Park, H. W., Chang, Y. Y., Kim, J., ... Guan, K. L. (2013). ULK1 induces autophagy by phosphorylating Beclin-1 and activating VPS34 lipid kinase. *Nature Cell Biology, 15*(7), 741–750.
Saari, H., Turunen, T., Lõhmus, A., Turunen, M., Jalasvuori, M., Butcher, S. J., ... Yliperttula, M. (2020). Extracellular vesicles provide a capsid-free vector for oncolytic adenoviral DNA delivery. *Journal of Extracellular Vesicles, 9*(1), 1747206.
Sieben, C., Sezgin, E., Eggeling, C., & Manley, S. (2020). Influenza A viruses use multivalent sialic acid clusters for cell binding and receptor activation. *PLoS Pathogens, 16*(7), e1008656.
Sims, B., Gu, L., Krendelchtchikov, A., & Matthews, Q. L. (2014). Neural stem cell-derived exosomes mediate viral entry. *International Journal of Nanomedicine, 9*, 4893–4897.
Snijder, E. J., Limpens, R. W. A. L., de Wilde, A. H., de Jong, A. W. M., Zevenhoven-Dobbe, J. C., Maier, H. J., ... Bárcena, M. (2020). A unifying structural and functional model of the coronavirus replication organelle: Tracking down RNA synthesis. *PLoS Biology, 18*(6), e3000715.
Song, X., Xi, Y., Dai, M., Li, T., Du, S., Zhu, Y., ... Liu, X. (2024). STING guides the STX17-SNAP29-VAMP8 complex assembly to control autophagy. *Cell Insight, 3*(2), 100147.
Strating, J. R., & van Kuppeveld, F. J. (2017). Viral rewiring of cellular lipid metabolism to create membranous replication compartments. *Current Opinion in Cell Biology, 47*, 24–33.
Takeuchi, O., & Akira, S. (2010). Pattern recognition receptors and inflammation. *Cell, 140*(6), 805–820.
Tinkov, A. A., Aschner, M., Santamaria, A., Bogdanov, A. R., Tizabi, Y., Virgolini, M. B., ... Skalny, A. V. (2023). Dissecting the role of cadmium, lead, arsenic, and mercury in non-alcoholic fatty liver disease and non-alcoholic steatohepatitis. *Environmental Research, 238*(Pt 1), 117134.
Wang, C. W., & Klionsky, D. J. (2003). The molecular mechanism of autophagy. *Molecular Medicine (Cambridge, Mass.), 9*(3–4), 65–76.
Wang, Y., Sharma, P., Jefferson, M., Zhang, W., Bone, B., Kipar, A., ... Wileman, T. (2021). Non-canonical autophagy functions of ATG16L1 in epithelial cells limit lethal infection by influenza A virus. *The EMBO Journal, 40*(6), e105543.
Wei, W., Bai, L., Yan, B., Meng, W., Wang, H., Zhai, J., ... Zheng, C. (2022). When liquid-liquid phase separation meets viral infections. *Frontiers in Immunology, 13*, 985622.
Wolff, G., Limpens, R. W. A. L., Zevenhoven-Dobbe, J. C., Laugks, U., Zheng, S., de Jong, A. W. M., ... Bárcena, M. (2020). A molecular pore spans the double membrane of the coronavirus replication organelle. *Science (New York, N. Y.), 369*(6509), 1395–1398.
Yang, Z., & Klionsky, D. J. (2009). An overview of the molecular mechanism of autophagy. *Current Topics in Microbiology and Immunology, 335*, 1–32.
Yang, Z., & Klionsky, D. J. (2010). Eaten alive: A history of macroautophagy. *Nature Cell Biology, 12*(9), 814–822.

Yin, Y., Zhao, Y., Chen, Q., Chen, Y., & Mao, L. (2022). Dual roles and potential applications of exosomes in HCV infections. *Frontiers in Microbiology, 13*, 1044832.

Yu, L., Chen, Y., & Tooze, S. A. (2018). Autophagy pathway: Cellular and molecular mechanisms. *Autophagy, 14*(2), 207–215.

Zhang, H., Freitas, D., Kim, H. S., Fabijanic, K., Li, Z., Chen, H., ... Lyden, D. (2018). Identification of distinct nanoparticles and subsets of extracellular vesicles by asymmetric flow field-flow fractionation. *Nature Cell Biology, 20*(3), 332–343.

Zhang, L., Duan, Y., Wang, W., Li, Q., Tian, J., Zhu, Y., ... Xie, Z. (2023). Autophagy induced by human adenovirus B7 structural protein VI inhibits viral replication. *Virologica Sinica, 38*(5), 709–722.

Zhang, Y., Xu, L., Zhang, Z., Su, X., Wang, Z., & Wang, T. (2023). Enterovirus D68 infection upregulates SOCS3 expression to inhibit JAK-STAT3 signaling and antagonize the innate interferon response of the host. *Virologica Sinica, 38*(5), 755–766.

CHAPTER SIX

Sex matters in macroautophagy: Sex-driven differences in autophagic function in aging and aging related-diseases

José Ignacio García-Muñoz[a], Alicia Domenech-Bendaña[a], Lucia Gimeno-Mallench[a], Alejandro Ponce-Mora[a], Álvaro F. Fernández[b,c,d], Daniel López-López[e,f,g], Jose Luis Nieto-Torres[a], and Eloy Bejarano[a,*]

[a]Department of Biomedical Sciences, School of Health Sciences, Universidad Cardenal Herrera-CEU, CEU Universities, Valencia, Spain
[b]Departamento de Bioquímica y Biología Molecular, Universidad de Oviedo, Oviedo, Spain
[c]Instituto Universitario de Oncología del Principado de Asturias (IUOPA), Oviedo, Spain
[d]Instituto de Investigación Sanitaria del Principado de Asturias (ISPA), Oviedo, Spain
[e]Clinical Bioinformatics Area, Fundación Progreso y Salud (FPS), CDCA, Hospital Virgen del Rocio, Sevilla, Spain
[f]Computational Systems Medicine, Institute of Biomedicine of Seville (IBIS), Hospital Virgen del Rocio, Sevilla, Spain
[g]Bioinformatics in Rare Diseases (BiER), Centro de Investigación Biomédica en Red de Enfermedades Raras (CIBERER), FPS, Hospital Virgen del Rocío, Sevilla, Spain
*Corresponding author. e-mail address: eloy.bejaranofernandez@uchceu.es

Contents

Abstract

Autophagy is a fundamental cellular process essential for maintaining homeostasis, particularly in the context of aging and age-related diseases. Increasing evidence highlights biological sex as a critical modulator of autophagy, influencing its basal

International Review of Cell and Molecular Biology, Volume 406
ISSN 1937-6448, https://doi.org/10.1016/bs.ircmb.2026.07.006

activity, regulatory pathways, and responsiveness to stress. Distinct autophagic profiles in males and females (across tissues, species and developmental stages) may underlie sex-specific vulnerabilities and divergent disease trajectories. These differences are evident in conditions such as neurodegeneration, cardiovascular disease, cancer, sarcopenia or chronic inflammation, all of which are commonly associated with aging. Indeed, aging is a major risk factor for the onset and progression of these pathologies. Importantly, sex-dependent variations in autophagy might also impact the efficacy and safety of therapeutic interventions, challenging the validity of uniform treatment strategies. Despite growing recognition of these disparities, significant knowledge gaps remain. This review summarizes current understanding of sex-related differences in autophagy, focusing on genetic and hormonal influences across the lifespan. The evidence supports the need for future research to systematically incorporate sex as a biological variable in experimental design and data analysis, utilize dynamic assessments of autophagy flux and investigate the interplay among genetic, hormonal, epigenetic, and post-translational regulatory mechanisms. Developing preclinical models that reflect human diversity—including genetic heterogeneity and relevant hormonal states—is imperative. Embracing the complexity of sex-dependent autophagy regulation is essential for translating mechanistic insights into effective, personalized interventions that improve health outcomes for both women and men.

1. Autophagy in healthspan and lifespan

Aging is a complex, multifactorial process marked by the progressive decline of physiological functions, ultimately increasing susceptibility to age-related diseases and culminating in death. In a landmark publication by López-Otín et al. the concept of the "hallmarks of aging" was coined, identifying nine key biological processes that drive aging (López-Otín et al., 2013). Since then, the field of longevity research has expanded considerably, deepening our understanding of the molecular and cellular mechanisms underlying aging. In 2023, the same group of researchers proposed an updated framework, incorporating three additional hallmarks. The current list includes genomic instability, telomere attrition, epigenetic alterations, loss of proteostasis, deregulated nutrient sensing, mitochondrial dysfunction, cellular senescence, stem cell exhaustion, altered intercellular communication, microbiome disturbances (dysbiosis), chronic inflammation, and— of particular relevance to this review—disabled autophagy (López-Otín et al., 2023).

Autophagy, from the Greek for "self-eating," is an evolutionarily conserved catabolic process essential for cellular homeostasis in eukaryotic organisms. This catabolic process enables the selective degradation and recycling of intracellular components, including damaged organelles, misfolded proteins, and other cytoplasmic debris in the lysosome. There are alternative molecular routes that allow the delivery of such cellular components to the lysosome, which define three main autophagic pathways: chaperone-mediated autophagy (CMA), microautophagy and macroautophagy (Khawaja et al., 2025; Manastireanu et al., 2024; Ponce-Mora et al., 2025). CMA allows the elimination of cellular proteins via their direct translocation through the lysosomal membrane, assisted by a chaperone and a receptor. Specifically, proteins containing a KFERQ, or a KFERQ-like amino acid sequence, are recognized by the chaperone Hsc70, which guides them to the lysosomal membrane. Once there, the proteins are translocated to the lysosomal lumen upon interaction with the LAMP2A receptor (Bejarano & Cuervo, 2010; Jafari et al., 2022). CMA is therefore a highly specific degradation pathway, estimated to control the turnover of at least 40% of the whole cellular proteome, which displays KFERQ-like canonical motifs (Jafari et al., 2022). Microautophagy not only degrades proteins, but also lipids and small fragments of organelles. In lysosomal microautophagy (lMi) this is achieved via the invagination of the lysosomal membrane, ultimately delivering the components to be degraded to the lysosomal lumen. Alternatively, cargo invagination takes place at the endosomal membrane, and subsequent fusion with lysosomes permits degradation of the endosomal content. This process is called endosomal microautophagy (eMi) and appears to be the prevalent molecular route over lMi in higher organisms (Manastireanu et al., 2024). Macroautophagy represents the most complex type of autophagy, as it facilitates the degradation of the widest array of cargo from proteins, lipids and organelle fragments to protein aggregates, complete organelles, and invading pathogens (Nieto-Torres & Hansen, 2021). Macroautophagy requires the coordinated action of a numerous group of proteins that identify and target cargo for degradation, create a specialized double-membrane vesicle, the autophagosome, in which cargo is encapsulated, and finally deliver the autophagosomes to the lysosome for cargo degradation (for details of the process see next section) (Fig. 1). Given their fundamental role in degrading critical cellular components and their crucial contribution to homeostasis, it is not surprising that accumulating reports highlight the profound implications of autophagy subtypes in healthspan and lifespan. CMA, for example, declines with age in a process that is conserved among different organisms and tissues, ultimately compromising cellular homeostasis (Dong et al., 2020; Kaushik & Cuervo, 2018; Madrigal-Matute et al., 2022). The molecular root of age-dependent CMA deficiency seems

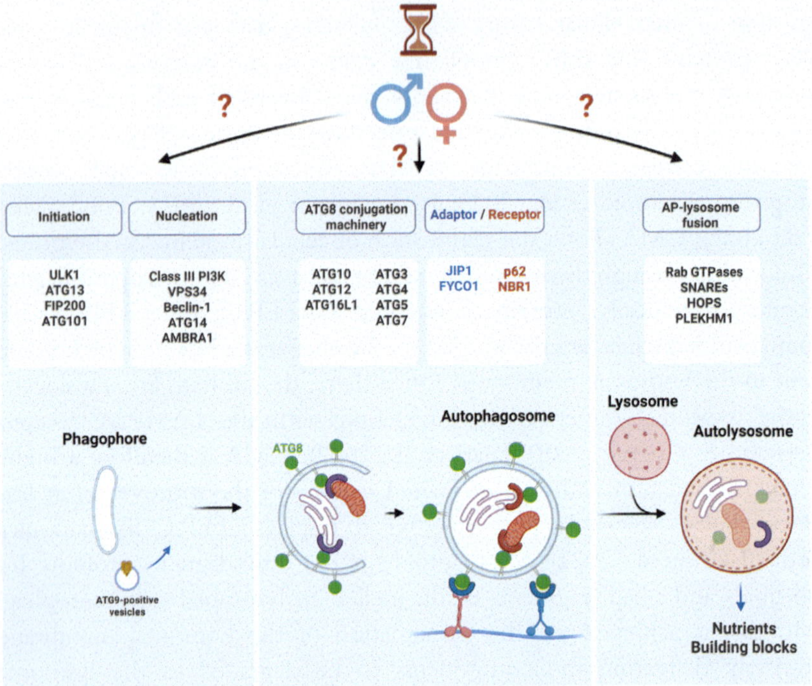

Fig. 1 Molecular machinery involved in the sequential phases of macroautophagy. This figure illustrates the key molecular complexes orchestrating the major phases of macroautophagy: initiation, nucleation, phagophore expansion, autophagosome closure, fusion with the lysosome and degradation. The ULK1 complex initiates autophagy in response to cellular stress and nutrient deprivation. The PI3K class III complex drives membrane nucleation, while the ATG5-ATG12 and ATG8 conjugation systems facilitate phagophore elongation. Different receptors are involved in the recruitment of cargo while different adaptors contribute to vesicle motility. SNARE proteins and Rabs mediate autophagosome-lysosome fusion, enabling cargo degradation and recycling. Together, these components ensure the dynamic regulation of autophagy, essential for cellular homeostasis and adaptation to stress.

to stem from reduced LAMP2A levels or stability (Kaushik et al., 2006; Kiffin et al., 2007; Valdor et al., 2014). Thus, unpaired CMA correlates with multiorgan malfunction. Interestingly, restoration of CMA is associated with enhanced resistance to stress and organ homeostasis (Li et al., 2025; Reynolds et al., 2024; Salinas et al., 2025a; Valdor et al., 2014; Zhang & Cuervo, 2008). Furthermore, CMA is increased in a hormone deficient longevity mouse model (Endicott et al., 2021). As abovementioned, there are two major protagonists: chaperone Hsc70, which recognizes the KFERQ-like motifs in molecular targets, and the lysosome-membrane-associated protein

LAMP2A, which mediates the direct translocation of that cargo through the membrane (Kaushik & Cuervo, 2018). Interestingly, Hsc70 also plays an important role in eMI along with the ESCRT system, with the latter participating in lMi, too (Krause et al., 2023; Kuchitsu & Taguchi, 2024; Sakai & Oku, 2024; Tekirdag & Cuervo, 2018).

The connection of microautophagy and aging is not as developed as for CMA, but there are recent reports that support its protective role, as well. Lysosome repair via microautophagy, a process that requires the action of the kinase STK38 and the ATG8 protein GABARAP, has important repercussions in aging. Elimination of STK38 or GABARAP leads to enhanced lysosomal damage and accelerated progression of cellular senescence, a process conserved in *Caenorhabditis elegans*, where gene knockdown also leads to lifespan reduction (Ogura et al., 2023). eMi is severely compromised during aging in mice liver, and such reduction correlates with a concomitant alteration in protein secretion (Krause et al., 2022). Remarkably, the connection between macroautophagy and aging is very well established. Longevity paradigms in *Saccharomyces cerevisiae*, *C. elegans* and *Drosophila melanogaster* show enhanced macroautophagy levels, which may be responsible for their lifespan benefits, as macroautophagy inhibition block such effects (Hansen et al., 2018). Furthermore, overexpression of autophagy-related proteins in model organisms extends lifespan, as demonstrated for HLH-30/TFEB and SQST-1/p62 in *C. elegans*, ATG8 and ATG1 in *D. melanogaster* (Simonsen et al., 2008; Ulgherait et al., 2014) and ATG5 in mice (Lapierre et al., 2013; Pyo et al., 2013). Furthermore, a gain-of-function mutation in the autophagy protein Beclin-1, which leads to enhanced macroautophagy, confers multiorgan health benefits to mice and prolongs the lifespan of a model of accelerated aging (Fernández et al., 2018). Overall, to different extents, the beneficial effects of the different subtypes of autophagy have been sufficiently proved in model organisms. In humans, information is so far limited to correlational studies on the age-related decline of CMA and macroautophagy components in certain tissues (Carosi et al., 2025; Jafari et al., 2022; Nieto-Torres & Hansen, 2021). Such studies are now being improved with the development of models that allow accurate measurement of autophagy efficiency (Moreno et al., 2025). Besides this, the direct effect of autophagy boosting approaches on extending healthspan and lifespan needs to be further investigated.

Given the limited literature available on the impact of biological sex on CMA and microautophagy (Khawaja et al., 2025; Luo, 2025)—and considering that macroautophagy is the most extensively studied subtype—the

Table 1

Publication year	Authors	Biological parameters	Cell line	Sex differences	Expression change	Autophagy effect
2024	Green et al.	Sphk2	Primary hepatocytes (♂♀, C57BL/6NJ)	↑ in males	↓ (KO or inhibition) alters lipid metabolism and p62 levels	↓ Sphk2 produce ↑ autophagy in males; ↓ autophagy in females
		p62/SQSTM1		↑ in males (WT)	↓ in males; ↑ in females	↓ autophagy in females; ↑ in males
		NQO1		↑ in males	↓ in males; ↑ in females	Correlates with p62 accumulation; sex-dependent oxidative stress
		LC3B		No sex differences	No changes	No autophagy effect
2023	Franconi et al.	Beclin-1 (BECN1)	Primary HUVECs (female & male)	↑ in males under basal conditions; ↑ in females after Ang II	↑ mRNA and protein in FHUVECs after Ang II; no change in MHUVECs	Promotes autophagosome nucleation; activates autophagy in females
		Parkin (PARK2)		↑ in females (basal and Ang II)	↑ mRNA and protein in FHUVECs; ↓ protein in MHUVECs after Ang II	Induces mitophagy (mitochondrial clearance) in females

	p62 (SQSTM1)	No basal differences	↑ in MHUVECs after Ang II	Accumulation indicates impaired autophagy flux in males
	LAMP-1	No basal differences	↑ in MHUVECs after Ang II	Suggests enhanced lysosomal fusion; late-stage activation in males
	LC3 (MAP1LC3A)	LC3-II/I ratio ↑ in males	No major change with Ang II	Indicates higher autophagosome formation in males
	miR-126-3p	↓ in males after Ang II	↓ only in MHUVECs	Regulates endothelial integrity; its decrease impairs autophagy and vascular repair
	miR-133a-3p	↑ in females after Ang II	↑ only in FHUVECs	Enhances Parkin and Beclin-1 expression → activates mitophagy
	miR-223-3p	↓ in both sexes (greater in females)	↓ with Ang II treatment	Suppresses autophagy and contributes to endothelial dysfunction

(continued)

Table 1 *(cont'd)*

Publication year	Authors	Biological parameters	Cell line	Sex differences	Expression change	Autophagy effect
		miR-29b-3p		No sex differences	No change with Ang II	No evident effect on autophagy
		miR-133b		No sex differences	No change with Ang II	No significant role in autophagy regulation
2022	Bazan et al.	Parkin (PARK2)	Primary mouse lung endothelial cells (male & female WT and Parkin$^{-}/^{-}$)	↑ in females under basal conditions	↓ after hypoxia ($CoCl_2$ or 1 % O_2) in both sexes	Promotes mitophagy; its reduction under hypoxia impairs mitochondrial quality control
		BNIP3		No basal difference	↑ after hypoxia in both sexes	Induces mitophagy through mitochondrial membrane permeabilization
		DRP1 (Dynamin-related protein 1)		↑ in females at baseline	↓ after hypoxia in both sexes	Regulates mitochondrial fission; reduction under hypoxia impairs mitophagy

		LC3B (MAP1LC3B)		Similar between sexes	↑ after hypoxia in both sexes	Marker of autophagic flux; supports ongoing autophagy during stress
		PGC1α		↑ in females	↑ after hypoxia in females	Stimulates mitochondrial biogenesis; compensates for mitophagy imbalance
2021	Mughal et al.	DAP1 (Death-associated protein 1)	Adult *S. mansoni* worms (♂♀) cultured in M199	↑ in females	Upregulated in females at day 1 of culture, then decreases	Inhibits excessive autophagy; limits cell death in female gonads
		DRAM (Damage-regulated autophagy modulator)		↑ in males	Higher transcription in males	Activates autophagy via p53 pathway; promotes autophagic activity in males
		LC3B (MAP1LC3B)		↓ in females	Decreased LC3B-I and LC3B-II during first 3 days of culture	Lower LC3B-II/LC3B-I ratio → reduced autophagy in females
		Beclin-1		No sex differences	Stable expression during culture	Essential for autophagosome initiation

(continued)

Table 1 *(cont'd)*

Publication year	Authors	Biological parameters	Cell line	Sex differences	Expression change	Autophagy effect
		Ambra1		No sex differences	Slightly reduced over time	Regulates early phagophore formation
		Vps34 (PIK3C3)		No sex differences	Stable expression	Required for autophagosome membrane formation
2021	Campesi et al.	LC3	Primary monocytes isolated from healthy men and women (18–37 years)	↑LC3-II/I ratio in males	Basal LC3-II/I ratio higher in males; LPS ↓ autophagy only in males	Indicates higher basal autophagy in males; LPS inhibits autophagy selectively in male cells
		TNF-α (release)		↑ in males after LPS	LPS induced TNF-α release in both sexes	Enhanced inflammatory response to LPS in males; linked to reduced autophagy
		ER-α (Estrogen receptor alpha)		↓ only in females after LPS	LPS significantly downregulated ER-α expression in female monocytes only	Suggests estrogen-mediated modulation of LPS signaling in females
		ER-β (Estrogen receptor beta)		↓ in both sexes, stronger in females	LPS reduced ER-β in males and females	Indicates sex-dependent ER regulation after LPS stimulation

2020	Liu et al.	LC3 (MAP1LC3B)	HepaRG cells (human hepatocytes)	Sex differences mediated by estrogen signaling	↓ with DHA + APAP; ↑ with DHA + E2 (estrogen) + APAP	Estrogen restores autophagy flux suppressed by n−3 PUFA (DHA) under APAP stress
		p62 (SQSTM1)		Estrogen-dependent regulation	↓ with DHA + APAP; ↑ with DHA + E2 + APAP	E2 reverses DHA-mediated autophagy inhibition, promoting clearance of damaged mitochondria
		β-catenin (CTNNB1)		↑ in males (E2 reduces its activation)	↑ with DHA + APAP; ↓ with DHA + E2 + APAP	β-catenin activation suppresses autophagy; estrogen inhibits β-catenin → autophagy restored
		p-JNK (MAPK8)		↑ activation in males	↑ with DHA + APAP; ↓ with DHA + E2 + APAP	JNK activation linked to oxidative stress → autophagy inhibition
		GSK3β		Differentially regulated by estrogen	↑ phosphorylation with DHA + APAP; reversed by E2	GSK3β–β-catenin signaling mediates sex-dependent autophagy control

Table 2

Publication Year	Authors	Biological Parameters	Sex differences	Expression change	Autophagy effect
2025	Zaheer et al.	Beclin-1	males = females	no changes	↓ (functional disruption impairs autophagy; effect only in males)
2025	Le et al.	LGALS3 (Galectin-3)	↑ in males	↑ in both sexes (higher in males)	↑ (stronger in males)
		CSF1R	males = females	↓ in both sexes	↓ (both sexes)
		TREM2	↑ in males	↑ in both sexes (higher in males)	↑ (stronger in males)
		CX3CR1	males = females	↓ in both sexes	↓ (both sexes)
2024	Jung et al.	Beclin-1	↑ in males	↑ in males	↑ (males)
		LC3B	males = females	no changes	No alteration detected in autophagy flux via LC3B levels.
2024	Uğuz et al.	LC3B	↑ in females	↑ after acute alcohol (females > males); ↓ during chronic exposure (both sexes)	Altered: ↑ acute, ↓ chronic

	BECN1 (Beclin-1)	males = females	Findings inconsistent with literature (no coordinated change with LC3B)	same in both sexes
	IBA1	↑ in females (hippocampus & cortex), ↑ in males (cortex only)	↑ with chronic alcohol exposure; acute alcohol ↑ only in males (cortex)	↑ (both sexes)
	Parvalbumin	males = females	↑ in hippocampus & cortex with chronic alcohol exposure	↑ (both sexes)
	GAD65	males = females	↑ in hippocampus & cortex with chronic alcohol exposure	↑ (both sexes)
	mGluR5	males = females	↑ in hippocampus & cortex with chronic alcohol exposure	↑ (both sexes)
	TRPM2	↓ in both sexes (higher in females)	↓ in both sexes (acute & chronic)	↓ (both sexes)

(continued)

Table 2 *(cont'd)*

Publication Year	Authors	Biological Parameters	Sex differences	Expression change	Autophagy effect
2024	Di Veroli et al.	NRF2	↑ in males	↑ total proteinin both sexes; ↓ nuclear NRF2 in both sexes; ↑ cytosolic NRF2 only in females	Impaired in females (nuclear translocation blocked)
		KEAP1	↑ in females	↑ in females	↑ KEAP1 prevents NRF2 nuclear translocation in females
		HO-1	↑ in males	↑ with NRF2 activity	↑ antioxidant defense (males stronger)
		NQO1	males = females	↓ in both sexes	↓ antioxidant defense
		p62/SQSTM1	males = females	no changes	same in both sexes
		LC3B	males = females	↑ in both sexes	↑ autophagosome formation
		LAMP2	↑ in males	↑ in males	↑ autophagic flux in males only
		XBP1	↑ in males	↑ in males	↑ ER-stress response in males

2024	Green et al.	Sphk2	↑ in males	↓ tumor incidence in males, ↑ in females	↓ Sphk2 → ↑ autophagy in males / ↓ autophagy in females
		p62/SQSTM1	↑ accumulation in females (WT)	↓ in males; ↑ in females	p62 accumulation → ↓ autophagy flux in females; opposite in males
		NQO1	↑ in females	↓ in males; ↑ in females	Correlates with p62: ↑ oxidative stress and ↓ autophagy in females
		LC3B	males = females	no changes	Basal autophagy marker; unchanged
		KEAP1/NRF2 pathway	More active in females	↑ NRF2 signaling in females	↑ antioxidant response but ↓ autophagy flux in females
2024	Baer et al.	Atg12	↑ in males	↑ with obesity in males; ↓ with caloric restriction in both sexes	↑ with obesity (males); ↓ with caloric restriction
		Atg5	males = females	no changes	same in both sexes
		Atg7	↑ in males	↑ with caloric restriction in control males	↑ (males, caloric restriction-dependent)

(continued)

Table 2 *(cont'd)*

Publication Year	Authors	Biological Parameters	Sex differences	Expression change	Autophagy effect
		Becn1 (Beclin-1)	males = females	no changes	same in both sexes
		SQSTM1 (p62)	males = females	no changes	same in both sexes
		Slc38a9	↑ in males	↑ with obesity in males	↑ nutrient sensing/ mTOR activation (males only)
		TFE3	↑ in females	↑ with caloric restriction in control females	↑ lysosomal/autophagy regulation (females only, caloric restriction-dependent)
		FNIP2	↑ in females; ↓ in males	↑ with caloric restriction in control females; ↓ with caloric restriction in obese males	↑ in females (caloric restriction); ↓ in males
		FLCN	↑ in females	↑ with obesity in females	↑ with obesity (females), ↓ with caloric restriction
		LAMP1	males = females	no changes	same in both sexes
		CTSB / CTSD	males = females	no changes	same in both sexes

2024	Panagaki et al.	Atg7	↑ in females	↓ in Down syndrome; ↑ with inhibitor in females	↓ in Down syndrome; rescued by inhibitor (females)
		Beclin-1 (Becn1)	↑ in females	↓ in Down syndrome; ↑ with inhibitor in females	↓ in Down syndrome; rescued by inhibitor (females)
		LC3B	↑ in females	↓ in Down syndrome; ↑ with inhibitor in females	↓ in Down syndrome; rescued by inhibitor (females)
		Atg3	males = females	no changes	same in both sexes
		CBS	↑ in both sexes (higher in females)	↑ (due to trisomy); ↓ with inhibitor	↑ CBS → disrupted autophagy; inhibition restores balance
		3-MST	↑ in both sexes (higher in females)	↑ in Down syndrome; ↓ with inhibitor in both sexes	↑ H_2S metabolism → indirect effect on autophagy
2024	Oliveira et al.	ATG7	↑in females	↑ in WT females	↑ autophagy initiation in females (TFE3-dependent)
		Beclin-1	↑ in females	↑ in both sexes	↑ autophagy activation

(continued)

Table 2 *(cont'd)*

Publication Year	Authors	Biological Parameters	Sex differences	Expression change	Autophagy effect
		Parkin	↑ in females	↑ with denervation in both sexes	↑ mitophagy (stronger in females, TFE3-dependent)
		TFEB	↑ in females	↑ in WT females	↑ lysosomal biogenesis in females (TFE3-dependent)
		TFE3	males = females	no changes	same in both sexes
		LC3-II flux	↑ in females	↓ with denervation in WT females	↓ mitophagy in females (TFE3-dependent)
		Cathepsin D (mature)	↑ in females	↑ with denervation in both sexes	↑ lysosomal function
		v-ATPase	↑ in females	↑ with denervation in both sexes	↑ lysosomal acidification
		Cathepsin B / MCOLN1 / LAMP1	males = females	↑ with denervation in both sexes	↑ lysosomal capacity

2024	Lopez-Lee et al.	LAMP2 (LAMP2A)	↑ in females	↓ with aging in both sexes	↓ CMA flux; contributes to aggregate buildup
		ATP6AP2	↑ in females	no changes	Involved in lysosomal acidification; potential minor ↑ efficiency in females
		Rab family (e.g., RAB9A)	↑ in females	no changes	Regulates macroautophagy trafficking; minor ↑ activity (females)
		PTEN	↑ in females (mouse cortex)	no changes	↑ autophagy via AKT inhibition (females)
		Klotho	↑ in males (mouse cortex)	no changes	↑ autophagy via AKT inhibition (males)
		Beclin-1 (BECN1)	↓ in females	↓ with age in both sexes (higher in females)	↓ autophagy initiation (especially in females)
		AKT (phosphorylated form)	males = females	no changes	Maintains balance; sex-specific regulators differ
		ERα/ERβ (estrogen receptors)	↑ in females	↓ with menopause	↓ autophagy gene expression under estrogen presence

(continued)

Table 2 *(cont'd)*

Publication Year	Authors	Biological Parameters	Sex differences	Expression change	Autophagy effect
2024	Kerr et al.	Sqstm1 (p62)	↓ in males	↓ with sarcopenia in males	↓ autophagy flux (males)
		Map1lc3a (LC3)	↓ in males	↓ with sarcopenia in males	↓ autophagosome formation (males)
		AMPK/p-AMPK (Thr172)	↓ in males	↓ in sarcopenic inmales	↓ autophagy activation via AMPK (males)
		PGC-1α	↓ in males	↓ with sarcopenia in males	↓ mitochondrial biogenesis and mitophagy (males)
		Atrogin-1 (FBXO32)	males = females	no changes	No effect on autophagy (ubiquitin–proteasome pathway)
		MuRF1 (TRIM63)	↓ in males	↓ with sarcopenia in males	↓ protein degradation signaling
2023	Zhazykbayeva et al.	p62/SQSTM1	↑ in both sexes (higher in males)	↑ in transgenic mices (higher in males)	↑ accumulation → ↓ autophagy flux
		LC3 (LC3A/B)	males = females	no changes	No change in autophagy initiation

		mTOR (phospho-mTOR)	↓ in males	↓ in transgenic females	↓ mTOR activity → ↑ autophagy signaling in females
		NF-κB (phospho-p65)	↑ in both sexes	↑ in transgenic mice	↑ inflammation-related autophagy modulation
		Cathepsin L	↓ in males; ↑ in females	↓ in transgenic males; ↑ in transgenic females	↑ lysosomal degradation in females
		Calpain	↓ in males; ↑ in females	↓ in transgenic males	Possible ↓ autophagy-linked proteolysis in males
		Caspase-3	↑ in males; ↓ in females	↑ in transgenic males; ↓ intransgenic females	↑ apoptotic turnover in males → impaired autophagy balance
		Caspase-1	↑ in females	↑ in transgenic females	↑ inflammasome activation → ↑ autophagy stress
2023	Zhang et al.	LAMP2A	↑ in females (brain); ↑ in liver (both sexes)	↑ with age in liver; = in kidney and brain	↑ or maintained CMA activity
		HSPA8 (HSC70)	↑ in males (kidney); ↑ in females (brain)	no changes with age (all tissues)	Maintained CMA substrate recognition

(continued)

Table 2 *(cont'd)*

Publication Year	Authors	Biological Parameters	Sex differences	Expression change	Autophagy effect
		LAMP1	↑ with age in females; ↓ in males	↑ with age females; ↓ in males	Possible sex-dependent lysosomal remodeling
		CTSD (Cathepsin D)	males = females	↑ with age in both sexes	↑ lysosomal degradation
		CTSB (Cathepsin B)	males = females	no changes	no effect on autophagy
		FASN	males = females	↑ with age; ↓ with fasting	Minor effect on CMA target turnover
		GAPDH, ACSS2, IDH1	males = females	↓ with fasting; = with age	Fasting activates CMA equally in both sexes
2023	Franconi et al.	Beclin-1 (BECN1)	↑ in females	↓ with angiotensin II in both sexes; stronger in females	↓ autophagy initiation (more in females)
		LC3B	↑ in females	↓ with angiotensin II in both sexes	↓ autophagosome formation
		p62/SQSTM1	↑ in males	↑ with angiotensin II in males	↓ autophagy flux (accumulation in males)

		ATG7	males = females	↓ with angiotensin II in both sexes	↓ autophagy initiation
		miR-30a	↑ in females	↑ with angiotensin II in females	↓ Beclin-1-mediated autophagy (via repression)
		miR-101	↑ in females	↑ with angiotensin II in females	↓ autophagy (targets ATG4 and STMN1)
		miR-204	↑ in males	↑ with angiotensin II in males	↓ autophagy via LC3 regulation
2023	Merino et al.	DARPP-32 (Ppp1r1b)	↓ in both sexes	↑ after treatment in males	↑ autophagy-mediated degradation of mHTT in males
		Nfl (Neurofilament light chain)	↓ in males	↑ after treatment in males	↑ neuronal integrity and autophagy recovery in males
		Mlh1	↓ in males	no changes	Possible involvement in DNA repair/autophagy regulation in males
		Fan1	↓ in males	no changes	Possible impact on genome stability; no autophagy change detected

(continued)

Table 2 *(cont'd)*

Publication Year	Authors	Biological Parameters	Sex differences	Expression change	Autophagy effect
		mHtt (mutant Huntingtin aggregates)	↑ in both sexes	↓ in males after treatment	↑ clearance via autophagy-proteasome system in males
		ERK1/2 (MAPK pathway)	↓ phosphorylation in males	↓ in males with tratment	↓ ERK1/2 → ↑ autophagy activation via TSC1/2–AMPK in males
2023	Baumgartner et al.	Cd68	↑ in females	↑ with aging in females	↑ lysosomal/autophagy activity in females
		Gusb	↑ in females	↑ with aging in females	↑ lysosomal degradation in females
		Man2b1	↑ in females	↑ with aging in females	↑ lysosomal/autophagy pathway in females
		Hexb	↑ in females	↑ with aging in females	↑ lysosomal/autophagy pathway in females
2022	Noh et al.	Beclin-1 (BECN1)	↑ in males, ↓ in females	↑ after ischemia in males	↑ autophagy initiation in males; ↓ in females
		LC3 (LC3-II/I)	↑ in males, = or ↓ in females	↑ after ischemia in males	↑ autophagosome formation in males

		p62/SQSTM1	↓ in males	↓ after ischemia in males	↑ autophagy flux in males
		ATG7	↑ in females	↑ after stroke in females	↑ autophagosome maturation in females
		HIF-1α	↑ in males	↑ after ischemia in males	↑ autophagy activation in males
		UCP-2	↑ in males	↑ after stroke in males	↑ mitophagy/oxidative stress adaptation in males
		Lamp2 / Atp6ap2 / Wdr45 / Hdac6 / Rab39b	↑ in females	↑ in aged females	↑ lysosomal and CMA-related autophagy in females
2022	Muehleman et al.	LC3 (LC3-II/ LC3-I)	↑ in both sexes	↓ during isoproterenol-induced hypertrophy in females; ↑ after isoproterenol removal in both sexes	↑ autophagy flux after Iso removal (both sexes)
		Proteasome activity	↑ in males	↑ in males with isoproterenol and angiotensin II	↑ protein degradation in males

(continued)

Table 2 *(cont'd)*

Publication Year	Authors	Biological Parameters	Sex differences	Expression change	Autophagy effect
		p-Smad2/3	↑ in females	↑ with hypertrophy	↓ Smad2/3 signaling linked to autophagy activation in Iso groups
		p-ERK1/2	↑ in males	↓ after removal of stimuli in both sexes	ERK1/2 inactivation associated with autophagy induction
		TGF-β1 (pathway activity)	↓ in males	↓ after Iso removal in males	Inhibition promotes autophagy and regression in males
2021	Svoboda et al.	Atg5	↑ in males	↑ in males	↓ autophagy (hypermethylation reduces activity)
		Lamp2	↑ in males	↑ in males	↑ lysosomal/autophagy activity (compensatory)
		Tlr4	↑ in females	no transcript data	↓ autophagy (inflammatory pathway suppression)
		Akap1	↑ in females	↑ in females	↑ mitochondrial autophagy/mitophagy regulation

		Grk5	↓ in females	↓ in females	↓ autophagy signaling
		Tgfb2	↑ in females	epigenetic change only	↓ autophagy via TGF-β pathway inhibition
2021	Kim et al.	PAR2 (F2RL1)	↑ in both sexes with high-fat diet	Induced by high-fat diet; absent in KO	Promotes β-arrestin-2 complex formation → inhibits AMPK
		β-arrestin-2	↑ in both sexes (high-fat diet)	↑ interaction with PAR2 and CAMKKβ	Inhibits AMPK-mediated autophagy
		AMPK (p-AMPK T172)	↓ in WT high-fat diet males; restored in PAR2 KO	High-fat diet reduces phosphorylation; PAR2 KO rescues it	Key autophagy activator; inhibited by PAR2
		mTOR (p-S2448)	↑ with high-fat diet; ↓ in PAR2 KO	High-fat diet activates mTOR signaling	Suppresses autophagy initiation
		LC3-II / LC3-I ratio	↓ in WT high-fat diet; ↑ in PAR2 KO high-fat diet	Indicates enhanced autophagy in KO	Reflects increased autophagosome formation
		p62 (SQSTM1)	↑ in WT high-fat diet; ↓ in PAR2 KO high-fat diet	Accumulates when autophagy is blocked	Reduced in KO → improved autophagic flux

(continued)

Table 2 *(cont'd)*

Publication Year	Authors	Biological Parameters	Sex differences	Expression change	Autophagy effect
		ATG12 conjugate	↓ in WT high-fat diet; ↑ in PAR2 KO high-fat diet	Restored in KO mice	Promotes autophagosome formation
2020	Prudente de Mello et al.	PTEN	↑ in females	no changes	↓ AKT activity → ↓ autophagy signaling in females
		Klotho	↑ in males	no changes	↓ insulin/AKT pathway → ↑ autophagy regulation in males
		mTOR (p-mTOR)	↑ in females	↑ activation (higher in females)	↓ autophagy induction in females
		FoxO3a (p-FoxO3a)	↑ phosphorylation (inactive) in females	↑ inactive form in females	↓ transcription of autophagy genes in females
		Beclin-1 (BECN1)	males = females	no changes	no effect on autophagy
		ATG7	males = females	no changes	no effect on autophagy

2020	George et al.	p38δ (MAPK13)	↑ in females	↑ with treatment in WT females; ↓ in KO females	↑ autophagy in KO females (↑ LC3-II, ↓ mTOR)
		mTOR (p-mTOR)	males = females	↓ with treatment in females	↓ mTOR → ↑ autophagy in females
		LC3-II	↑ in females	↑ after treatment in KO females	↑ autophagy flux in females
		p38γ (MAPK12)	males = females	no changes with treatment	no effect on autophagy
		Akt (p-Akt)	males = females	↓ with treatment in both sexes	no effect on autophagy

Table 3

Biological parameters	Sex differences	Expression change	Autophagy effect	Study population
LC3B (MAP1LC3B)	↑ in females vs males	↑ in female Laryngeal Cancer tissues	↑ autophagy activity in females	Human vocal fold tissues from Reinke Edema and Laryngeal Cancer patients (Asan Medical Center, Korea)
p62 (SQSTM1)	↑ in females vs males	↑ in Reinke Edema; ↓ in Laryngeal Cancer	↑ autophagy flux in female Reinke Edema	
HAS3 (Hyaluronan synthase 3)	↑ in Reinke Edema (female smokers)	↑ with cigarette smoke exposure	Promotes extracellular matrix remodeling and contributes to protective autophagy in Reinke Edema	
Atg5 / Beclin-1 / p62	No sex difference	No changes	No changes	The Cancer Genome Atlas - Lung Cancer (TCGA)
RASD1	↓ in females	↓ in female HCM hearts	↓ autophagy signaling via RASD1–AMPK axis	106 human cardiac samples (52 ♀, 54 ♂, GEO GSE36961)
NAMPT	↓ in females	↓ with RASD1 reduction	↓ NAD^+ synthesis → ↓ autophagy flux	
MYH6	↓ in females	↓ in female HCM vs. male	Structural downregulation linked to impaired autophagy	

Sphk2	↑ in males	↑ in male Hepatocellular Carcinoma tissues	↑ p62 accumulation → ↓ autophagy in males	TCGA-LIHC (Liver Hepatocellular Carcinoma) + Tokyo Hepatocellular Carcinoma cohort
p62/SQSTM1	↑ in males	↑ in male Hepatocellular Carcinoma (correlated with Sphk2 and Nqo1)	↓ autophagy flux; linked to poor survival in males	
NQO1	↑ in males	↑ in male Hepatocellular Carcinoma tissues	↑ oxidative stress and ↓ autophagy efficiency in males	
LC3B	No sex difference	No changes	No changes	
ATG7	↓ in women vs. men	↓ in men with HFrEF (↑ catabolic activation)	↑ autophagy signaling in men (linked to muscle atrophy)	Muscle biopsies from 22 ♂ + 16 ♀ HeartFailure and reduced EjectionFactor patients; 18 ♂ + 16 ♀ controls (Leeds Teaching Hospitals, UK)
LC3B (MAP1LC3B)	↓ in HFrEF (both sexes)	↓ autophagy flux	↓ autophagy flux	
MuRF1 (TRIM63)	↑ in men vs. women	↑ proteasomal degradation	↑ autophagy-proteasome system	

(continued)

Table 3 *(cont'd)*

Biological parameters	Sex differences	Expression change	Autophagy effect	Study population
MAFbx/Atrogin-1 (FBXO32)	↑ in men vs. women	↑ protein breakdown linked to aurophagy	↑ protein breakdown, linked to autophagy-related wasting	
UBB (Ubiquitin B)	↑ in men vs. women	↑ autophagy-proteasome activity	↑ autophagy-proteasome activity	
IGF1 (Insulin-like growth factor 1)	↑ in women vs. men (interaction with disease)	No changes	No changes	
MSTN (Myostatin)	↑ in men vs. women	↓ autophagy and regeneration	↓ autophagy and regeneration	
HIF1A	↑ in men vs. women	↑ in men with HFrEF	No changes	
PGC1A	No sex difference	↑ mitochondrial biogenesis and autophagy support	No changes	
ESR1 (Estrogen receptor 1)	↓ in women vs. men	↓ in female and HFrEF	↓ estrogen-driven autophagy	

VEGF	↑ in women vs. men	↓ in muscle; ↑ in serum of women	↑ angiogenesis and protective autophagy	
AMPK (Thr172 phosphorylation)	↑ in females only	↑ activation in female schizophrenia DLPFC; ↔ in males	↑ beclin-1 activation and pro-autophagy signaling	Human postmortem samples (n = 80; 20♂ + 20♀ control, 20♂ + 20♀ Schizoprenia)
Beclin-1 (Ser93 phosphorylation)	↑ in females only	↑ AMPK-mediated phosphorylation in female schizophrenia; ↔ in males	↑ PI3K-III complex activity produce ↑ autophagy induction	
Beclin-1 (Ser15 phosphorylation)	No sex difference	↔ in all groups	No change; AMPK-independent site	
ULK1 (Ser317/467/555/638 phosphorylation)	No sex difference	↔ in schizophrenia	No changes	
LC3B-II / LC3B-I ratio	↑ in females (mouse model)	↑ after Beclin-1 S91/94D activation in females; ↔ in males	↑ autophagosome formation only in females	
LC3 (MAP1LC3A)	No sex difference	↑ LC3-II/I ratio in female amniotic fluid cells	↑ basal autophagy in females	42 pregnant women (22 female fetuses, 20 male

(continued)

Table 3 *(cont'd)*

Biological parameters	Sex differences	Expression change	Autophagy effect	Study population
p62 (SQSTM1)	↓ protein in females; ↑ mRNA in females	Male amniotic fluid cells: higher protein; female amniotic fluid cells: higher mRNA	↑ autophagy flux; male → autophagy blockage	fetuses; 16–18 weeks gestation)
LAMP1	↑ mRNA in males; protein similar between sexes	↑ LAMP1 mRNA in males	↑ lysosomal gene activation but not increased protein levels	
ERα (ESR1)	↑ in females	↑ protein levels	↑ autophagy in an mTOR-independent pathway	
ERβ (ESR2)	No sex difference	No changes	No changes	
Caspase-9	↑ in females	↑ activity	Linked to intrinsic apoptosis and possible autophagy–apoptosis crosstalk	
Caspase-3	↑ in females	No changes	No changes	
TERT	↑ in females	↑ activity	↓ replicative senescence in female amniotic fluid cells	

MTOR	↑ in females vs. males in Alzheimer Desease	↑ in female; no changes in males	↑ mTOR activation produce ↓ autophagy induction	78 postmortem brains (16 controls, 28 asymAD, 34 AD; 37 ♀, 41 ♂)
RAPTOR (MTORC1 component)	↑ in females	↑ in female AD	↑ mTORC1 signaling produce ↓ autophagy flux	
ULK1	↓ in females	↓ in female AD	↓ autophagy initiation (mTOR-dependent)	
BECN1 (Beclin-1)	↓ in females	↓ in female AD vs. asymAD	↓ autophagosome nucleation	
ATG5 / ATG7	↓ in females	↓ in female AD	↓ autophagy machinery efficiency	
SQSTM1 (p62)	↑ in females	↑ in female AD	↑ accumulation due to impaired autophagy degradation	
MAP1LC3B (LC3B)	↓ in females	↓ in female AD	↓ autophagic flux (LC3-II depletion)	
TFEB	↓ in females	↓ in female AD	↓ lysosomal biogenesis and autophagy regulation	

(continued)

Table 3 *(cont'd)*

Biological parameters	Sex differences	Expression change	Autophagy effect	Study population
NFE2L2 (NRF2)	↓ in females	↓ in female AD	↓ oxidative stress defense and mitophagy activation	
SIRT1	↓ in females	↓ in female AD; ↔ in males	↓ AMPK–autophagy signaling; reduced Aβ clearance	
PRKN (Parkin)	↓ in females	↓ in female AD	↓ mitophagy efficiency and mitochondrial quality control	
FOXO3	↑ in women	↑↑ at 12 h and 24 h in women	↑ transcriptional initiation of autophagy, but not translated to flux	8 ♂ (22 ± 3 y) and 8 ♀ (20 ± 1 y), untrained
BECN1	↑ in women	↑ at 24 h in women; transient ↓ then ↑ in men	↑ initiation signaling; no increase in flux	
ULK1, ATG5	No sex difference	No changes	No change in initiation	
LC3-I	↓ in men; ↑ in women (at 24 h)	↓ in men; ↑ in women	↑ flux in men; ↓ flux in women	
LC3-II	↓ in men	↓ in men	↑ autophagosome degradation in men	

focus of this review will be on sex-driven differences in macroautophagy during aging and in aging-related diseases. Tables 1–3 provide a comprehensive summary of studies that have taken biological sex into account as a key parameter in the assessment of macroautophagy in the context of aging and aging-related diseases (Suppl Material).

2. Molecular machinery of autophagy: Sex differences in key proteins across distinct pathways

A complex network of cellular effectors is involved in the autophagic response (Galluzzi et al., 2017) (Fig. 1). In fact, it is difficult to establish the boundaries that define what an autophagy protein is (usually termed as "ATG", from "autophagy-related"), as many of them show additional roles in other cellular processes, or only participate in autophagy in very restricted circumstantial contexts. Still, it is possible to identify important proteins and complexes that participate in this catabolic pathway. In this regard, the lysosome (and its components, including membrane proteins like LAMP1 or degradative enzymes such as cathepsins) is the only structure that is shared by all three autophagy modes, as these routes only differ in how the substrates are transported into this organelle (Yim & Mizushima, 2020).

Intense research has been focused on unveiling the molecular machinery core of macroautophagy, responsible for the formation of autophagosomes (Noda, 2024; Shatz & Elazar, 2024). In this regard, many effectors interplay to initiate or block the biogenesis of these vesicles, from repressors (such as mTOR or DAP1) (Koren et al., 2010; Wang & Zhang, 2019) to inducers (such as AMPK or DRAMs proteins) (Barthet & Ryan, 2022; Tamargo-Gómez & Mariño, 2018) and transcription factors (including TFEB, members of the FOXO family or HIF-1α) (Cheng, 2019; Settembre et al., 2011; Wang et al., 2017). Once the process is triggered, different complexes act to begin the nucleation and elongation of the phagophore, a nascent structure that will sequester the cytosolic cargo (Nakatogawa, 2020) (Fig. 1). First, the ULK1/2 complex (which also includes ATG13, ATG101 or FIP200/RB1CC1) translocates to formation sites for the recruitment and activation of the class III phosphatidylinositol 3-kinase (PI3KC3) complex (Zachari & Ganley, 2017). The latter is a group of proteins (namely, Beclin-1, VPS34/PIK3C3, VPS15/p150/PIK3R4 and Barkor/ATG14L) that generates phosphatidylinositol 3-phosphate (PI(3)P), a signaling lipid that recruits further downstream factors involved in autophagosome biogenesis (Hurley & Young, 2017). Specifically, PI(3)P-binding proteins like DFCP1, ALFY or the WIPI family, that in turn facilitate the action of ATG9 and ATG2. ATG9 is the only transmembrane protein directly involved in autophagosome biogenesis, which drives the transportation of vesicles to the phagophore so they can be used as membrane sources. ATG2 translocates lipids from the aforementioned vesicles and other membranes to nascent autophagosomes (Noda, 2021). While the phagophore evolves to form the autophagosome, two major ubiquitin-like conjugation systems get activated to promote phagophore elongation and drive cargo sequestering (Mizushima, 2020; Noda, 2023). First, ATG7 and ATG10 mediate the covalent conjugation of ATG12 to ATG5, a conjugate that will finally bind to ATG16. This complex, ATG12-ATG5-ATG16, then participates in the second conjugation system along with ATG3 and, again, ATG7, to transfer molecules of ATG8 to the membrane of the phagophore. ATG8 proteins (that are classified into two different subfamilies in mammals: LC3 and GABARAP) need to be first activated by the ATG4 proteases, which cleave the C-terminal of ATG8 to expose a glycine residue that is essential for their conjugation to the lipid phosphatidylethanolamine and transfer to the membrane (Fernández & López-Otín, 2015). Interestingly, ATG8 proteins are also necessary for cargo sequestering, as they bind to "autophagy receptors" that are labelling the material that will be degraded (Johansen & Lamark, 2020). This is the case, for example, of SQSTM1/p62, which can bind to both ATG8 proteins and ubiquitinated substrates, establishing a link between the autophagosome and the target (Lamark et al., 2017). More importantly, these autophagy receptors add specificity to autophagic degradation, allowing for the definition of different types of selective autophagy depending on the nature of the molecules or structures that will be transferred to the lysosome (Adriaenssens et al., 2022; Lamark & Johansen, 2021; Vargas et al., 2023). These include mitophagy (the selective elimination of mitochondria, mediated by effectors like Parkin and autophagy receptors such as p62, OPTN, NDP52, BNIP3, BNIP3L/NIX, FUNDC1, FKBP8, PHB2 or NIPSNAP1/2); ER-phagy (which targets endoplasmic reticulum by the action of FAM134B, SEC62, RTN3, CCPG1, ATL3 or TEX264); lipophagy (selective autophagic degradation of lipid droplets that involves interactions with lipolytic enzymes such as ATGL and HSL and the participation of VPS4A and the mTORC1–Plin3 signaling axis); glycophagy (eliminating glycogen that is recognized by STBD1); ribophagy (with NUFIP1 acting as a autophagy receptor for ribosomes) or aggrephagy (targeting protein aggregates, thanks to p62 or TOLLIP). Once the autophagosome is closed, it can be transported through the cytoplasm by motor proteins like dynein and kinesin/FYCO1, maturing during this process (Bejarano et al., 2018; Kast & Dominguez, 2017; Nieto-Torres et al., 2021). Finally, the autophagosome fuses with the lysosome, a process that is mediated by proteins families such as SNAREs or RABs, and complexes like HOPs or ESCRT (Barz et al., 2021; Jiang et al., 2021; Tian et al., 2021; van der Beek & Klumperman, 2025; Vietri et al., 2020; Zhao et al., 2021; Das et al., 2024; Garcia-Macia et al. 2021).

Accumulating evidence robustly demonstrates that both the baseline state of autophagy and its responsiveness to various challenges are significantly influenced by biological sex across different species, tissues, and developmental stages. These differences suggest intrinsic variations in cellular turnover, quality control capacity and stress adaptation strategies between males and females (Shang et al., 2021; Tower et al., 2020). There is, for example, increasing literature pointing out to sex differences in basal autophagy. Comparative studies of basal autophagy markers between sexes reveal a complex and tissue-specific landscape. For example, male peripheral human blood monocytes exhibit higher LC3-II/I ratios (which might point out accumulation of autophagosomes, as LC3-II is the lipidated form of this protein) whereas, conversely, female human umbilical vein endothelial cells (HUVECs) display increased levels of lysosomal markers and the mitophagy regulator Parkin, although male HUVECs express higher levels of Beclin-1 protein (Campesi et al., 2023; Campesi et al., 2022; Franconi et al., 2023). In rodent models, young female skeletal muscle shows enhanced basal expression of key autophagy and lysosomal proteins (ATG7, TFEB, LAMP1 and Cathepsin D) as well as greater mitophagy flux compared to males (Oliveira et al., 2024; Triolo et al., 2022). In contrast, female neurons may exhibit lower basal autophagy than their male counterparts, and male mouse liver lysosomes demonstrate superior CMA substrate uptake (Patrizz et al., 2021; Zhang et al., 2023). Invertebrate models further support the existence of sex-specific autophagic regulation: young male *C. elegans* display constitutive tubular lysosomes in the gut, a feature absent in hermaphrodites; *Schistosoma mansoni* exhibits sex-dependent expression of autophagy-related genes such as *DAP1* and *DRAM*; and male *Bombyx mori* larvae show higher basal levels of LC3-II/I and p62 (Mughal et al., 2021; Punyavathi & Manjunatha, 2020; Ramos et al., 2022). Collectively, these findings highlight the sexually dimorphic nature of basal autophagy, which varies significantly depending on species, tissue type, and the specific markers assessed.

Autophagy responds dynamically to physiological challenges and to stress and injury, and this triggered autophagic response is markedly influenced by biological sex. Acute exhaustive exercise induces autophagy in a sex-dependent manner, as young male humans exhibit a significant increase in autophagic parameters—such as p62 degradation and nuclear TFEB translocation—in skeletal muscle post-exercise, whereas females primarily show upregulation of autophagy initiation genes without corresponding flux activation (Luk et al., 2021; Triolo et al., 2022). Similarly, in peripheral blood mononuclear cells, LC3-II levels rise only in males, suggesting a more complete autophagic response to exercise in males (Escobar et al., 2021).

Nutritional interventions also reveal sex-specific autophagic regulation, although the differences can vary depending on the nature of the dietary strategy. Early-life calorie restriction preserves muscle mass in both sexes in young rats, but does not significantly alter canonical autophagy markers, although downstream mitochondrial responses diverge (Torrens-Mas et al., 2022). However, chronic undernutrition impairs autophagy through PPARα depletion, predominantly in male mice due to a higher baseline expression (Suh et al., 2022). Of note, high-fat diets elicit distinct effects. PAR2 deletion restores autophagy and prevents hepatic steatosis in males, while in females, impaired autophagy flux is linked to NRF2 pathway dysfunction (Di Veroli et al., 2024; Kim et al., 2021). In the context of acute liver injury, omega-3 fatty acids enhance protective autophagy in females but suppress the same activity in males via estrogen-dependent GSK3β/β-catenin signaling (Liu et al., 2020). Interestingly, maternal diet during gestation impacts offspring cardiac autophagy pathways in a sex-dependent manner, with females often showing greater alterations (Liu et al., 2022). Finally, it was showed that protein restriction impacts mTOR signaling and reduces Aβ burden only in male Alzheimer's model mice, though the link to autophagy remains unclear (Bensalem et al., 2023).

Sex also modulates autophagic responses to other stressors, including inflammation, environmental toxins and psychological or physical stress. For example, lipopolysaccharide exposure that mimics bacterial infection suppresses autophagy in male human monocytes but not in females (Campesi et al., 2022). In alignment with these findings, lipopolysaccharide-induced neuroinflammation reduces hippocampal autophagy, with resveratrol reversing this effect via SIRT1/YAP inhibition in males (Tian et al., 2020). However, viral infection in fish leads to enhanced autophagy-mediated degradation of MITA in females, increasing susceptibility through Cyp19a1a signaling (Lu et al., 2022). Also, environmental stress impacts autophagic function in a sex dependent manner. For example, ionizing radiation preferentially induces autophagy in female endothelial cells, prenatal alcohol exposure alters autophagy markers in the developing brain in a sex-specific manner, maternal smoking leads to persistent LC3-II elevation in male offspring lungs, and mild cold exposure enhances mitophagy signaling more strongly in male voles (Campesi et al., 2023; Uğuz et al., 2025; Wang et al., 2020a; Wang et al., 2020b). Finally, psychological and physical stress in early life, such as maternal separation, induces long-term, sex-specific changes in hippocampal autophagy and mitophagy pathways, and social stress models increase Beclin-1 expression selectively in the male amygdala (Jung et al., 2024; Sierra-Fonseca et al., 2021).

In summary, although the molecular scenario behind the sex-dependent phenotype is unknown, these findings collectively demonstrate that both basal autophagy and its activation are subject to significant sexual dimorphism. The differences extend beyond magnitude to involve distinct regulatory mechanisms and functional outcomes, suggesting that males and females engage autophagy differently to maintain homeostasis and adapt to physiological and pathological challenges. These sex-specific patterns might contribute to differential aging trajectories and disease susceptibility.

3. Sex as a biological variable in autophagy: Implications for health and disease

Autophagy is essential for maintaining cellular integrity throughout life, and its dysregulation is increasingly recognized as a contributor to aging and the development of age-related diseases. Evidence indicates that both basal autophagy and its responsiveness to stressors are modulated by biological sex, influencing aging trajectories and disease susceptibility (Hägg & Jylhävä, 2021). Thus, given its central role in cellular physiology, it is not surprising that autophagy dysregulation is implicated in both normal and pathological aging. Additionally, alterations of autophagic function have been described in a wide spectrum of human diseases, including neurodegenerative disorders (e.g. Alzheimer's, Parkinson's or Huntington's disease), cardiovascular diseases (e.g. heart failure and atherosclerosis), inflammatory disorders (e.g. inflammatory bowel disease or ankylosing spondylitis), etc.

Although autophagy generally declines with age, this process does not follow a uniform pattern across sexes. In skeletal muscle, aging is associated with accumulation of autophagy and mitophagy proteins (e.g.: Beclin1, ATG7, Parkin, TFEB, p62), suggesting increased stress or impaired flux. Notably, young females exhibit higher baseline levels of these proteins, and aging blunts exercise-induced autophagy differently in males and females (Moradi et al., 2024; Triolo et al., 2022). Thus, mitochondrial dynamics also diverge with age: males tend to lose mitochondrial content, while females preserve it, albeit with functional decline—patterns that correlate with sex-specific autophagy and mitophagy regulation (Triolo et al., 2022). In the brain, increased autophagy gene expression in the ventral hippocampus of aged female mice is linked to socioemotional changes, whereas other tissues, such as the Harderian gland in striped dwarf hamsters (*Cricetulus barabensis*), show age-related increases in autophagy markers in both sexes, being consistently higher in males (Baumgartner et al., 2023; Kong et al., 2021). Nevertheless, these findings diverge from previous reports in other hamster models (Vega-Naredo et al., 2009), suggesting that the conclusions should be interpreted with caution. Overall, while aging impacts autophagy across sexes, females often exhibit higher baseline markers or different adaptive/maladaptive responses compared to males, potentially linked to hormonal changes and inherent genetic differences.

Sex differences are well established in cardiovascular disease, and growing evidence implicates autophagy dysregulation as a contributing factor. Studies on isoproterenol-induced cardiac hypertrophy reveal that autophagy responses are both stimulus- and sex-dependent. For instance, regression from isoproterenol-induced hypertrophy involves rapid autophagy activation in both sexes—though females exhibit higher baseline levels—whereas regression from angiotensin II-induced hypertrophy is limited, particularly in females, who fail to mount an early autophagic response (Muehleman et al., 2022). Similarly, therapeutic induction of autophagy with trehalose, an autophagy inducer via TFEB, promotes atherosclerosis regression effectively only in female mice, likely due to their lower basal autophagy flux in plaque foam cells (Robichaud et al., 2024). In heart failure with preserved ejection fraction, p62 accumulation is observed in both sexes but appears to arise from distinct upstream mechanisms; specifically, heightened NF-κB signaling in males versus reduced mTOR inhibition in females (Kaur & Lau, 2022). In human endothelial cells, angiotensin II elicits divergent autophagic responses: females upregulate Beclin-1 and Parkin, suggesting enhanced autophagy and mitophagy, while males accumulate p62 and LAMP-1 despite increased

LC3-II/I, pointing to impaired flux. These effects may be mediated by sex-specific miRNA profiles (Zhazykbayeva et al., 2023). Although direct evidence of sex differences in lysosomal dysfunction in diabetic cardiomyopathy remains limited, the hormonal regulation of key autophagy components suggests plausible sex-specific effects. Drugs like metformin showed sex-dependent effects on kidney pathology possibly linked to differential autophagy regulation (Luetić et al., 2021). Additional models, such as pulmonary hypertension, also point to sex-dependent involvement of mitophagy regulators like Parkin. Moreover, environmental exposures such as perinatal lead have been shown to induce persistent, sex-specific epigenetic changes in autophagy-related genes like *ATG5*, predominantly affecting males (Svoboda et al., 2021).

Neurodegenerative disorders frequently exhibit sex-specific patterns in prevalence, progression, and pathology—differences that may be partially explained by divergent autophagy regulation. In Alzheimer's disease, which disproportionately affects women, lower basal autophagy levels throughout life in females may contribute to increased vulnerability (Cui et al., 2023; Lopez-Lee et al., 2024). Although transcriptomic analyses reveal shared disruptions in autophagy pathways, the surrounding molecular networks differ markedly between sexes. In contrast, Parkinson's disease, more common in men, shows significant alterations in autophagy and lysosomal gene expression specifically in the female substantia nigra, suggesting sex-specific compensatory or pathological mechanisms (López-Cerdán et al., 2022; Santiago et al., 2022). In Huntington's disease, the autophagy-enhancing compound 4QMn slows disease progression only in male zQ175 mice, potentially due to sex-dependent modulation of MAPK signaling (Merino et al., 2023). Similarly, in the male-specific condition spinal and bulbar muscular atrophy, impaired autophagy flux is directly linked to androgen-driven accumulation of mutant androgen receptor aggregates (Chivet et al., 2020). Other rare neurodevelopmental disorders, such as ADNP syndrome, also show sex-specific brain pathology associated with autophagy dysregulation via FOXO3 signaling (Karmon et al., 2022). Of note, ischemic stroke models consistently demonstrate opposing outcomes of autophagy modulation: administration of 3-Methyladenine (3-MA) is neuroprotective in males and low-estrogen females, but detrimental in estrogen-replete females, aligning with greater stroke-induced autophagy activation in males (Noh et al., 2023; Patrizz et al., 2021). Caution is warranted, however, as although 3-MA has traditionally been used as an autophagy inhibitor—primarily due to its suppressive effect on class III PI3K activity—this compound may exhibit a dual role depending on the duration of treatment (Wu et al., 2010). While microglial autophagy is essential for resolving neuroinflammation via NLRP3 clearance, sex-specific regulation of this axis remains underexplored. Finally, stress-related models reveal long-lasting, sex-dependent alterations in autophagy: early-life stress impairs hippocampal autophagy flux, while social stress selectively increases Beclin-1 expression in the male amygdala (Jung et al., 2024; Sierra-Fonseca et al., 2021).

Age-related muscle wasting (sarcopenia) and responses to muscle disuse exhibit marked sex differences, many of which are closely linked to autophagy and mitophagy regulation. In male mice, sarcopenia progression is strongly associated with a decline in AMPK signaling and downstream autophagy/mitophagy pathways. In contrast, these pathways appear more stable in aging females, potentially contributing to their relatively milder sarcopenic phenotype (Kerr et al., 2024; Tian et al., 2023). In models of acute muscle disuse, such as hindlimb unloading, females initiate a more rapid and robust autophagy and mitophagy response—evidenced by elevated Bnip3 and LC3-II/I levels—suggesting a protective mechanism against mitochondrial stress (Rosa-Caldwell et al., 2021). However, following denervation, females paradoxically suppress mitophagy flux through a TFE3-dependent mechanism, resulting in the accumulation of dysfunctional mitochondria. In contrast, males maintain mitophagy flux but experience a loss of mitochondrial content (Oliveira et al., 2024). These findings underscore the complexity of sex-specific autophagic regulation in skeletal muscle and highlight the need to consider sex as a critical variable in both the study and treatment of muscle aging and disuse-related atrophy.

Autophagy plays a dual role in cancer, acting as both a tumor suppressor and promoter depending on context—a complexity further modulated by biological sex. In hormone-dependent cancers, sex hormones regulate the expression of proteoglycans, which in turn modulate autophagy, suggesting an indirect but potentially significant sex-specific influence. More directly, sex-biased microRNA expression profiles can target autophagy-related genes, as illustrated by the miR-22/ERβ axis, which modulates autophagy and chemotherapy response in colorectal cancer (Caserta et al., 2023). Sex differences also emerge in cancer incidence and progression. The male predominance in laryngeal cancer has been linked to a protective autophagic response to smoking in females, potentially diverting epithelial transformation toward benign outcomes such as Reinke's edema (Kim et al., 2025). In contrast, in MASH-associated hepatocellular carcinoma, impaired autophagy driven by SphK2 promotes tumorigenesis specifically in males, while SphK2 appears protective in females (Green et al., 2024). In glioblastoma, secreted proteins associated with CMA—such as Lumican and Osteopontin—exhibit sex-dependent prognostic value, further underscoring the relevance of sex in autophagy-related cancer biology (Salinas et al., 2025b).

Autophagy modulation is also emerging as relevant in pain and inflammatory conditions, often with sex specificity. In neuropathic pain models, functional Beclin-1-mediated autophagy limits pain hypersensitivity in males but is less critical in females; conversely, it facilitates pain-associated anxiety only in males (Tam et al., 2024; Zaheer et al., 2025). The mTOR/autophagy pathway in the insular cortex is implicated in neuropathic pain, but specific sex differences here need study (Wang et al., 2021). Ankylosing Spondylitis involves sex-specific gene signatures for programmed cell death, including the autophagy regulator RHEB in females (Dong et al., 2024). In addition, in inflammatory bowel disease, autophagy is altered, and sex hormones influence disease via the "microgenderome," but a direct sex-dependent autophagy link requires explicit investigation (Hassan et al., 2023).

As abovementioned, autophagy is a highly conserved cellular process and the sex-dependet response can be found beyond mammals. In fact, the influence of sex on autophagy regulation extends across diverse taxonomic groups. In *Danio rerio*, for instance, increased susceptibility of females to viral infections has been associated with elevated expression of Cyp19a1a, which promotes the selective autophagic degradation of the antiviral adaptor MITA (Lu et al., 2022). In sea trout (*Salmo trutta*), sex-specific responses in lysosomal and autophagy markers have been documented in relation to life cycle stressors (Kurhaluk & Tkachenko, 2021). Similarly, while the fungicide difenoconazole induces autophagy and exhibits sex-specific toxicity, a direct causal relationship between these phenomena remains to be established. Even in parasites such as *S. mansoni*, basal expression levels of key autophagy-related genes differ between males and females (Mughal et al., 2021).

4. Molecular determinants of biological sex-associated differences in autophagy

It has become increasingly evident that human biological sex—defined by chromosomal complement (XX in females, XY in males) and the physiological differences primarily mediated by sex hormones—is a fundamental variable influencing virtually every aspect of health and disease (Hägg & Jylhävä, 2021; Tower et al., 2020). Sex differences are observed in the incidence, prevalence, clinical presentation, progression and therapeutic response of numerous conditions, many of which are closely linked to autophagy, including cardiovascular diseases, Alzheimer's disease, Parkinson's disease, osteoporosis, autoimmune disorders, certain cancers and metabolic syndromes. As previously mentioned, women exhibit a higher prevalence of Alzheimer's disease and osteoporosis, whereas men show increased rates of early-onset cardiovascular disease and Parkinson's disease (Cui et al., 2023; Hägg & Jylhävä, 2021; Lopez-Lee et al., 2024; Noh et al., 2023; Ortona et al., 2023; Tower et al., 2020). Despite growing awareness and the mandated inclusion of sex as a biological variable in biomedical research, the specific molecular mechanisms underlying these disparities remain poorly understood. A critical and still underexplored area lies at the intersection of these two biological dimensions: the influence of sex on the autophagy pathway itself (Shang et al., 2021). These observed sex differences suggest that males and females may employ distinct intracellular strategies to maintain homeostasis. Understanding the sex-specific regulation of autophagy carries profound implications for both basic biological research and clinical applications. Unraveling how genetic and physiological factors differentially shape autophagic capacity and responsiveness could offer critical insights into the mechanisms underlying sex-based disparities in disease susceptibility, progression and therapeutic outcomes. Furthermore, given that autophagy is increasingly recognized as a promising therapeutic target across a wide range of diseases, acknowledging potential sex-based differences in pathway activity and responsiveness is essential for the development of effective and safe interventions (Green et al., 2024; Merino et al., 2023; Tam et al., 2024). Thus, a "one-size-fits-all" approach to autophagy modulation may be suboptimal—or even harmful—if the underlying regulatory mechanisms differ significantly between sexes (Liu et al., 2020; Noh et al., 2023; Patrizz et al., 2021). In the following section, we will summarize the current knowledge on the biological aspects that influence autophagy both directly—by modulating the expression and activity of autophagy-related genes and proteins—and indirectly, through their effects on signaling pathways, cellular stress responses and metabolic states that converge on the autophagic machinery (Fig. 2).

This figure schematic illustrates key regulatory nodes through which biological sex influences autophagic activity. For example, hormones modulate autophagy via receptor-mediated signaling cascades, sex chromosome-linked

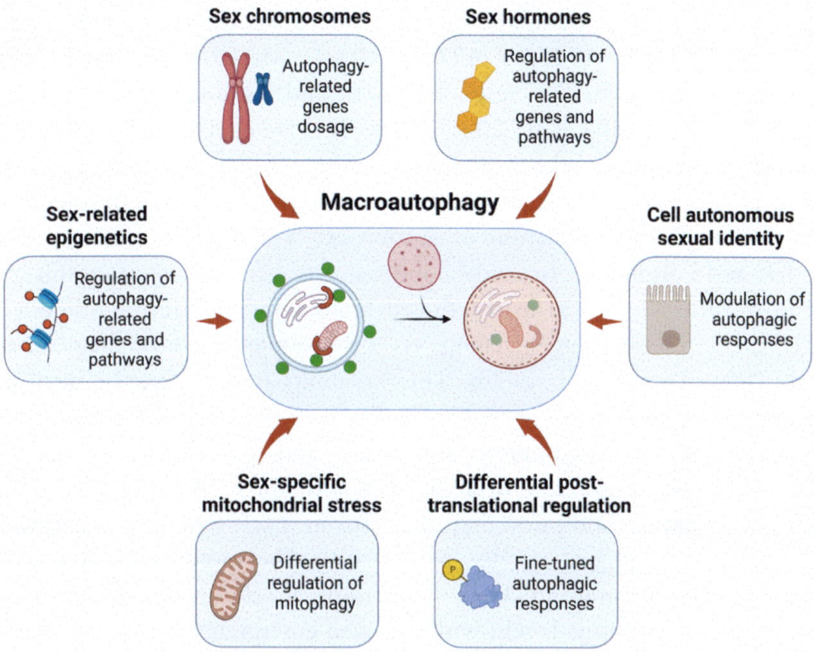

Fig. 2 Molecular determinants of biological sex-associated differences in autophagy.

genes and epigenetic regulators further contribute to sex-specific autophagic responses. These mechanisms interact with age-related changes in cellular homeostasis, shaping differential susceptibility to aging-related diseases in males and females.

4.1 The endocrine influence: Sex hormones and autophagy

Sex hormones—primarily estrogens (e.g., 17β-estradiol, E2) and androgens (e.g., testosterone, T)—exert profound and widespread effects on cellular physiology, including the regulation of autophagy. These hormones act predominantly through their nuclear receptors: estrogen receptors alpha and beta (ERα/ESR1, ERβ/ESR2) and the androgen receptor (AR) (Shang et al., 2021; Tower et al., 2020). A key mechanism involves the direct transcriptional regulation of autophagy-related genes. Hormone-receptor complexes can bind to specific hormone response elements located in promoter or intronic regions of core autophagy genes, thereby modulating their expression. These target genes span the entire autophagic process, including those involved in phagophore initiation (e.g., *ULK1/2*,

BECN1, *PIK3C3*, *AMBRA1*), membrane expansion and cargo recognition (e.g., *ATG3*, *ATG4B*, *ATG5*, *ATG7*, *MAP1LC3B*, *SQSTM1*), autophagosome maturation (e.g., *FYCO1*), and lysosomal function (e.g., *TFEB*, *CTSD*) (Shang et al., 2021). However, the effects of sex hormones on autophagy are complex and highly context-dependent. Estrogen, for instance, is generally considered neuroprotective and cardioprotective, partly through its modulation of autophagy. Yet, its effects vary across models and conditions. In some contexts, estrogen enhances autophagy, while in others—such as iron-induced brain injury or during adipocyte differentiation—estrogen signaling via ERα suppresses autophagy (Tao et al., 2021; Tower et al., 2020). This variability underscores the intricate integration of hormonal cues with cellular state. Conversely, testosterone has been shown to suppress autophagy and mitophagy in skeletal muscle (Tian et al., 2023). In alignment with these findings, a growing body of literature indicates that the estrous cycle exerts a significant influence on autophagic activity, particularly in hormone-sensitive organs such as the Harderian gland. These fluctuations are closely linked to cyclical changes in sex hormone levels, with estrogen emerging as a key modulator of autophagy dynamics (Xiang et al., 2019; García-Macia et al., 2019).

Beyond direct gene regulation, sex hormones also influence autophagy indirectly by modulating key upstream signaling pathways. These include the PI3K/AKT/mTOR axis (with mTORC1 as a central autophagy inhibitor), the AMPK pathway (a cellular energy sensor that activates autophagy) and MAPK cascades (Franconi et al., 2023; Tian et al., 2020). For example, estrogen's protective effect against acetaminophen-induced liver injury in female mice involves modulation of the GSK3β/β-catenin pathway, leading to enhanced autophagy—an effect not observed in males (Liu et al., 2020). Additionally, complex crosstalk exists between ERα and metabolic regulators such as Sirt1 in adipocytes, where ERα induces Sirt1 expression, and Sirt1 is required for ERα-mediated suppression of autophagy and adipogenesis (Tao et al., 2021). The differential expression and activity of hormone receptors across tissues and between sexes adds yet another layer of regulatory complexity (Noh et al., 2023).

4.2 Sex chromosomes: Genetic drivers of sex-based differences in autophagy

The disparate sex chromosome complements (XX in females, XY in males) provide a cell-intrinsic foundation for sex differences in autophagy regulation. In both mammals and *Drosophila*, the X chromosome is substantially larger than the Y chromosome and encodes numerous genes essential for cellular homeostasis, including several directly involved in autophagy and lysosomal function. Notable examples include *LAMP2* (encoding a lysosomal membrane protein critical for CMA and autophagosome–lysosome fusion), WDR45/WIPI4 (essential for autophagosome formation), ATP6AP2 (a component of the V-ATPase proton pump required for lysosomal acidification), HDAC6 (a deacetylase involved in autophagosome maturation and aggrephagy) or TFE3 (transcription factor regulating lysosomal biogenesis and autophagy gene expression). Additional X-linked genes such as RAB39B, FUNDC1 (a mitophagy receptor), and DHRSX also contribute to autophagic processes (Moradi et al., 2024; Noh et al., 2023; Oliveira et al., 2024; Shang et al., 2021; Tower et al., 2020).

Although X-chromosome inactivation in females serves to balance gene dosage with males, approximately 15–25 % of human X-linked genes escape inactivation (Navarro-Cobos et al., 2023). This escape might result in higher expression levels of certain autophagy-related genes in XX cells compared to XY cells. Such dosage differences may underline baseline disparities in autophagic capacity or regulatory flexibility between sexes. Conversely, males—being hemizygous for the X chromosome—are more vulnerable to the phenotypic consequences of deleterious mutations in X-linked autophagy genes. This concept, known as the "unguarded X" hypothesis, may help explain the male bias observed in certain X-linked disorders involving autophagy dysfunction. In addition, although the Y chromosome contains relatively few genes, some—such as SRY—may influence cellular stress responses and indirectly modulate autophagy-related pathways (Tower et al., 2020).

4.3 Other contributing factors

While sex chromosomes and hormones are primary determinants of autophagy regulation, additional factors contribute to sex-specific differences in autophagic activity. One such factor is cell-autonomous sexual identity, which can influence baseline autophagy levels and responsiveness to stimuli independently of systemic hormone levels. This has been demonstrated in *Drosophila* gut enterocytes, where intrinsic sexual identity modulates autophagic responses to mTOR inhibition (Regan et al., 2022). Evolutionary pressures may also play a role. The "Mother's Curse" hypothesis posits that mitochondria, being maternally inherited, may accumulate mutations that are neutral or beneficial in females but deleterious in males. This could result in sex-specific mitochondrial stress and differential reliance on mitophagy for mitochondrial quality control (Hägg & Jylhävä, 2021; Tower et al., 2020).

Post-translational modifications add another layer of complexity. For example, palmitoylation can regulate both autophagy-related proteins and sex hormone receptors, and emerging evidence suggests that this modification exhibits sex-specific patterns (Ramzan et al., 2023). These modifications may fine-tune autophagic responses in a sex-dependent manner, further diversifying the regulatory landscape. Thus, a key area demanding further investigation would be the identification of sex-specific epigenetic landscapes (DNA methylation, histone modifications) and epitranscriptomic marks (e.g., mRNA m6A methylation) involved in differential autophagy gene expression programs.

Collectively, these genetic, hormonal, and cell-intrinsic factors interact within a complex network that shapes autophagy differently in males and females, with important implications for physiology and disease. A deeper understanding of these sex-specific mechanisms may unveil novel therapeutic targets unique to males or females. Pathways that are disproportionately reliant on hormonal input in one sex, or specific autophagy regulators predominantly active in males versus females under certain conditions, could offer opportunities for targeted drug development (Caserta et al., 2023)

5. Concluding remarks and future directions

The evidence compiled in this review underscores the pivotal role of biological sex as a fundamental modulator of autophagy. Sex influences not only the basal regulation of autophagic activity but also its responsiveness to physiological and pathological stimuli, ultimately shaping cellular and organismal outcomes across species and biological systems (Hägg & Jylhävä, 2021; Shang et al., 2021). This modulation arises from the complex interplay between genetic determinants encoded by the sex chromosomes and hormonal influences mediated by sex steroids. On the genetic level, sex-specific differences are partly driven by two major factors: the localization of autophagy-related genes on the X chromosome and the X-chromosome inactivation escape, which may result in differential gene dosage between sexes (Noh et al., 2023; Oliveira et al., 2024; Tower et al., 2020). Superimposed on this genetic framework, sex hormones—particularly estrogens and androgens—exert transcriptional control over core autophagy genes via nuclear receptors. These hormonal effects dynamically modulate autophagic capacity throughout the lifespan, with critical transitions occurring during development, puberty, reproductive aging and menopause (Lopez-Lee et al., 2024; Tao et al., 2021; Tian et al., 2023; Mughal et al., 2021).

The differential regulatory mechanisms described above translate into measurable sex-based disparities in basal autophagy activity and the expression of pathway components, which vary according to tissue type, species, and age (Campesi et al., 2023; Mughal et al., 2021; Ramos et al., 2022; Triolo et al., 2022; Zhang et al., 2023). More importantly, these baseline differences appear to condition the sexes for distinct autophagic responses when exposed to physiological or pathological stressors. Numerous studies have documented sex-specific autophagic modulation in response to exercise, nutritional challenges (e.g. undernutrition, over-nutrition, and exposure to specific fatty acids) and environmental stressors (e.g. smoking, radiation or cold exposure) (Di Veroli et al., 2024; Escobar et al., 2019; Kim et al., 2025; Liu et al., 2020; Liu et al., 2022; Moradi et al., 2024; Suh et al., 2022; Wang et al., 2020b). Furthermore, sex has emerged as a critical variable in autophagic responses to disease-specific insults, including ischemic injury and peripheral nerve damage (Noh et al., 2023; Patrizz et al., 2021; Tam et al., 2024; Zaheer et al., 2025). Given those tissue and cell-type specificities, future work should investigate sex differences with greater cellular resolution, potentially using single-cell omics or cell-type-specific manipulations.

These findings collectively highlight the necessity of incorporating sex as a biological variable in autophagy research, particularly in the context of aging and disease. The sex-specific regulation of autophagy is thus far from a mere physiological nuance. It may hold profound implications for our understanding of disease mechanisms and the observed sex biases in numerous pathological conditions. Variations in autophagic regulation and functional capacity between males and females likely contribute to differences in disease susceptibility, progression, and clinical presentation. This may be particularly relevant in neurodegenerative disorders where sex-linked disparities in autophagic efficiency may influence neuronal vulnerability and resilience (Cui et al., 2023; López-Cerdán et al., 2022; Merino et al., 2023). Sex-dependent differences in autophagic flux may underlie divergent trajectories of tissue degeneration, immune response or tumor progression, and similar patterns are observed in cardiovascular diseases, sarcopenia, cancer or chronic inflammatory diseases (Caserta et al., 2023; Green et al., 2024; Kerr et al., 2024; Muehleman et al., 2022; Robichaud et al., 2024; Rosa-Caldwell et al., 2021; Zhazykbayeva et al., 2023). Importantly, impaired or maladaptive autophagy—whether due to intrinsic sex-linked mechanisms or differential responses to external stressors—can lead to the accumulation of toxic aggregates, dysfunctional organelles, or unresolved inflammation, thereby amplifying sex-specific disease vulnerability (Noh et al., 2023; Zhazykbayeva et al., 2023). Given the sex-specific regulation of autophagy, therapeutic strategies aimed at modulating this pathway must be critically evaluated through a sex-informed lens. Evidence from preclinical models of stroke, chronic pain, hepatocellular carcinoma or Huntington's disease already demonstrated that interventions targeting autophagy can yield markedly different outcomes depending on sex, including variations in efficacy, optimal dosing and safety profiles (Green et al., 2024; Merino et al., 2023; Noh et al., 2023; Patrizz et al., 2021).

Despite growing recognition of sex as a critical modulator of autophagy, substantial knowledge gaps remain. Future research must systematically integrate sex as a biological variable in both experimental design and data interpretation, across preclinical models and human studies alike. Documented sex differences in autophagy demand a shift in research methodology, moving beyond male-biased paradigms to incorporate sex as a primary biological variable in design, analysis, and interpretation (Lopez-Lee et al., 2024; Tower et al., 2020). Future preclinical studies investigating autophagy should routinely include both male and female subjects in sufficient numbers to enable adequately powered sex-based comparisons, and report outcomes should be stratified by sex as standard practice, facilitating meta-analyses and cross-study comparisons. This methodological rigor must extend to clinical trials evaluating autophagy-modulating therapies. Stratification of participants by sex—and, where relevant, by hormonal status (e.g., pre- vs. post-menopause)—will be essential for identifying differential responders, optimizing therapeutic efficacy and minimizing sex-specific adverse effects.

Last but not least, accurate assessment of autophagic activity will also require moving beyond static markers (e.g., LC3-II/I ratio, p62 levels) toward dynamic measurements of autophagy flux, which better reflect the functional state of the pathway. Rigorous assessment of autophagy flux (the dynamic rate of degradation) using methods like lysosomal inhibitors (e.g., Bafilomycin A1, chloroquine) or flux reporter systems (e.g., tandem fluorescent reporters) is essential to truly understand pathway activity, as marker levels alone can be misleading (Bensalem et al., 2025). Moreover, a deeper understanding of the interplay between sex chromosome-linked genetic factors, hormonal signaling across the lifespan—including critical transitions such as puberty and menopause—and additional regulatory layers such as epigenetic modifications and post-translational control is essential. Advancing this field will also depend on the development and use of preclinical models that more accurately reflect human diversity, including genetic heterogeneity and physiologically relevant hormonal states. Ultimately, embracing the complexity of sex-dependent autophagy regulation will be key to translating mechanistic insights into effective, personalized interventions that improve health outcomes for both women and men.

In conclusion, the substantial evidence for sex differences in autophagy across diverse biological contexts necessitates a paradigm shift in both research and therapeutic development. Continued investigation, employing rigorous, sex-stratified methodologies and focusing on dynamic flux measurements and underlying mechanisms, is crucial. These findings underscore the potential risks of overlooking sex as a biological variable in therapeutic development. To maximize clinical benefit and minimize harm, the design of autophagy-targeted interventions must align with the principles of precision medicine, explicitly incorporating sex-based differences in autophagic regulation and function. Integrating this sex-specific understanding of autophagy is indispensable for moving beyond "one-size-fits-all" approaches and realizing the potential of precision medicine to improve health outcomes for both men and women.

Acknowledgments

This work was supported by RYC2021–032836-I, Ayuda a la Investigación en Ciencias de la Vida y de la Materia- Fundación Ramón Areces (to J.N-T); CNS2024–154954 and PID2021–127534OB-I00, funded by MICIU/AEI/10.13039/501100011033 and by ERDF/EU (Á.F.F.) and RYC2018–024434-I and MINECO PID2023–151873OB-I00 (to EBF).

Competing interests

The authors declare no competing or financial interest.

References

Adriaenssens, E., Ferrari, L., & Martens, S. (2022). Orchestration of selective autophagy by cargo receptors. *Current Biology: CB, 32*, R1357–r1371.

Barthet, V. J. A., & Ryan, K. M. (2022). DRAMs and autophagy: A family affair. *Autophagy Reports, 1*, 170–174.

Barz, S., Kriegenburg, F., Sánchez-Martín, P., & Kraft, C. (2021). Small but mighty: Atg8s and Rabs in membrane dynamics during autophagy. *Biochimica et Biophysica Acta. Molecular Cell Research, 1868*, 119064.

Baumgartner, N. E., Biraud, M. C., & Lucas, E. K. (2023). Sex differences in socio-emotional behavior and changes in ventral hippocampal transcription across aging in C57Bl/6J mice. *Neurobiology of Aging, 130*, 141–153.

Bejarano, E., & Cuervo, A. M. (2010). Chaperone-mediated autophagy. *Proceedings of the American Thoracic Society, 7*, 29–39.

Bejarano, E., Murray, J. W., Wang, X., Pampliega, O., Yin, D., Patel, B., ... Cuervo, A. M. (2018). Defective recruitment of motor proteins to autophagic compartments contributes to autophagic failure in aging. *Aging Cell, 17*, e12777.

Bensalem, J., Hein, L. K., Hassiotis, S., Trim, P. J., Proud, C. G., Heilbronn, L. K., & Sargeant, T. J. (2023). Modifying dietary protein impacts mTOR signaling and brain deposition of amyloid β in a knock-in mouse model of Alzheimer disease. *The Journal of Nutrition, 153*, 1407–1419.

Bensalem, J., Teong, X. T., Hattersley, K. J., Hein, L. K., Fourrier, C., Dang, L. V. P., ... Sargeant, T. J. (2025). Intermittent time-restricted eating may increase autophagic flux in humans: An exploratory analysis. *The Journal of Physiology, 603*, 3019–3032.

Campesi, I., Brunetti, A., Capobianco, G., Galistu, A., Montella, A., Ieri, F., & Franconi, F. (2023). Sex differences in X-ray-induced endothelial damage: Effect of taurine and N-acetylcysteine. *Antioxidants, 12*, 77.

Campesi, I., Montella, A., & Franconi, F. (2022). Human monocytes respond to lipopolysaccharide (LPS) stimulation in a sex-dependent manner. *Journal of Cellular Physiology, 237*, 580–588.

Carosi, J. M., Martin, A., Hein, L. K., Hassiotis, S., Hattersley, K. J., Turner, B. J., ... Sargeant, T. J. (2025). Autophagy across tissues of aging mice. *PLoS One, 20*, e0325505.

Caserta, S., Gangemi, S., Murdaca, G., & Allegra, A. (2023). Gender differences and miRNAs expression in cancer: Implications on prognosis and susceptibility. *International Journal of Molecular Sciences, 24*, 11544.

Cheng, Z. (2019). The foxO-autophagy axis in health and disease. *Trends in Endocrinology and Metabolism: TEM, 30*, 658–671.

Chivet, M., Marchioretti, C., Pirazzini, M., Piol, D., Scaramuzzino, C., Polanco, M. J., ... Pennuto, M. (2020). Polyglutamine-expanded androgen receptor alteration of skeletal muscle homeostasis and myonuclear aggregation are affected by sex, age and muscle metabolism. *Cells, 9*, 325.

Cui, S.-S., Jiang, Q.-W., & Chen, S.-D. (2023). Sex difference in biological change and mechanism of Alzheimer's disease: From macro- to micro-landscape. *Ageing Research Reviews, 87*, 101918.

Das, D., Sharma, M., Gahlot, D., Nia, S. S., Gain, C., Mecklenburg, M., ... Singh, R. (2024). VPS4A is the selective receptor for lipophagy in mice and humans. *Molecular Cell, 84*(22), 4436–4453.e8. https://doi.org/10.1016/j.molcel.2024.10.022.

Di Veroli, B., Bentanachs, R., Roglans, N., Alegret, M., Giona, L., Profumo, E., ... Buttari, B. (2024). Sex differences affect the NRF2 signaling pathway in the early phase of liver steatosis: A high-fat-diet-fed rat model supplemented with liquid fructose. *Cells, 13*, 1247.

Dong, S., Aguirre-Hernandez, C., Scrivo, A., Eliscovich, C., Arias, E., Bravo-Cordero, J. J., & Cuervo, A. M. (2020). Monitoring spatiotemporal changes in chaperone-mediated autophagy in vivo. *Nature Communications, 11*, 645.

Dong, T., Li, X., Yu, W., Liu, Y., & Yang, J. (2024). Identification of sex-specific biomarkers related to programmed cell death and analysis of immune cells in ankylosing spondylitis. *Scientific Reports, 14*, 15358.

Endicott, S. J., Boynton, D. N., Jr, Beckmann, L. J., & Miller, R. A. (2021). Long-lived mice with reduced growth hormone signaling have a constitutive upregulation of hepatic chaperone-mediated autophagy. *Autophagy, 17*, 612–625.

Escobar, K. A., Cole, N. H., Mermier, C. M., & VanDusseldorp, T. A. (2019). Autophagy and aging: Maintaining the proteome through exercise and caloric restriction. *Aging Cell, 18*, e12876.

Escobar, K. A., Welch, A. M., Wells, A., Fennel, Z., Nava, R., Li, Z., ... Amorim, F. T. (2021). Autophagy response to acute high-intensity interval training and moderate-intensity continuous training is dissimilar in skeletal muscle and peripheral blood mononuclear cells and is influenced by sex. *Human Nutrition & Metabolism, 23*, 200118.

Fernández, Á. F., & López-Otín, C. (2015). The functional and pathologic relevance of autophagy proteases. *The Journal of Clinical Investigation, 125*, 33–41.

Fernández, Á. F., Sebti, S., Wei, Y., Zou, Z., Shi, M., McMillan, K. L., ... Levine, B. (2018). Disruption of the beclin 1–BCL2 autophagy regulatory complex promotes longevity in mice. *Nature, 558*, 136–140.

Franconi, F., Capobianco, G., Diana, G., Lodde, V., De Donno, A., Idda, M. L., ... Campesi, I. (2023). Sex influence on autophagy markers and miRNAs in basal and angiotensin II-treated human umbilical vein endothelial cells. *International Journal of Molecular Sciences, 24*, 14929.

García-Macia, M., Santos-Ledo, A., Caballero, B., Rubio-González, A., de Luxán-Delgado, B., Potes, Y., ... Coto-Montes, A. (2019). Selective autophagy, lipophagy and mitophagy, in the Harderian gland along the oestrous cycle: a potential retrieval effect of melatonin. *Scientific Reports, 9*(1), 18597. https://doi.org/10.1038/s41598-019-54743-5.

Garcia-Macia, M., Santos-Ledo, A., Leslie, J., Paish, H. L., Collins, A. L., Scott, R. S., ... Mann, D. A. (2021). A mammalian target of rapamycin-perilipin 3 (mTORC1-Plin3) pathway is essential to activate lipophagy and protects against hepatosteatosis. *Hepatology (Baltimore, Md.), 74*(6), 3441–3459. https://doi.org/10.1002/hep.32048.

Galluzzi, L., Baehrecke, E. H., Ballabio, A., Boya, P., Bravo-San Pedro, J. M., Cecconi, F., ... Kroemer, G. (2017). Molecular definitions of autophagy and related processes. *The EMBO Journal, 36*, 1811–1836.

Green, C. D., Brown, R. D. R., Uranbileg, B., Weigel, C., Saha, S., Kurano, M., ... Spiegel, S. (2024). Sphingosine kinase 2 and p62 regulation are determinants of sexual dimorphism in hepatocellular carcinoma. *Molecular Metabolism, 86*, 101971.

Hägg, S., & Jylhävä, J. (2021). Sex differences in biological aging with a focus on human studies. *eLife, 10*, e63425.

Hansen, M., Rubinsztein, D. C., & Walker, D. W. (2018). Autophagy as a promoter of longevity: Insights from model organisms. *Nature Reviews. Molecular Cell Biology, 19*, 579–593.

Hassan, A., Sara, S.-Z., Ehsan, R.-A., Hossein Kargar, J., Payman Raise, A., Karim, D., ... Ali Reza, S. (2023). The role of gut microbiota in inflammatory bowel disease-current state of the art. *Mini-Reviews in Medicinal Chemistry, 23*, 1376–1389.

Hurley, J. H., & Young, L. N. (2017). Mechanisms of autophagy initiation. *Annual Review of Biochemistry, 86*, 225–244.

Jafari, M., McCabe, M., & Cuervo, A. M. (2022). Chaperone-mediated autophagy: Mechanisms and physiological relevance. *Current Opinion in Physiology, 30*, 100597.

Jiang, W., Chen, X., Ji, C., Zhang, W., Song, J., Li, J., & Wang, J. (2021). Key regulators of autophagosome closure. *Cells, 10*.

Johansen, T., & Lamark, T. (2020). Selective autophagy: ATG8 family proteins, LIR motifs and cargo receptors. *Journal of Molecular Biology, 432*, 80–103.

Jung, J. T. K., Marques, L. S., Brambila, C. A., da Cruz Weber Fulco, B., Nogueira, C. W., & Zeni, G. (2024). Social-single prolonged stress affects contextual fear conditioning in male and female Wistar rats: Molecular insights in the amygdala. *Progress in Neuro-Psychopharmacology and Biological Psychiatry, 133*, 111021.

Karmon, G., Sragovich, S., Hacohen-Kleiman, G., Ben-Horin-Hazak, I., Kasparek, P., Schuster, B., ... Gozes, I. (2022). Novel ADNP syndrome mice reveal dramatic sex-specific peripheral gene expression with brain synaptic and tau pathologies. *Biological Psychiatry, 92*, 81–95.

Kast, D. J., & Dominguez, R. (2017). The cytoskeleton-autophagy connection. *Current Biology: CB, 27*, R318–r326.

Kaur, G., & Lau, E. (2022). Sex differences in heart failure with preserved ejection fraction: From traditional risk factors to sex-specific risk factors. *Women's Health, 18*, 17455057221140209.

Kaushik, S., & Cuervo, A. M. (2018). The coming of age of chaperone-mediated autophagy. *Nature Reviews. Molecular Cell Biology, 19*, 365–381.

Kaushik, S., Massey, A. C., & Cuervo, A. M. (2006). Lysosome membrane lipid microdomains: Novel regulators of chaperone-mediated autophagy. *The EMBO Journal, 25*, 3921–3933.

Kerr, H. L., Krumm, K., Anderson, B., Christiani, A., Strait, L., Li, T., ... Garcia, J. M. (2024). Mouse sarcopenia model reveals sex- and age-specific differences in phenotypic and molecular characteristics. *The Journal of Clinical Investigation, 134*.

Khawaja, R. R., Martín-Segura, A., Santiago-Fernández, O., Sereda, R., Lindenau, K., McCabe, M., ... Cuervo, A. M. (2025). Sex-specific and cell-type-specific changes in chaperone-mediated autophagy across tissues during aging. *Nature Aging, 5*, 691–708.

Kiffin, R., Kaushik, S., Zeng, M., Bandyopadhyay, U., Zhang, C., Massey, A. C., ... Cuervo, A. M. (2007). Altered dynamics of the lysosomal receptor for chaperone-mediated autophagy with age. *Journal of Cell Science, 120*, 782–791.

Kim, B. M., Kim, D. H., Park, Y. J., Ha, S., Choi, Y. J., Yu, H. S., ... Chung, H. Y. (2021). PAR2 promotes high-fat diet-induced hepatic steatosis by inhibiting AMPK-mediated autophagy. *The Journal of Nutritional Biochemistry, 95*, 108769.

Kim, M. J., Hong, J., Chang, H. W., Lee, Y.-H., Myong, J.-P., Jung, A. R., ... Kwon, M. (2025). Autophagy activation in response to cigarette smoke: Exploring the disparity in laryngeal cancer incidence and outcomes between sexes in South Korea. *Translational Oncology, 52*, 102229.

Kong, X.-T., Wang, Z., Mou, J.-J., Li, C.-S., Xue, H.-L., Wu, M., ... Xu, L.-X. (2021). Change on apoptosis, autophagy and mitochondria of the Harderian gland in Cricetulus barabensis during age. *Comparative Biochemistry and Physiology Part B: Biochemistry and Molecular Biology, 253*, 110547.

Koren, I., Reem, E., & Kimchi, A. (2010). DAP1, a novel substrate of mTOR, negatively regulates autophagy. *Current Biology: CB, 20*, 1093–1098.

Krause, G. J., Diaz, A., Jafari, M., Khawaja, R. R., Agullo-Pascual, E., Santiago-Fernández, O., ... Cuervo, A. M. (2022). Reduced endosomal microautophagy activity in aging associates with enhanced exocyst-mediated protein secretion. *Aging Cell, 21*, e13713.

Krause, G. J., Kirchner, P., Stiller, B., Morozova, K., Diaz, A., Chen, K. H., ... Cuervo, A. M. (2023). Molecular determinants of the crosstalk between endosomal microautophagy and chaperone-mediated autophagy. *Cell Reports, 42*, 113529.

Kuchitsu, Y., & Taguchi, T. (2024). Lysosomal microautophagy: An emerging dimension in mammalian autophagy. *Trends in Cell Biology, 34*, 606–616.

Kurhaluk, N., & Tkachenko, H. (2021). Antioxidants, lysosomes and elements status during the life cycle of sea trout Salmo trutta m. trutta L. *Scientific Reports, 11*, 5545.

Lamark, T., & Johansen, T. (2021). Mechanisms of selective autophagy. *Annual Review of Cell and Developmental Biology, 37*, 143–169.

Lamark, T., Svenning, S., & Johansen, T. (2017). Regulation of selective autophagy: The p62/SQSTM1 paradigm. *Essays in Biochemistry, 61*, 609–624.

Lapierre, L. R., De Magalhaes Filho, C. D., McQuary, P. R., Chu, C.-C., Visvikis, O., Chang, J. T., ... Hansen, M. (2013). The TFEB orthologue HLH-30 regulates autophagy and modulates longevity in Caenorhabditis elegans. *Nature Communications, 4*, 2267.

Li, Z., Yu, Y., Zhao, X., Qu, Y., Wang, J., & Zhang, D. (2025). Chaperone-mediated autophagy reactivation protects against severe acute pancreatitis-associated liver injury through upregulating keap1/Nrf2 signaling pathway and inhibiting NLRP3 inflammasome activation. *Cell Biochemistry and Biophysics*.

Liu, Y., Chen, Y., Xie, X., Yin, A., Yin, Y., Liu, Y., ... Zuo, D. (2020). Gender difference on the effect of omega-3 polyunsaturated fatty acids on acetaminophen-induced acute liver failure. *Oxidative Medicine and Cellular Longevity, 2020*, 8096847.

Liu, Y., Ding, Q., Halderson, S. J., Arriola Apelo, S. I., Jones, A. K., Pillai, S. M., ... Guo, W. (2022). Maternal overnutrition during gestation in sheep alters autophagy associated pathways in offspring heart. *Frontiers in Genetics, 12*, 2021.

López-Cerdán, A., Andreu, Z., Hidalgo, M. R., Grillo-Risco, R., Català-Senent, J. F., Soler-Sáez, I., ... García-García, F. (2022). Unveiling sex-based differences in Parkinson's disease: a comprehensive meta-analysis of transcriptomic studies. *Biology of Sex Differences, 13*, 68.

Lopez-Lee, C., Torres, E. R. S., Carling, G., & Gan, L. (2024). Mechanisms of sex differences in Alzheimer's disease. *Neuron, 112*, 1208–1221.

López-Otín, C., Blasco, M. A., Partridge, L., Serrano, M., & Kroemer, G. (2013). The hallmarks of aging. *Cell, 153*, 1194–1217.

López-Otín, C., Blasco, M. A., Partridge, L., Serrano, M., & Kroemer, G. (2023). Hallmarks of aging: An expanding universe. *Cell, 186*, 243–278.

Lu, L.-F., Jiang, J.-Y., Du, W.-X., Wang, X.-L., Li, Z.-C., Zhou, X.-Y., ... Li, S. (2022). Fish female-biased gene cyp19a1a leads to female antiviral response attenuation between sexes by autophagic degradation of MITA. *PLoS Pathogens, 18*, e1010626.

Luetić, M., Kretzschmar, G., Grobe, M., Jerčić, L., Bota, I., Ivić, V., ... Filipović, N. (2021). Sex-specific effects of metformin and liraglutide on renal pathology and expression of connexin 45 and pannexin 1 following long-term high-fat high-sugar diet. *Acta Histochemica, 123*, 151817.

Luk, H.-Y., Appell, C., Levitt, D. E., Jiwan, N. C., & Vingren, J. L. (2021). Differential autophagy response in men and women after muscle damage. *Frontiers in Physiology, 12*, 2021.

Luo, R. (2025). Unraveling the complexity of chaperone-mediated autophagy in aging: Insights into sex-specific and cell-type-specific regulation. *Autophagy, 21*, 1848–1849.

Madrigal-Matute, J., de Bruijn, J., van Kuijk, K., Riascos-Bernal, D. F., Diaz, A., Tasset, I., ... Cuervo, A. M. (2022). Protective role of chaperone-mediated autophagy against atherosclerosis. *Proceedings of the National Academy of Sciences, 119*, e2121133119.

Manastireanu, D. M., Salazar, N. A., Bejarano, E., & Nieto-Torres, J. L. (2024). Selective autophagy: A therapeutic target for healthy aging? *Aging Advances, 1*.

Merino, M., González, S., Tronch, M. C., Sánchez-Sánchez, A. V., Clares, M. P., García-España, A., ... Mullor, J. L. (2023). Small molecule pytren-4QMn metal complex slows down Huntington's disease progression in male zQ175 transgenic mice. *International Journal of Molecular Sciences, 24*, 15153.

Mizushima, N. (2020). The ATG conjugation systems in autophagy. *Current Opinion in Cell Biology, 63*, 1–10.

Moradi, N., Sanfrancesco, V. C., Champsi, S., & Hood, D. A. (2024). Regulation of lysosomes in skeletal muscle during exercise, disuse and aging. *Free Radical Biology and Medicine, 225*, 323–332.

Moreno, T. M., Nieto-Torres, J. L., & Kumsta, C. (2025). Monitoring autophagy in human aging: Key cell models and insights. *FBL, 30*.

Muehleman, D. L., Crocini, C., Swearingen, A. R., Ozeroff, C. D., & Leinwand, L. A. (2022). Regression from pathological hypertrophy in mice is sexually dimorphic and stimulus specific. *American Journal of Physiology-Heart and Circulatory Physiology, 322*, H785–H797.

Mughal, M. N., Grevelding, C. G., & Haeberlein, S. (2021). First insights into the autophagy machinery of adult Schistosoma mansoni. *International Journal for Parasitology, 51*, 571–585.

Nakatogawa, H. (2020). Mechanisms governing autophagosome biogenesis. *Nature Reviews. Molecular Cell Biology, 21*, 439–458.

Navarro-Cobos, M. J., Bhangu, K., & Brown, C. J. (2023). Chapter 14 - Human X-chromosome inactivation: Complexity and clinical implications. In P. C. K. Leung, & J. Qiao (Eds.). *Human reproductive and prenatal genetics* (pp. 355–385)(Second ed.). Academic Press.

Nieto-Torres, J. L., & Hansen, M. (2021). Macroautophagy and aging: The impact of cellular recycling on health and longevity. *Molecular Aspects of Medicine, 82*, 101020.

Nieto-Torres, J. L., Shanahan, S. L., Chassefeyre, R., Chaiamarit, T., Zaretski, S., Landeras-Bueno, S., ... Hansen, M. (2021). LC3B phosphorylation regulates FYCO1 binding and directional transport of autophagosomes. *Current Biology: CB, 31*, 3440–3449.e3447.

Noda, N. N. (2021). Atg2 and Atg9: Intermembrane and interleaflet lipid transporters driving autophagy. *Biochimica et Biophysica Acta. Molecular and Cell Biology of Lipids, 1866*, 158956.

Noda, N. N. (2023). Structural biology of the Atg8 and Atg12 conjugation systems. *Autophagy Reports, 2*, 2277582.

Noda, N. N. (2024). Structural view on autophagosome formation. *FEBS Letters, 598*, 84–106.

Noh, B., McCullough, L. D., & Moruno-Manchon, J. F. (2023). Sex-biased autophagy as a potential mechanism mediating sex differences in ischemic stroke outcome. *Neural Regeneration Research, 18*, 31–37.

Ogura, M., Kaminishi, T., Shima, T., Torigata, M., Bekku, N., Tabata, K., ... Nakamura, S. (2023). Microautophagy regulated by STK38 and GABARAPs is essential to repair lysosomes and prevent aging. *EMBO Reports, 24*, e57300.

Oliveira, A. N., Memme, J. M., Wong, J., & Hood, D. A. (2024). Dimorphic effect of TFE3 in determining mitochondrial and lysosomal content in muscle following denervation. *Skeletal Muscle, 14*, 7.

Ortona, E., Pagano, M. T., Capossela, L., & Malorni, W. (2023). The role of sex differences in bone health and healing. *Biology, 12*, 993.

Patrizz, A. N., Moruno-Manchon, J. F., O'Keefe, L. M., Doran, S. J., Patel, A. R., Venna, V. R., ... McCullough, L. D. (2021). Sex-specific differences in autophagic responses to experimental ischemic stroke. *Cells, 10*, 1825.

Ponce-Mora, A., Salazar, N. A., Domenech-Bendaña, A., Locascio, A., Bejarano, E., & Gimeno-Mallench, L. (2025). Interplay between polyphenols and autophagy: Insights from an aging perspective. *Frontiers in Bioscience (Landmark edition), 30*, 25728.

Punyavathi, & Manjunatha, H. B. (2020). Comprehensive analysis of differentially expressed proteins in the male and female Bombyx mori larval instars exposed to thermal stress. *Archives of Insect Biochemistry and Physiology, 105*, e21719.

Pyo, J.-O., Yoo, S.-M., Ahn, H.-H., Nah, J., Hong, S.-H., Kam, T.-I., ... Jung, Y.-K. (2013). Overexpression of Atg5 in mice activates autophagy and extends lifespan. *Nature Communications, 4*, 2300.

Ramos, C. D., Bohnert, K. A., & Johnson, A. E. (2022). Reproductive tradeoffs govern sexually dimorphic tubular lysosome induction in Caenorhabditis elegans. *Journal of Experimental Biology, 225*.

Ramzan, F., Abrar, F., Mishra, G. G., Liao, L. M. Q., & Martin, D. D. O. (2023). Lost in traffic: Consequences of altered palmitoylation in neurodegeneration. *Frontiers in Physiology, 14*, 1166125.

Regan, J. C., Lu, Y.-X., Ureña, E., Meilenbrock, R. L., Catterson, J. H., Kißler, D., ... Partridge, L. (2022). Sexual identity of enterocytes regulates autophagy to determine intestinal health, lifespan and responses to rapamycin. *Nature Aging, 2*, 1145–1158.

Reynolds, C. A., Pelka, S., Gjergjova, F., Tasset, I., Khawaja, R. R., Lindenau, K., ... Macian, F. (2024). Restoration of LAMP2A expression in old mice leads to changes in the T cell compartment that support improved immune function. *Proceedings of the National Academy of Sciences, 121*, e2322929121.

Robichaud, S., Rochon, V., Emerton, C., Laval, T., & Ouimet, M. (2024). Trehalose promotes atherosclerosis regression in female mice. *Frontiers in Cardiovascular Medicine, 11*, 2024.

Rosa-Caldwell, M. E., Lim, S., Haynie, W. S., Brown, J. L., Lee, D. E., Dunlap, K. R., ... Greene, N. P. (2021). Mitochondrial aberrations during the progression of disuse atrophy differentially affect male and female mice. *Journal of Cachexia, Sarcopenia and Muscle, 12*, 2056–2068.

Sakai, Y., & Oku, M. (2024). ATG and ESCRT control multiple modes of microautophagy. *FEBS Letters, 598*, 48–58.

Salinas, M. D., Martínez, C. M., Roca, F. J., García-Bernal, D., Martínez-Morga, M., Rodríguez-Madoz, J. R., ... Valdor, R. (2025a). Chaperone-mediated autophagy sustains pericyte stemness necessary for brain tissue homeostasis. *Journal of Advanced Research*.

Salinas, M. D., Rodriguez, P., Rubio, G., & Valdor, R. (2025b). Expression of lumican and osteopontin in perivascular areas of the glioblastoma peritumoral niche and its value for prognosis. *International Journal of Molecular Sciences, 26*, 192.

Santiago, J. A., Quinn, J. P., & Potashkin, J. A. (2022). Sex-specific transcriptional rewiring in the brain of Alzheimer's disease patients. *Frontiers in Aging Neuroscience, 14*, 2022.

Settembre, C., Di Malta, C., Polito, V. A., Garcia Arencibia, M., Vetrini, F., Erdin, S., ... Ballabio, A. (2011). TFEB links autophagy to lysosomal biogenesis. *Science (New York, N. Y.), 332*, 1429–1433.

Shang, D., Wang, L., Klionsky, D. J., Cheng, H., & Zhou, R. (2021). Sex differences in autophagy-mediated diseases: Toward precision medicine. *Autophagy, 17*, 1065–1076.

Shatz, O., & Elazar, Z. (2024). Autophagy in a nutshell. *FEBS Letters, 598*, 7–8.

Sierra-Fonseca, J. A., Hamdan, J. N., Cohen, A. A., Cardenas, S. M., Saucedo, S., Lodoza, G. A., & Gosselink, K. L. (2021). Neonatal maternal separation modifies proteostasis marker expression in the adult hippocampus. *Frontiers in Molecular Neuroscience, 14*, 2021.

Simonsen, A., Cumming, R. C., Brech, A., Isakson, P., Schubert, D. R., & Finley, K. D. (2008). Promoting basal levels of autophagy in the nervous system enhances longevity and oxidant resistance in adult Drosophila. *Autophagy, 4*, 176–184.

Suh, J. H., Kim, K. H., Conner, M. E., Moore, D. D., & Preidis, G. A. (2022). Hepatic PPARα is destabilized by SIRT1 deacetylase in undernourished male mice. *Frontiers in Nutrition, 9*, 831879.

Svoboda, L. K., Wang, K., Jones, T. R., Colacino, J. A., Sartor, M. A., & Dolinoy, D. C. (2021). Sex-specific alterations in cardiac DNA methylation in adult mice by perinatal lead exposure. *International Journal of Environmental Research and Public Health, 18*, 577.

Tam, T. H., Zhang, W., Tu, Y., Hicks, J. L., Farcas, S., Kim, D., & Salter, M. W. (2024). Pain hypersensitivity is dependent on autophagy protein Beclin 1 in males but not females. *Cell Reports, 43*, 114293.

Tamargo-Gómez, I., & Mariño, G. (2018). AMPK: Regulation of metabolic dynamics in the context of autophagy. *International Journal of Molecular Sciences, 19*.

Tao, Z., Shi, L., Parke, J., Zheng, L., Gu, W., Dong, X. C., ... Cheng, Z. (2021). Sirt1 coordinates with ERα to regulate autophagy and adiposity. *Cell Death Discovery, 7*, 53.

Tekirdag, K., & Cuervo, A. M. (2018). Chaperone-mediated autophagy and endosomal microautophagy: Jointed by a chaperone. *Journal of Biological Chemistry, 293*, 5414–5424.

Tian, Q., Fan, X., Ma, J., Han, Y., Li, D., Jiang, S., ... Fan, J. (2020). Resveratrol ameliorates lipopolysaccharide-induced anxiety-like behavior by attenuating YAP-mediated neuro-inflammation and promoting hippocampal autophagy in mice. *Toxicology and Applied Pharmacology, 408*, 115261.

Tian, X., Lou, S., & Shi, R. (2023). From mitochondria to sarcopenia: Role of 17β-estradiol and testosterone. *Frontiers in Endocrinology, 14*, 1156583.
Tian, X., Teng, J., & Chen, J. (2021). New insights regarding SNARE proteins in autophagosome-lysosome fusion. *Autophagy, 17*, 2680–2688.
Torrens-Mas, M., Navas-Enamorado, C., Wahl, D., Sanchez-Polo, A., Picca, A., Oliver, J., ... Gonzalez-Freire, M. (2022). Sex specific differences in response to calorie restriction in skeletal muscle of young rats. *Nutrients, 14*, 4535.
Tower, J., Pomatto, L. C. D., & Davies, K. J. A. (2020). Sex differences in the response to oxidative and proteolytic stress. *Redox Biology, 31*, 101488.
Triolo, M., Oliveira, A. N., Kumari, R., & Hood, D. A. (2022). The influence of age, sex, and exercise on autophagy, mitophagy, and lysosome biogenesis in skeletal muscle. *Skeletal Muscle, 12*, 13.
Uğuz, A. C., Okan, A., Doğanyiğit, Z., Yilmaz, S., Ateş, Ş., Arikan Söylemez, E. S., ... Espino, J. (2025). Evaluation of TRPM2 channel-mediated autophagic signaling pathway in hippocampus and cortex tissues of rat offspring following prenatal exposure to elevated alcohol levels. *Environmental Toxicology, 40*, 222–244.
Ulgherait, M., Rana, A., Rera, M., Graniel, J., & Walker, D. W. (2014). AMPK modulates tissue and organismal aging in a non-cell-autonomous manner. *Cell Reports, 8*, 1767–1780.
Valdor, R., Mocholi, E., Botbol, Y., Guerrero-Ros, I., Chandra, D., Koga, H., ... Macian, F. (2014). Chaperone-mediated autophagy regulates T cell responses through targeted degradation of negative regulators of T cell activation. *Nature Immunology, 15*, 1046–1054.
van der Beek, J., & Klumperman, J. (2025). Trafficking to the lysosome: HOPS paves the way. *Current Opinion in Cell Biology, 94*, 102515.
Vargas, J. N. S., Hamasaki, M., Kawabata, T., Youle, R. J., & Yoshimori, T. (2023). The mechanisms and roles of selective autophagy in mammals. *Nature Reviews. Molecular Cell Biology, 24*, 167–185.
Vega-Naredo, I., Caballero, B., Sierra, V., Huidobro-Fernández, C., de Gonzalo-Calvo, D., García-Macia, M., ... Coto-Montes, A. (2009). Sexual dimorphism of autophagy in Syrian hamster Harderian gland culminates in a holocrine secretion in female glands. *Autophagy, 5*(7), 1004–1017. https://doi.org/10.4161/auto.5.7.9610.
Vietri, M., Radulovic, M., & Stenmark, H. (2020). The many functions of ESCRTs. *Nature Reviews. Molecular Cell Biology, 21*, 25–42.
Wang, B., Chan, Y. L., Zhou, S., Saad, S., Chen, H., & Oliver, B. G. (2020a). Offspring sex affects the susceptibility to maternal smoking-induced lung inflammation and the effect of maternal antioxidant supplementation in mice. *Journal of Inflammation, 17*, 24.
Wang, N., Zhang, Y.-H., Wang, J.-Y., & Luo, F. (2021). Current understanding of the involvement of the insular cortex in neuropathic pain: A narrative review. *International Journal of Molecular Sciences, 22*, 2648.
Wang, P., Long, M., Zhang, S., Cheng, Z., Zhao, X., He, F., ... Ming, L. (2017). Hypoxia inducible factor-1α regulates autophagy via the p27-E2F1 signaling pathway. *Molecular Medicine Reports, 16*, 2107–2112.
Wang, Y., & Zhang, H. (2019). Regulation of autophagy by mTOR signaling pathway. *Advances in Experimental Medicine and Biology, 1206*, 67–83.
Wang, Z., Xu, J.-H, Mou, J.-J, Kong, X.-T, Zou, J.-W, Xue, H.-l, ... Xu, L.-X (2020b). Novel ultrastructural findings on cardiac mitochondria of huddling Brandt's voles in mild cold environment. *Comparative Biochemistry and Physiology Part A: Molecular & Integrative Physiology, 249*, 110766.
Wu, Y. T., Tan, H. L., Shui, G., Bauvy, C., Huang, Q., Wenk, M. R., ... Shen, H. M. (2010). Dual role of 3-methyladenine in modulation of autophagy via different temporal patterns of inhibition on class I and III phosphoinositide 3-kinase. *The Journal of Biological Chemistry, 285*, 10850–10861.

Xiang, J., Liu, X., Ren, J., Chen, K., Wang, H. L., Miao, Y. Y., & Qi, M. M. (2019). How does estrogen work on autophagy? *Autophagy, 15*(2), 197–211. https://doi.org/10.1080/15548627.2018.1520549.

Yim, W. W.-Y., & Mizushima, N. (2020). Lysosome biology in autophagy. *Cell Discovery, 6*, 6.

Zachari, M., & Ganley, I. G. (2017). The mammalian ULK1 complex and autophagy initiation. *Essays in Biochemistry, 61*, 585–596.

Zaheer, F., Levine, G. J., Simal, A. L., Fatemi Tabatabaei, S. R., Martino, T. A., & Descalzi, G. (2025). Sex-specific contrasting role of BECLIN-1 protein in pain hypersensitivity and anxiety-like behaviors. *eNeuro, 12*, ENEURO.0244-0224.2024.

Zhang, C., & Cuervo, A. M. (2008). Restoration of chaperone-mediated autophagy in aging liver improves cellular maintenance and hepatic function. *Nature Medicine, 14*, 959–965.

Zhang, K. K., Zhang, P., Kodur, A., Erturk, I., Burns, C. M., Kenyon, C., ... Endicott, S. J. (2023). LAMP2A, and other chaperone-mediated autophagy related proteins, do not decline with age in genetically heterogeneous UM-HET3 mice. *Aging, 15*, 4685–4698.

Zhao, Y. G., Codogno, P., & Zhang, H. (2021). Machinery, regulation and pathophysiological implications of autophagosome maturation. *Nature Reviews. Molecular Cell Biology, 22*, 733–750.

Zhazykbayeva, S., Hassoun, R., Herwig, M., Budde, H., Kovács, Á., Mannherz, H. G., ... Hamdani, N. (2023). Oxidative stress and inflammation distinctly drive molecular mechanisms of diastolic dysfunction and remodeling in female and male heart failure with preserved ejection fraction rats. *Frontiers in Cardiovascular Medicine, 10*, 1157398.

Printed and bound by CPI Group (UK) Ltd, Croydon, CR0 4YY
25/09/2026
02222752-0001